Viral Gastroenteritis

Edited by
Shunzo Chiba, Mary K. Estes,
Shuji Nakata, and Charles H. Calisher

Archives of Virology
Supplement 12

Springer-Verlag Wien GmbH

Prof. Dr. Shunzo Chiba
Sapporo Medical University, Sapporo, Japan

Prof. Dr. Mary K. Estes
Baylor College of Medicine, Houston, USA

Ass. Prof. Dr. Shuji Nakata
Sapporo Medical University, Sapporo, Japan

Prof. Dr. Charles H. Calisher
Colorado State University, Fort Collins, USA

© 1996 Springer-Verlag Wien
Originally published by Springer-Verlag/Wien in 1996
Softcover reprint of the hardcover 1st edition 1996
Typesetting: Thomson Press (India) Ltd., New Delhi

Graphic design: Ecke Bonk
Printed on acid-free and chlorine-free bleached paper

With 62 partly coloured Figures

ISSN 0939-1983
ISBN 978-3-211-82875-5   ISBN 978-3-7091-6553-9 (eBook)
DOI 10.1007/978-3-7091-6553-9

Preface

Over twenty years have passed since the monumental discoveries of Norwalk virus and rotavirus as etiologic agents of gastroenteritis. During these years, many new findings on viral gastroenteritis have accumulated.

In 1986 the Symposium on "Novel Diarrhea Viruses" was held at the Ciba Foundation in London. In that Symposium, mainly novel rotaviruses, enteric adenoviruses, and small round viruses were discussed. Since then there have been rapid developments in the field of viral gastroenteritis. The Norwalk virus genome was cloned, which has made a great impact on further characterization of other small round viruses. The first rotavirus vaccine will be licensed before long.

In this Symposium, the Sapporo International Symposium on Viral Gastroenteritis, held at the Hotel-Nidom, Tomakomai Hokkaido, Japan, June 28–30, 1995, the fundamental and practical aspects of the latest developments in the field of viral gastroenteritis were discussed by distinguished scientists from around the world. Of the various agents which can cause viral gastroenteritis, this conference focused on the group A rotaviruses and small round structured viruses (represented by Norwalk virus and now classified as caliciviruses) because of the clinical impact and accumulating new findings about these two groups. Recent advances in enteric adenoviruses were also discussed.

I am very grateful to the members of the International Advisory Board for their participation and excellent suggestions in preparing the program. I would like to extend my gratitude to all of the invited speakers and participants, the World Health Organization and many other organizations and enterprises that economically supported this Symposium.

I would like to express special thanks to Drs. S. Urasawa, K. Taniguchi, and S. Nakata for spending much time in organizing this Symposium, and to Dr. M. K. Estes who didn't spare herself to help me at every step in preparing for the Symposium. Without their personal support, my wishes to hold this Symposium could not have been realized. I can never thank them enough.

Finally, I hope that international symposia on viral gastroenteritis, such as this one, will be held regularly hereafter somewhere in the world.

Shunzo Chiba

Contents

Arch Virol (1996) [Suppl] 12: 1–6

Introduction

S. Urasawa

Department of Hygiene, Sapporo Medical University School of Medicine,
Chuoku, Sapporo, Japan

At the beginning of this special issue on viral gastroenteritis, a brief introduction on the viruses concerned may help readers better understand the papers that follow. In this international symposium, viruses from families causing gastroenteritis in humans, rotaviruses (*Reoviridae*), caliciviruses (*Caliciviridae*), astroviruses (*Astroviridae*) and adenoviruses (*Adenoviridae*), were highlighted. Some basic characters and epidemiologic features of these viruses are summarized in Table 1, and structural features are illustrated in Figs. 1 and 2.

Rotaviruses, family *Reoviridae*, genus *Rotavirus*, are the most important etiologic agents of severe gastroenteritis afflicting young children and animals wordwide [2, 8]. Severe rotavirus gastroenteritis, if not treated properly, may lead to a fatal outcome. Rotavirus gastroenteritis occurs in a clearly repetitive pattern in the temperate countries with peaks in the cooler months of each year. Mature rotavirus particles measure about 75 nm in diameter and possess a triple-layered protein capsid (outer capsid, inner capsid and core). The virus genome comprises 11 segments of double stranded RNA, each of which encode a single protein (six structural proteins VP1–4, 6 and 7, and five non-structural proteins NSP 1–5). Treatment of virus with protease enhances its infectivity by cleaving VP4 into VP5* and VP8*. Different electrophoretic patterns in polyacrylamide gels (electropherotypes) of the segmented genomic RNA of strains have been successfully employed for the identification of strains. The segmented nature of the genome facilitates genetic reassortment when cells are infected with different viruses. Genome rearrangement also occurs within specific segments by recombination. Double-layered particles contain an RNA-dependent RNA polymerase and other enzymes capable of producing mRNA.

Six distinct groups (A through F) of rotaviruses have been reported to date [11]. The group specificity of a virus can usually be determined in serologic assays by the immunodominant group antigens on the inner capsid protein VP6 encoded by RNA segment 6. Only group A, B and C rotaviruses have been found in humans. Among them, group A virus is by far the most important based on its high prevalence and pathogenicity in humans and animals [8]. The outermost protein layer (or outer capsid) of group A rotaviruses consists of VP4 and VP7, which are independent neutralization antigens [4]. VP7, which is encoded by

Table 1. Gastroenteritis viruses in humans

Virus	Epidemiologic features	Clinical symptoms	Characteristics of Virus					
			Size (nm)	Buoyant density (g/ml)	Nucleic acid	Pathogenic in experimental animals	Replication in cell culture	Antigenic type (or genogroup)
Rotavirus	Sporadic seasonal occurrence; Infants and young children	Severe diarrhea (occasionally fatal)	ca 75	1.36	dsRNA segmented	Yes / Very young animals of various species	Yes	3 groups (A, B and C) group A: 7 G-serotypes and 4 P-serotypes
Calicivirus Norwalk Virus Hawaii Agent	Epidemic gastroenteritis	Generally mild diarrhea	SRSV (amorphous surface, ragged outline) 30–35	1.36–1.41	ssRNA nonsegmented	No	No	3 genogroups Genogroup I Genogroup II
Snow Mountain Agent	All age groups	Self-limited						Genogroup II
Human Calicivirus Sapporo			"Star of David" configuration 30–38					Genogroup III
Other Morphologically Typical Caliciviruses	Epidemic or sporadic							Many strains not yet typed
Astrovirus	Epidemic or sporadic	Generally mild	5- or 6-pointed surface star 27–30		ssRNA nonsegmented		Yes	At least 7 serotypes
Adenovirus	Epidemic or sporadic; No seasonality	Generally mild	ca 80	1.34	dsDNA nonsegmented		Yes	2 serotypes

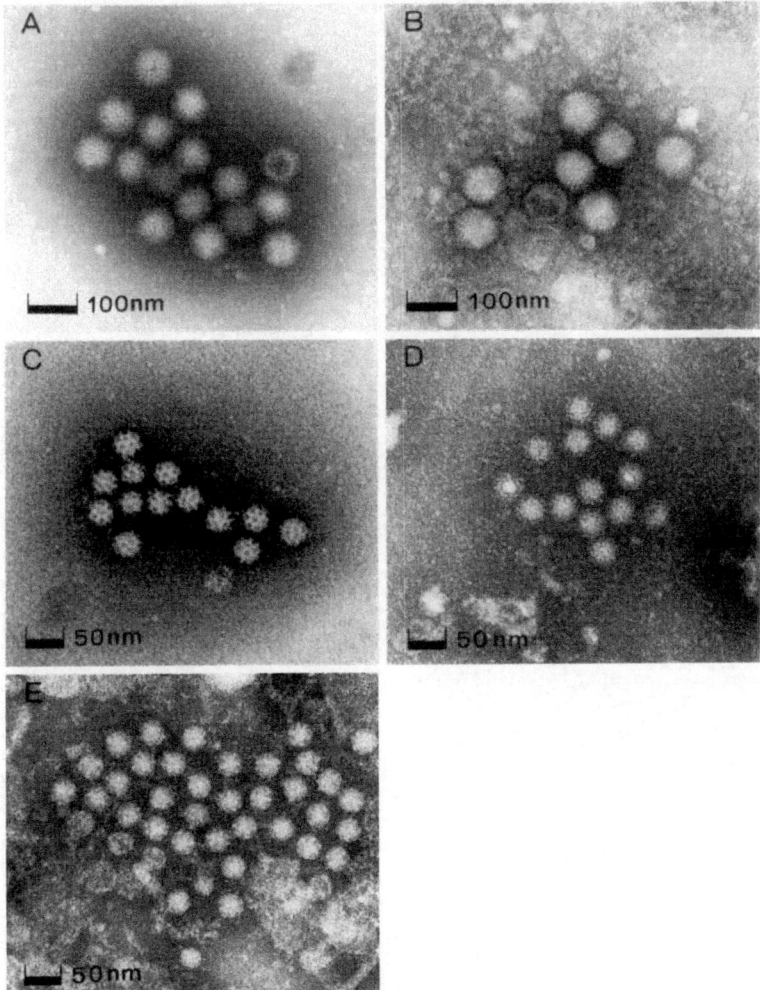

Fig. 1. Electron micrographs of five gastroenteritis viruses. A) Group A rotaviruses, B) Enteric adenoviruses, C) Classical caliciviruses (a "Star of David" configuration is apparent), D) Norwalk-like viruses or small round structured viruses (SRSVs) and E) Astroviruses. The samples were negatively stained with phosphotungstic acid. Bar = 100 nm for A) and B) and 50 nm for C), D) and E) (courtesy of S. Nakata)

RNA segment 7, 8 or 9 depending on the strain, determines the specificity of G (for glycoprotein) serotype. Fourteen G serotypes have been identified by cross-neutralization tests using type-specific hyperimmune serum; seven (G1–4, G8, G9 and G12) G serotypes have been found in humans. VP4, which is encoded by RNA segment 4, determines the specificity of P (for protease-sensitive protein) serotype. Four P serotypes and a pair of subtypes (P1A, P1B, P2, P3 and P4) have been found in prototype human rotaviruses [2].

Norwalk virus and other enteric caliciviruses [3] are the major causes of epidemic gastroenteritis, which usually occurs in family or community-wide

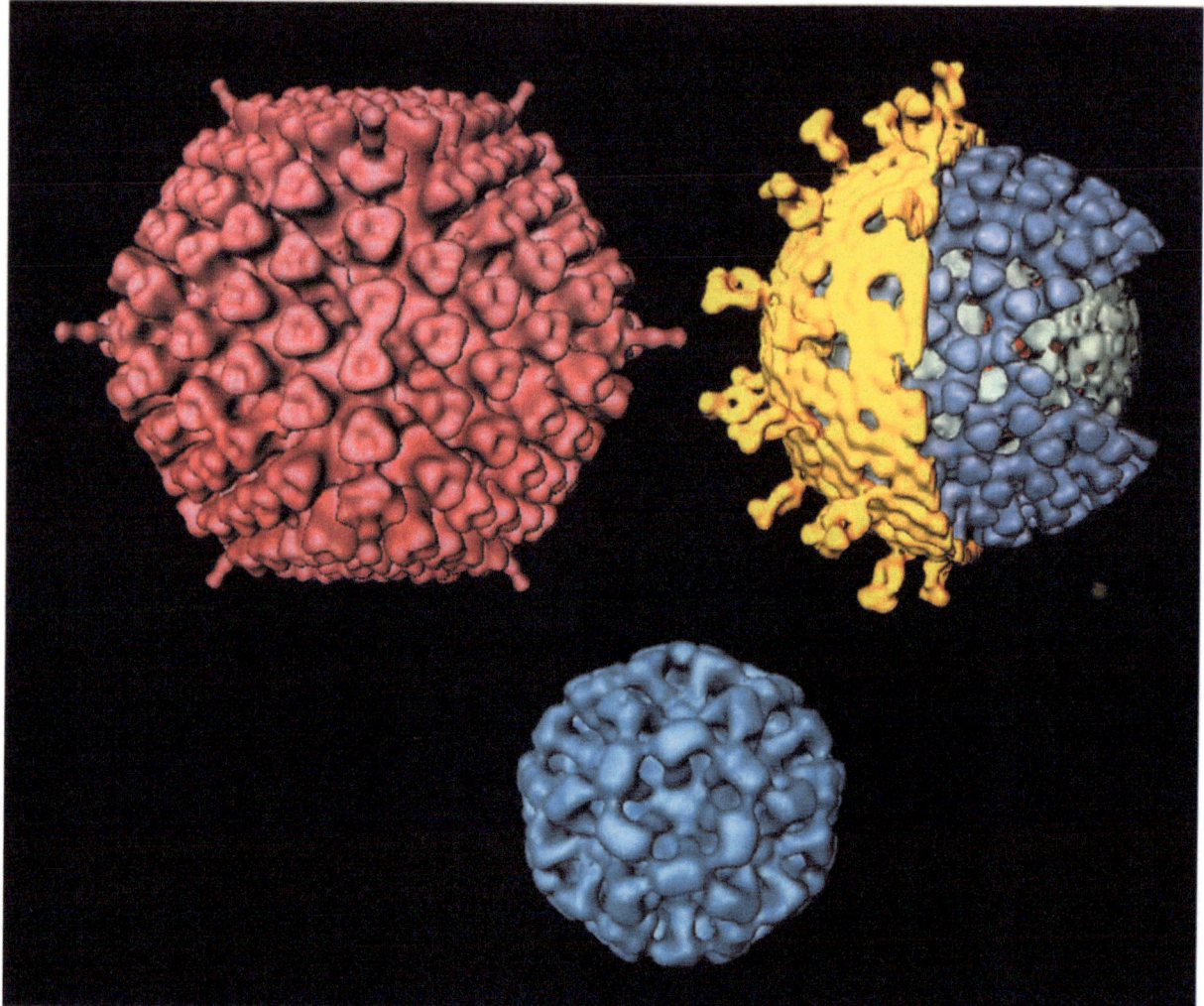

Fig. 2. Surface representation of the structure of three of the four gastroenteritis viruses discussed at the Sapporo International Symposium on Viral Gastroenteritis. These structures were obtained by electron cryo-microscopy and computer image processing of an adenovirus (top left, courtesy of R Burnett), a rotavirus (top right, courtesy of BVV Prasad) and Norwalk virus (bottom center, courtesy of BVV Prasad). The rotavirus image shows a cutaway of the triple layered particle and each protein shell is in a different color. The images shown are at a resolution of 35 angstroms (adenovirus), 28 angstroms (rotavirus) and 22 angstroms (Norwalk virus). The adenovirus structure was reported by Steward PL, Burnett RM, Cyrklaff M and Fuller SD (1991). Image reconstruction reveals the complex molecular organization of adenovirus. Cell 67: 145–154. See Chapters 3 and 25 in this Proceeding for details about rotavirus and Norwalk virus structure

outbreaks. Adults, school-age children and young children are affected. Clinical symptoms are usually mild and self-limited. Infections with these viruses do not show clear-cut seasonal patterns. Seroprevalence studies have shown that the caliciviruses are distributed worldwide; antibody acquisition begins slowly in childhood and continues throughout adulthood. The significance of these viruses in infantile diarrhea is being reevaluated by improved assays using recombinant Norwalk virus capsid antigen [5].

Since the detection of Norwalk virus in stools by immunolectron microscopy [7], a vast number of morphologically similar agents, including the Hawaii agent, the Snow Mountain agent, the Taunton agent, Minireovirus, and the Otofuke agent have been reported. These Norwalk-like viruses were initially classified based on morphology as a small round structured virus (SRSV). Besides these, virus particles having typical calcivirus morphology ("Star of David configuration") were also reported [3]. Until now, however, neither cell culture systems nor animal models have been found which allow the replication of these viruses. Recently, the Norwalk virus genome was successfully cloned from virus purified from stool and sequenced [6]. The results showed the virus genome to be a positive sense single-stranded RNA of 7 642 nucleotides which encodes three predicted open reading frames. A comparison of the overall genome organ- ization of Norwalk virus with those reported for established animal caliciviruses demonstrated Norwalk virus is a member of the family Caliciviridae [1].

Calicivirus particles measure 30–35 nm in diameter; a particle consists of a capsid made up of a single polypeptide and a single-stranded RNA genome [3]. Once the Norwalk virus sequence had been reported, it became possible to amplify numerous Norwalk-like viruses previously characterized as SRSVs with primers based on the Norwalk virus sequence (mainly of the polymerase region). Comparative sequence analysis of Norwalk, Norwalk-like viruses, and morphologically typical human caliciviruses disclosed that these viruses fall into one of three genogroups called genogroups I, II and III [3]. The availability of nucleotide sequence information that accumulated and baculovirus-expressed Norwalk virus capsid antigens enabled the development of highly sensitive diagnostic assays to detect both viruses and antibodies to them. These improved assays and the nucleotide sequence database are enabling rapid progress in epidemiologic and phylogenetic studies on human caliciviruses.

Astroviruses have been visualized by electron microscopy in fecal specimens from humans and animals[9]. Human astroviruses measure 27–30 nm in diameter and have a characteristic star-shaped outline, hence their name. While astroviruses usually cause epidemic or sporadic gastroenteritis in infants and young children, they are also found during outbreaks of gastroenteritis in adults. Astroviruses can be propagated in cell culture, and seven distinct serotypes have been reported[9]. The astrovirus genome is a positive-sense single-stranded RNA. Based on the recently acquired knowledge of the unique genome organization of this virus, astroviruses were classified in a new family, the Astroviridae [10]. Although astrovirus gastroenteritis had been regarded as relatively unimportant because of its relatively mild clinical symptoms and the low detection rate of virions in diarrhea specimens, recent epidemiologic studies using sensitive diagnostic assays with monoclonal antibody or RT-PCR appear to indicate that the illness is more clinically severe and prevalent than was presumed before.

Thirty-nine serotypes of human adenoviruses classified into five subgenera (A-E) have been isolated from patients with respiratory disease, keratoconjunctivitis and persistent kidney infection since 1983. There was also a group of adenoviruses which, though present in large number in stool specimens from diarrhea patients, could not grow in cell cultures. These noncultivatable viruses were designated enteric, or fastidious adenoviruses. Later, these viruses were found to grow in selected cell cultures and to consist of two new serotypes 40 and 41; restriction endonuclease analysis of their genomes indicated they represented a new subgenus F[12, 13]. Adenovirus serotype 31 has also been frequently isolated from pediatric diarrhea, implicating it as a causative agent of acute gastroenteritis. Enteric adenoviruses are associated with sporadic cases or outbreaks of acute gastroenteritis and are regarded as the most important viral pathogen, second to rotavirus, in pediatric gastroenteritis.

References

1. Cubitt D, Bradley DW, Carter MJ, Chiba S, Estes MK, Saif LJ, Schaffer FL, Smith AW, Studdert MJ, Thiel HJ (1995) Family Caliciviridae. Arch Virol Supp 10 (Virus Taxonomy) : 359–363
2. Estes MK (1996) Rotaviruses and their replication. In: Fields BN, Knipe DM, Howley PM, Chanock RM, Melnick JL, Monath TP, Roizman B, Straus SE (eds) Fields Virology, 3rd edn. Lippincott-Raven Publishers, Philadelphia, pp 1625–1655
3. Estes MK, Hardy ME (1995) Norwalk virus and other enteric caliciviruses. In: Blaser MJ, Smith PD, Ravdin JI, Greenberg HB, Guerrant RL (eds) Infections of the Gastrointestinal Tract. Raven Press, New York, pp 1009–1034
4. Hoshino Y, Sereno MM, Midthun K, Flores J, Kapikian AZ, Chanock RM (1985) Independent segregation of two antigenic specificities (VP3 and VP7) involved in neutralization of rotavirus infectivity. Proc Natl Acad Sci USA 82: 8701–8704
5. Jiang X, Wang M, Graham DY, Estes MK (1992) Expression, self-assembly, and antigenicity of the Norwalk virus capsid protein. J Virol 66: 6527–6532
6. Jiang X, Wang M, Wang KM, Estes MK (1993) Sequence and genomic organization of Norwalk virus. Virology 195: 51–61
7. Kapikian AZ, Wyatt RG, Dolin R, Thornhill TS, Kalica AR, Chanock RM (1972) Visualization by immune electron microscopy of a 27 nm particle associated with acute infectious nonbacterial gastroenteritis. J Virol 10: 1075–1081
8. Kapikian AZ, Chanock RM (1996) Rotaviruses. In: Fields BN, Knipe DM, Howley PM, Chanock RM, Melnick JL, Monath TP, Roizman B, Straus SE (eds) Fields Virology, 3rd edn. Lippincott-Raven Publishers, Philadelphia, pp 1657–1708
9. Mastui SM, Greenberg HB (1996) Astroviruses. In: Fields BN, Knipe DM, Howley PM, Chanock RM, Melnick JL, Monath TP, Roizman B, Straus SE (eds) Fields Virology, 3rd edn. Lippincott-Raven Publishers, Philadelphia, pp 811–824
10. Monroe SS, Carter MJ, Herrmann JE, Kurtz JB, Mastui SM (1995) Family Astroviridae. Arch Virol Suppl 10 (Virus Taxonomy): 364–367
11. Saif LJ, Jiang B (1994) Nongroup A rotaviruses of humans and animals. Curr Top Microbiol Immunol 185: 339–371
12. Van Loon A E, Rozijn TH, de Jong JC, Sussenbach JS (1985) Physico-chemical properties of the DNAs of the fastidious adenovirus species 40 and 41. Virology 140: 197–200
13. Wadell G (1984) Molecular epidemiology of adenoviruses. Curr Top Microbiol Immunol 110: 191–220

Author's address: Dr. Shozo Urasawa, Department of Hygiene, Sapporo Medical University School of Medicine, South-1 West-17, Chuoku, Sapporo 060, Japan.

Arch Virol (1996) [Suppl] 12: 7–19

Overview of viral gastroenteritis

A. Z. Kapikian

Epidemiology Section, Laboratory of Infectious Diseases, National Institute
of Allergy and Infectious Diseases, National Institutes of Health,
Bethesda, MD

Summary. Diarrheal illnesses in humans have been recognized since antiquity. Such illnesses continue to take a great toll of lives, with a disproportionately high mortality in infants and young children in developing countries. Bacteriologic and parasitologic advances made during the past century led to the discovery of the etiology of some of the diarrheal illnesses, but the etiology of the major portion remained unknown. It was assumed that viruses caused most of these illnesses because: (i) bacteria were recovered from only a small proportion of episodes, and (ii) bacteria-free filtrates were found to induce gastroenteritis in adult volunteer studies. However, an etiologic agent could not be recovered despite the "golden age" of virology in the 1950's and 1960's when tissue culture technology enabled the discovery of numerous cultivatable enteric viruses, none of which emerged as an important etiologic agent of gastroenteritis. The discoveries of the Norwalk virus in 1972, and of rotaviruses in 1973, both without the benefit of *in vitro* tissue culture systems, ushered in a new era in the study of the etiology of viral gastroenteritis. The Norwalk virus was found to be an important cause of non-bacterial epidemic gastroenteritis in adults and older children, and rotaviruses were shown to be the single most important etiologic agents of severe diarrheal illnesses of infants and young children in both developed and developing countries. With the major advances in the study of rotaviruses, there is a high degree of optimism that in the not-too-distant future, a rotavirus vaccine will be available. In addition, the recent molecular biologic advances in the study of the Norwalk and Norwalk-like viruses, now firmly established as caliviviruses, represent a major new horizon in the study of these viruses.

Introduction

Although diarrheal diseases have been recognized since ancient times, having been recorded on papyrus in hieroglyphic form about 3300 B.C., their etiology has remained elusive for the most part until recently [30]. This brief overview will describe the importance of diarrheal diseases in general, and present some of the advances made in recognition of viruses as major etiologic agents of diarrheal illnesses of all age groups, from a historical, epidemiological, virological, and vaccinological perspective.

Diarrheal diseases are important cause of morbidity in developed countries and a major cause of morbidity and mortality in developing countries world-wide. For example, during the Cleveland Family Study in the United States, infectious gastroenteritis (considered nonbacterial) was the second most common disease experience, accounting for 16% of ~ 25 000 illnesses over a period of about ten years [8]. This was translated into an average of 1.52 episodes of diarrhea per person per year, a figure that is remarkably consistent with that observed in two U.S. family studies 20 to 30 years later [12, 32]. However, deaths from diarrheal illnesses occur relatively infrequently in developed countries. For example, in the U.S., a review of vital statistics from 1973 to 1983 revealed that an average of 504 children one month to four years of age died annually from diarrheal illnesses [13].

In constrast, diarrheal diseases are a major cause of death among infants and young children in developing countries. Various estimates continue to stress the major impact of diarrheal diseases: (i) 15% to 34% of all deaths in certain countries are associated with diarrheal diseases, with the greatest toll in infants and young children [38]; (ii) ~ 450 million diarrheal episodes in infants and young children less than five years of age in Asia, Africa and Latin America, with a fatality rate of 1% to 4%, resulting in five to 18 million deaths annually [34]; (iii) three to five billion cases of diarrhea and five to ten million deaths from diarrhea during a one year period in these same regions, ranking diarrhea first

> **During the two days of The World Summit for Children, this is what will happen.**
>
> 2,800 children will die from whooping cough.
>
> 8,000 children will die from measles.
>
> 4,300 children will die from tetanus.
>
> 5,500 children will die from malaria.
>
> 22,000 children will die from diarrhoea.
>
> 12,000 children will die from pneumonia.
>
> **Now you know why there's a summit.**

Fig. 1. Adaptation of UNICEF poster for The World Summit for Children, United Nations, New York, September 29–30, 1990

among infectious diseases with regard to frequency and mortality [37]; (iv) 744 million to 1 billion diarrheal episodes and 4.6 million deaths due to diarrhea annually in children less than five years of age in Africa, Latin America and Asia (excluding China) [35]; (v) in an update of the latter study the mortality estimates were lowered to 3.3 million deaths with a range of 1.5 to 5.1 million per year [3]; and (vi) a UNICEF poster heralding a two-day World Summit for children in 1990, at which strategies were presented to improve the health of children globally, showed in dramatic fashion the importance of diarrheal diseases, as shown in Fig. 1. Although the figures in the various estimates cited above vary widely with regard to absolute numbers, each of them lead to the same conclusion, that diarrheal diseases are a major scourge of infancy and early childhood in developing countries.

History

Despite major bacteriologic, parasitologic, and virologic discoveries made during the past century, the etiology of most diarrheal illnesses remained unknown until the early 1970's. From the 1940's to the early 1970's, it was assumed by exclusion that viruses were the major cause because: (i) bacteria could be isolated from only a small proportion of such illnesses, and (ii) bacteria-free filtrates derived from community outbreaks of gastroenteritis induced diarrheal illness in adult volunteers in studies carried out 1945–1947 in the U.S. and Japan [21, 23]. Although bacteria-free filtrates were transmitted serially and characterized antigenically in cross-challenge studies, an etiologic agent could not be identified in any *in vitro* system. This was especially difficult to understand in the 1950's, the "golden age" of virology, when hundreds of new viruses were discovered using newly described tissue culture technology. Initially, the isolation of many distinct enteric viruses (*e.g.*, echoviruses) from infants with diarrhea was particularly encouraging, but when carefully controlled epidemiologic studies revealed that children without diarrhea shed these agents with equal frequency, their role as important etiologic agents of diarrheal illness was voided [21].

Further efforts were made in the 1960's when newer techniques such as organ culture were used successfully for the discovery of fastidious respiratory viruses, such as coronaviruses [6]. Following this lead, fecal specimens derived from outbreaks of diarrheal illness were studied in human embryonic intestinal organ culture (with the rationale that such cells were less dedifferentiated than cells passaged in tissue culture) but once again a viral agent could not be recovered [6]. It was considered that perhaps this failure was due to the absence of an infectious agent in the outbreak specimens tested. Known infectious stool filtrates from the volunteer studies of the 1940's and 1950's described above had been exhausted or were not available for study.

Therefore, adult volunteer studies were revived in the 1970's to generate known infectious stool filtrates for further study. Several gastroenteritis outbreaks were evaluated in challenge studies but one deserves special attention. A sharp outbreak of acute gastroenteritis which affected 50% of the students and

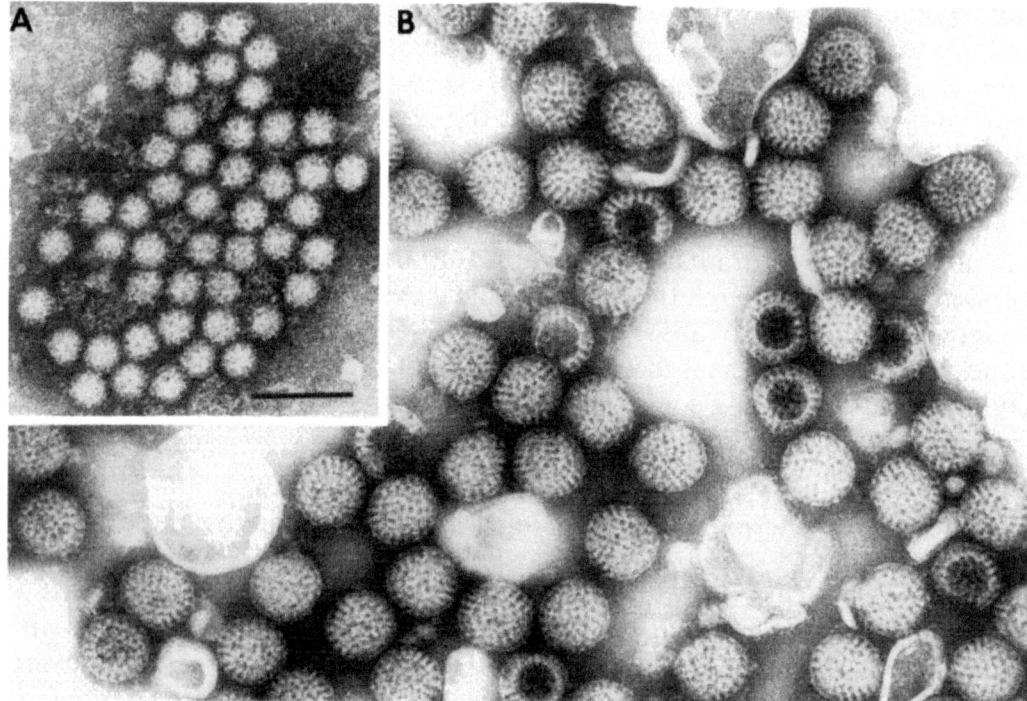

Fig. 2. **A** The 27-nm Norwalk virus observed by immune electron microscopy in the stool of an adult volunteer with gastroenteritis (from Kapikian *et al.* [22] [bar added]). **B** The 70-nm rotavirus observed by electron microsopy in the stool of an 11-month-old infant with gastroenteritis (from Kapikian *et al.* [23]. Bar: 100 nm (applies to A and B)

teachers in an elementary school in Norwalk, Ohio in 1968 failed to yield an etiologic agent [1]. Thus, with the aforementioned goal in mind, a bacteria-free filtrate which was prepared from a rectal swab specimen obtained from a secondary case was given orally to three adult volunteers, two of whom developed gastroenteritis one to two days after challenge (one had vomiting without diarrhea and the other had diarrhea without vomiting) [9]; serial passages in human volunteers were successful and the infectious Norwalk stool filtrate was studied for size (by filtration), acid and heat stability, and was evaluated in cross-challenge studies with other stool filtrates from other outbreaks [6, 9]. Because all efforts to propagate a viral agent *in vitro* were still unsuccessful, the indicator of infectivity remained the health-status of the volunteer following challenge. Thus, despite the more recent round of volunteer studies, the etiology of viral gastroenteritis was essentially at the same juncture as in the 1940's and 1950's.

However, in 1972, an important breakthrough was made when 27 × 32 nm (shortest × longest diameter) virus-like particles were discovered directly in an infectious stool filtrate derived from the Norwalk outbreak by the use of the technique of immune electron microscopy (IEM) (Fig. 2A) [26]. This method permits the direct visualization of antigen-antibody interaction, thus facilitating or enabling the recognition and identification of a particle which: (i) does not have a distinctive morphologic appearance; (ii) is present in low titer, and (iii) is among the smallest known viruses [24]. This is accomplished by the reaction of a test specimen (*e.g.*, stool filtrate) which may or may not contain virus, with a putative specific antiserum. If the homotypic virus is present, it will be coated by specific antibodies and, depending on the antigen-antibody ratio, the immune complex will appear in the form of large or small viral aggregates or as individual entities so heavily coated with antibody that they fail to aggregate because of steric hindrance. The Norwalk virus was discovered after incubation of the infectious Norwalk stool filtrate (designated 8FIIa for laboratory purposes) with a volunteer's convalescent phase serum. Serologic evidence of infection with the Norwalk virus was demonstrated by IEM in paired pre- and post-challenge sera of four of the ill volunteers and in acute and convalescent phase sera from certain individuals from the original outbreak; from this and other evidence it was suggested that the 27 nm virus-like particles were the etiologic agents of the outbreak in Norwalk, Ohio [26]. The 27 nm Norwalk virus is considered to be the first virus to be identified as an important etiologic agent of viral gastroenteritis in humans. It represents the prototype strain of a group of 27 nm viruses that have still defied cultivation in any cell culture system. The technique of IEM has led to the discovery of other Norwalk-related gastroenteritis viruses such as the Montgomery County, Hawaii and Snow Mountain viruses [21, 23]. These gastroenteritis viruses are usually named after the location of the outbreak from which they were derived.

One year later, another important breakthrough occurred when the 70 nm rotaviruses were discovered by thin section electron microscopy in biopsies of duodenal mucosa in two of four children under three years of age with acute gastroenteritis [4]. It was suggested that the virus-like particles were the cause of the gastroenteritis in these two children. These EM studies were prompted by prior light microscopic investigations which had demonstrated histologic changes in duodenal mucosa of children hospitalized with acute nonbacterial gastroenteritis [2]. Shortly after the initial virus positive biopsies were reported, rotaviruses were visualized by EM in additional duodenal biopsies and soon thereafter also in feces of infants and young children hospitalized with diarrhea [5, 22]. In a relatively short time, laboratories around the world reported the presence of rotavirus particles in feces of a major segment of infants and young children with diarrheal illness and it became apparent from cross-sectional studies that rotaviruses were indeed the long sought-after major etiologic agents of severe diarrhea of this age group (Fig. 2B) [21, 23]. Thus, following a long fallow period, within a span of one year (reports in Nov., 1972 and Nov., 1973, respectively) the Norwalk virus, a major cause of epidemic gastroenteritis of adults and older children, and rotaviruses, the major cause of severe gastroenteritis of infants and young children, were discovered, paving the way for many of the studies presented in this issue.

Norwalk and Norwalk-like viruses

The Norwalk virus is the prototype strain of a group of nonenveloped 27 nm viruses associated with epidemic gastroenteritis. These viruses: (i) have resisted all attempts to propagate them in tissue culture; (ii) are shed in feces; (iii) produce illness only in humans; (iv) infect only chimpanzees among numerous animal models studied; (v) have a positive-sense single-stranded RNA genome; and (vi) have a buoyant density of 1.33 to 1.41 g/cm^3 in CsCl$_2$ [23]. By EM, they display a feathery outer edge and do not have a definitive surface substructure, although in certain orientations they have suggestive surface indentations, somewhat reminiscent of caliciviruses (Fig. 2A). There are at least four distinct serotypes as demonstrated by IEM or SPIEM: these are designated Norwalk, Hawaii, Snow Mountain, and Taunton viruses according to the location from which each is derived [23]. Related "27 nm" viruses include the Montgomery County, Southampton, Desert Storm, Toronto, and Otofuke viruses [23].

Another major advance was achieved in 1990 when the Norwalk virus genome was cloned and the capsid protein expressed in a baculovirus system, forming virus-like particles [17]. Subsequently, as expression of other members of this group, followed by sequencing studies gained momentum, knowledge regarding the genetic relatedness of these viruses became available [23]. Although the Norwalk group does not display the distinctive cup-like surface indentations of the classical caliciviruses (calix = cup in Latin), on the basis of their genome organization, the Norwalk and related viruses are classified in the genus *Calicivirus* in the family *Caliciviridae*. It should be noted that other noncultivatable small round structured viruses (SRSVs) have been visualized in stools of pediatric or elderly patients with gastroenteritis, some of which have the "classical" calicivirus morphologic appearance by EM; their role as important etiologic agents of epidemic or sporadic gastroenteritis has not been established [23].

The Norwalk and related viruses are transmitted by the fecal-oral route; it is possible that airborne transmission occurs under special circumstances [21, 23]. The virus has also been visualized in vomitus [10]. The Norwalk and related viruses are major etiologic agents of acute epidemic nonbacterial gastroenteritis that affects adults, school-aged children and family contacts but they do not have an important role in the etiology of diarrhea of infants and young children [21, 23]. These outbreaks occur in all seasons of the year in various settings such as cruise ships, camps, schools, institutions, and families. Foods such as raw oysters, cake frosting, salads (as well as water and commercial ice) have been incriminated as the vehicle of transmission. In systematic studies of outbreaks, most of which were selected because they were non-bacterial, the Norwalk virus was associated with 42% of 74 that occurred from 1976 to 1980 [18]. In addition, it was estimated that almost 10% of 558 unselected outbreaks of gastroenteritis (occurring from 1975–1981) were likely caused by the Norwalk virus [27]. The prevalence of serum antibody to Norwalk virus is quite different in developed and developing countries: antibody acquisition proceeds more

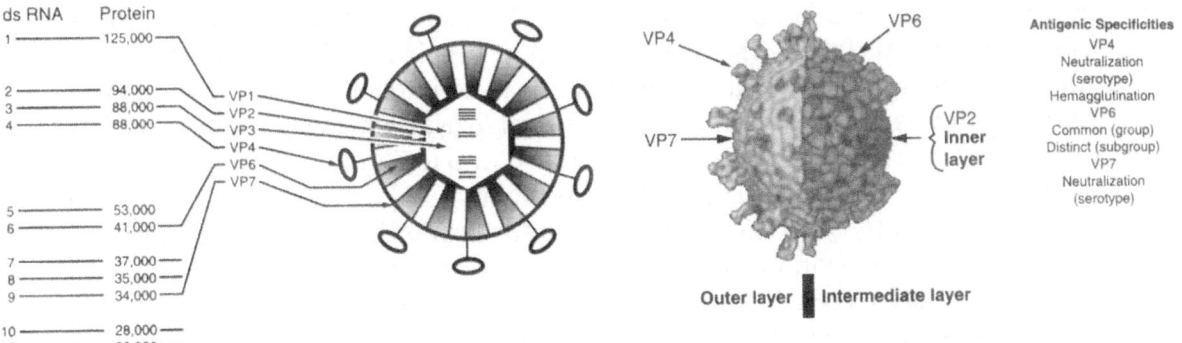

Fig. 3. Left: Schematic representation of the rotavirus particle. Right: Surface representations of the three-dimensional structures of the outer layer of the complete particle (left) and a particle (right) in which the outer layer and a small triangular portion of the intermediate layer have been removed exposing the inner layer. Modified from Kapikian and Chanock, [20]. The three-dimensional figure on the right is from courtesy of B.V.V. Prasad

slowly in former than in the latter countries, probably reflecting differences in sanitary conditions [21, 23]. Immunity to Norwalk virus remains a perplexing issue because susceptibility correlates inversely with serum or local jejunal antibodies [21, 23].

The Norwalk and related viruses do not play an important role in severe infantile diarrhea. They may be associated with mild gastroenteritis in infants and young children in certain settings [21, 23]. Unexplained at this time is the surprisingly high rate of Norwalk virus infection observed in infants and young children (49% of 154) in Finland over a period of almost two years, when sequential sera were studied by an enzyme immunoassay using the Norwalk virus outer capsid virus-like particle as antigen. However, these infections could not be associated retrospectively with illness [31].

Rotaviruses

Rotaviruses are classified as genus *Rotavirus* in the family *Reoviridae*. They are 70 nm in diameter, non-enveloped, and possess a distinctive double-shelled icosahedral outer shell (Fig. 2B). Within the inner shell is a third layer, the core, which encloses the virus genome, comprised of eleven segments of double-stranded RNA (Fig. 3) [22]. Rotaviruses possess three important antigenic specificities: group, subgroup and serotype. Group specificity is mediated predominantly by VP6 (encoded by gene 6); seven groups (A–G) are recognized with groups A, B and C found in humans and all seven groups in animals [14, 22]. Subgroup specificity is also determined by VP6 (encoded by gene 6); most human rotaviruses belong to subgroup I or II. Serotype specificity has been defined until recently exclusively by VP7 (encoded by gene 7, 8 or 9 depending on the strain) [14]. VP7, a glycoprotein, is one of two major neutralization antigens located on the outer shell. The other outer shell protein VP4 (encoded by gene 4) protrudes from the outer surface in the form of 60 spikes each 10–12 nm in length [33]. Antibodies to VP7 or VP4 are independently associated with protection against rotavirus illness [14]. The human group A rotaviruses, which are the most widely distributed rotaviruses, are further subdivided into ten human serotypes according to VP7 (or "G" [for glycoprotein]) specificity [14]. G serotypes 1, 2, 3, and 4 have consistently been shown to be the only strains of epidemiologic importance and for this reason vaccines are aimed at preventing severe diarrhea caused by these four serotypes. A serotyping system which also recognizes VP4 (or "P" [for protease sensitivity]) has been described recently and different numbers have been assigned according to neutralization specificity or genotype, giving rise to considerable confusion. A unified numbering system is under development but serotype numbers according to neutralization should have precedence to conform to standards established for other viruses [14].

Rotaviruses are the single most important etiologic agents of severe diarrhea of infants and young children world-wide [21, 22]. Infections with these agents are widespread in both developed and developing countries as evidenced by the prevalence of serum antibodies in ∼90% of infants and young children by three years of age. Although rotavirus infections are known to be transmitted by the fecal-oral route, the extremely high rate of infection in both developed and developing countries, regardless of sanitary conditions, has led to speculation that respiratory transmission might also occur [21, 22]. The consequences of rotavirus infections are dramatically different in these different settings. For example, it is estimated that: (i) three million infants and young children develop rotavirus diarrhea yearly in the U.S.; (ii) 82 000 are admitted to the hospital with rotavirus diarrhea; and (iii) rotavirus diarrhea causes about 150 deaths [15]. In contrast, in developing countries, the toll is staggering. Estimates indicate that each year 18 million cases of moderately severe or severe rotavirus diarrhea occur in children under five years of age and that more than 870 000 children in this age group die because of rotavirus diarrhea [16].

In cross sectional studies in developed countries, rotaviruses are responsible for ∼35 to 52% of acute diarrheal illnesses that require hospital admission of infants and young children. For example, at the Children's Hospital National Medical Center in Washington, DC, in a period of > 8 years, 34.5% of 1537 infants and young children hospitalized with diarrhea shed rotavirus in stool specimens (Fig. 4) [7]. In a similar study in Japan, 45% of 1910 infants and young children admitted with diarrhea shed rotavirus in a period > 6 years [29]. Rotavirus illnesses occur almost exclusively during the cooler seasons of the year in temperate climates. In the tropics, rotavirus diarrhea occurs throughout the year with less pronounced peaks. Rotavirus diarrhea occurs most frequently in children between six and 24 months of age, with the next highest frequency being in the < 6 month age group [21, 22]. Neonates experience a low rate of rotavirus diarrhea, even though in certain nurseries infections may occur frequently. Adults develop subclinical infections frequently, usually without clinical manifestations [22]. Rotavirus gastroenteritis has been observed in elderly residents

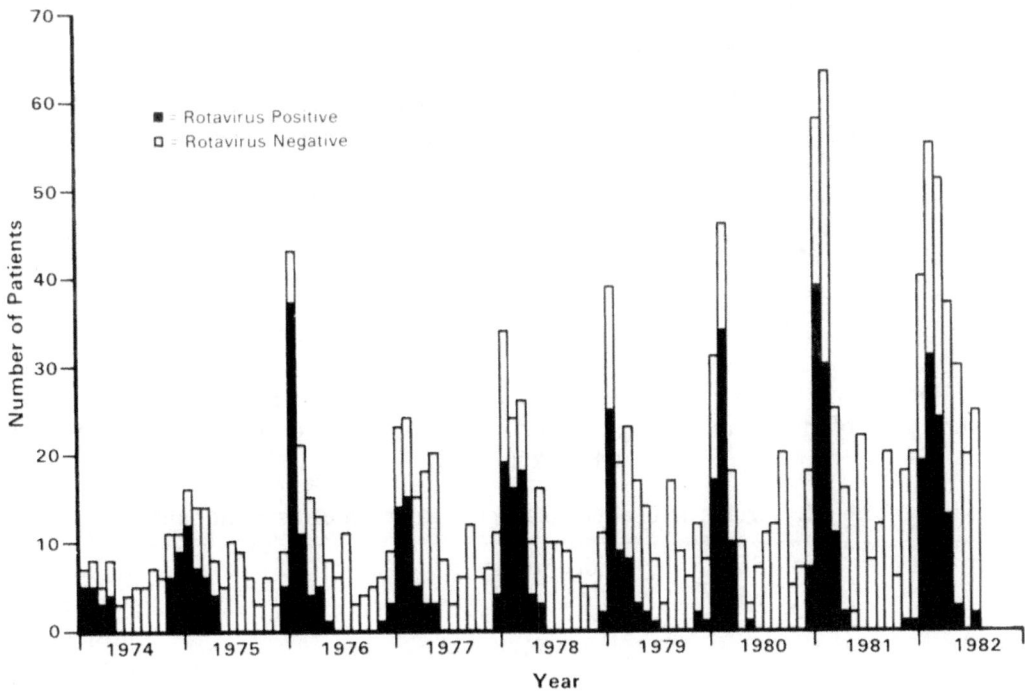

Fig. 4. Rotavirus infections in patients with gastroenteritis January 1974–July 1982 (as demonstrated by EM, IEM and rotavirus confirmatory ELISA) (from Brand et al. [7])

of nursing homes with several fatalities reported, but these viruses do not appear to be a major cause of morbidity in such settings [22]. The mechanism of immunity to human rotaviruses remains controversial, although in certain studies serum antibodies correlate with protection [21, 22].

Because rotaviruses are an important cause of severe illness and hospitalization in infants and young children in developed countries and a major cause of mortality in this same group in developing countries, the development of a rotavirus vaccine has received high priority from the international community [14, 21, 22]. Major efforts are aimed at developing a live, attenuated oral vaccine for administration during early infancy when the clinical manifestations of rotavirus illness are most severe. The most widely studied approach has been the "Jennerian" and "modified Jennerian" strategy in which an animal rotavirus that is antigenically related to group A rotaviruses, and human rotavirus-animal rotavirus reassortants are combined to form a four component vaccine that has VP7 antigenic specificity for each of the four epidemiologically important serotypes [19]. This quadrivalent vaccine is comprised of a rhesus rotavirus representing VP7 serotype 3, and three reassortant rotaviruses each containing ten rhesus rotavirus genes and a single human rotavirus gene that codes for VP7 serotype 1, 2, or 4 specificity [14, 20, 22]. Another vaccine, which is bovine rotavirus-based, is also under evaluation [36]. These vaccines are described later in this issue.

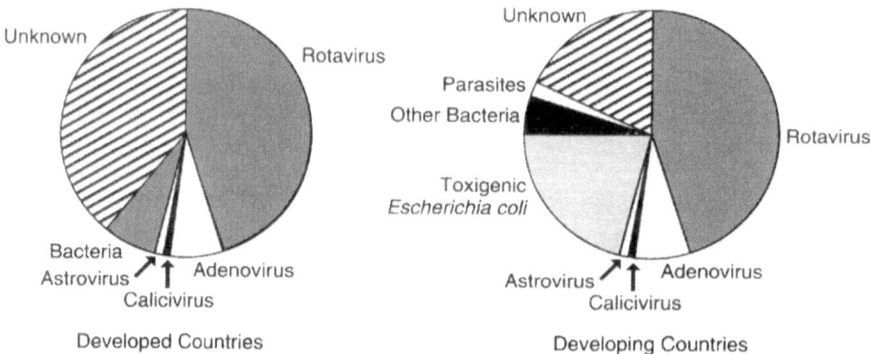

Fig. 5. An estimate of the role of etiological agents in severe diarrheal illnesses requiring hospitalization of infants and young children in developed and developing countries (from Kapikian [18])

Table 1. Viruses associated with acute gastroenteritis in humans[a]

Virus	Size, nm	Epidemiology	Important as a cause of hospitalization
Rotavirus Group A	70	Single most important cause (viral or bacterial) of endemic severe diarrheal illness in infants and young children worldwide (in cooler months in temperate climates)	Yes
Group B	70	Outbreaks of diarrheal illness in adults and children in China	No
Group C	70	Sporadic cases and occasional outbreaks of diarrheal illness in children	No
Enteric adenovirus	70–80	Second most important viral agent of endemic diarrheal illness of infants and young children worldwide	Yes
Norwalk virus and Norwalk-like viruses	27–32	Important cause of outbreaks of vomiting and diarrheal illness in older children and adults in families, communities, and institutions, frequently associated with ingestion of food	No
Caliciviruses	28–40	Sporadic cases and occasional outbreaks of diarrheal illness in infants, young children, and the elderly	No
Astroviruses	28	Sporadic cases and occasional outbreaks of diarrheal illness in infants, young children, and the elderly	No

[a] From Kapikian, Ref. 18

Other agents

Various other viruses, such as adenoviruses, astroviruses, and classical caliciviruses, have been described as etiologic agents of diarrhea of infants and young children [21, 23]. Enteric adenoviruses are considered to be the second most important group of viruses associated with severe diarrheal illness of infants and young children requiring hospitalization, with estimates ranging from 3–10% [11]. The morphologically classical caliciviruses, and the astroviruses have not been shown to be important causes of severe gastroenteritis of infants and young children [21]. Toxigenic *E. coli* are an important cause of diarrhea of infants and young children in developing countries [18]. As determined from published data, an estimate of the role of various microbial agents in severe diarrhea in developed and developing countries is shown in Fig. 5 [18]. In addition, Table 1 summarizes various salient points described in this brief review of etiologic agents of viral gastroenteritis. Other viral agents, such as the toroviruses and coronavirus-like objects observed by EM in stools, await further study before they can be evaluated as etiologic agents of diarrhea in humans [21].

Conclusion

Major advances have been made over the past two decades in elucidating the etiologic agents of viral gastroenteritis. For one of these, the rotaviruses, an effective vaccine may be available soon. However, as seen in Fig. 5, there is still a fairly substantial void in our knowledge regarding the etiology of severe diarrhea of infants and young children. Perhaps, by the time of the next international symposium on viral gastroenteritis, the etiologic "pie" will have neared completion.

References

1. Adler I, Zickl R (1969) Winter vomiting disease. J Infect Dis 119: 668–673
2. Barnes GL, Townley RRW (1973) Duodenal mucosal damage in 31 infants with gastroenteritis. Arch Dis Child 48: 343–349
3. Bern C, Martines J, De Zoysa I, Glass RI (1992) The magnitude of the problem of diarrhoeal disease: A ten-year update. Bull WHO 70: 705–714
4. Bishop RF, Davidson GP, Holmes IH, Ruck BJ (1973) Evidence for viral gastroenteritis. N Engl J Med 289: 1096–1097
5. Bishop RF, Davidson GP, Holmes IH, Ruck BJ (1973) Virus particles in epithelial cells of duodenal mucosa from children with viral gastroenteritis. Lancet 2: 1281–1283
6. Blacklow NR, Dolin R, Fedson DL, Dupont H, Northrup RS, Hornick RB, Chanock RM (1972) Acute infectious nonbacterial gastroenteritis: etiology and pathogenesis. Ann Intern Med 76: 993–1008
7. Brandt CD, Kin HW, Rodriguez WJ, Arrobio JO, Jeffries BC, Stallings EP, Lewis C, Miles AJ, Chanock RM, Kapikian AZ, Parrott RH (1983) Pediatric viral gastroenteritis during eight years of study. J Clin Microbiol 18: 71–78
8. Dingle JH, Badger GF, Jordan WS (1964) Illness in the home: a study of 25 000 illnesses in a group of Cleveland Families. Western Reserve University Press (1964), Cleveland, Ohio, pp 19–32

9. Dolin R, Blacklow NR, DuPont H, Formal S, Buscho RF, Kasel JA, Chames RP, Hornick R, Chanock RM (1971) Transmission of acute infectious nonbacterial gastroenteritis to volunteers by oral administration of stool filtrates. J Infect Dis 123: 307–312

10. Greenberg HB, Wyatt RG, Kapikian AZ (1979) Norwalk virus in vomitus. Lancet 2: 55

11. Grimwood K, Carzino R, Barnes GL, Bishop RF (1995) Patients with enteric adenovirus gastroenteritis admitted to an Australian pediatric teaching hospital from 1981 to 1992. J Clin Microbiol 33: 131–136

12. Guerrant RL, Hughes JM, Lima NL, Crane J (1990) Diarrhea in developed and developing countries: magnitude, special settings, and etrologies. Rev Inf Dis 12: S41–S50

13. Ho M-S, Glass RI, Pinsky PF, Young-Okoh N, Sappenfield WM, Buehler JW, Gunter N, Anderson LJ (1988) Diarrheal deaths in American children – are they preventable? JAMA 260: 3281–3285

14. Hoshino Y, Kapikian AZ (1994) Rotavirus vaccine development for the prevention of severe diarrhea in infants and young children. Trends in Virology 2: 242–249

15. Institute of Medicine. Prospects for immunizing against rotavirus. In: New Vaccine Development. Establishing Priorities. Diseases of Importance in the United States, vol 1. National Academy Press, Washington, DC, pp 410–423, 1985

16. Institute of Medicine. The prospects for immunizing against rotavirus. Establishing Prioritis. Diseases of Importance in Developing Countries, vol 2. National Academy Press, Washington, DC, pp 308–316, 1985

17. Jiang X, Graham DY, Wang K, Estes MK (1990) Norwalk virus genome cloning and characterization. Science 250: 1580–1583

18. Kapikian AZ (1993) Viral gastroenteritis. JAMA 269: 627–630

19. Kapikian AZ (1994) Jennerian and modified Jennerian approach to vaccination against rotavirus diarrhea in infants and young children: an introduction. In: Kapikian AZ (ed) Viral Infections of the Gastrointestinal Tract. Marcel Dekker, NY, pp 409–417

20. Kapikian AZ (1994) Rhesus rotavirus-based human rotavirus vaccines and observations on selected non-Jennerian approaches to rotavirus vaccination. In: Kapikian AZ (ed) Viral Infections of the Gastrointestinal Tract. Marcel Dekker, NY, pp 443–470

21. Kapikian AZ, Viral gastroenteritis. In: Evans AS, Kaslow R (eds) Viral infections of Humans, 4th edn (in press)

22. Kapikian AZ, Chanock RM (1996) Rotaviruses. In: Fields BN, Knipe DM, Howley PM, Chanock RM, Melnick JL, Monath TP, Roizman B, Straus SE (eds) Virology, vol 2. Lippincott-Raven Publishers, Philadelphia, pp 1657–1708

23. Kapikian AZ, Estes MK, Chanock RM (1996) Norwalk group of viruses. In: Fields BN, Knipe DM, Howley PM, Chanock RM, Melnick JL, Monath TP, Roizman B, Straus SE (eds) Virology pp 783–810

24. Kapikian AZ, Feinstone SM, Purcell RH, Wyatt RG, Thornhill TS, Kalica AR, Chanock RM (1975) Detection and identification by immune electron microscopy of fastidious agents associated with respiratory illness, acute nonbacterial gastroenteritis, and hepatitis A. Perspect Viral 9: 9–47

25. Kapikian AZ, Kim HW, Wyatt RG, Rodriguez WJ, Ross S, Cline WL, Parrott RH, Chanock RM (1974) Reovirus-like agent in stools: association with infantile diarrhea and development of serologic tests. Science 185: 1049–1053

26. Kapikian AZ, Wyatt RG, Dolin R, Thornhill TS, Kalica AR, Chanock RM (1972) Visualization by immune electron microscopy of a 27 nm particle associated with acute infectious non-bacterial gastroenteritis. J Virol 10: 1075–1081

27. Kaplan JE, Feldman R, Campbell DS, Lookabaugh C, Gary GW (1982) The frequency of a Norwalk-like pattern of illness in outbreaks of gastroenteritis. Am J Public Health 72: 1329–1332

28. Kaplan JE, Gary GW, Baron RC, Singh N, Schonberger LB, Feldman R, Greenberg HB (1982) Epidemiology of Norwalk gastroenteritis and the role of Norwalk virus in outbreaks of acute nonbacterial gastroenteritis. Ann Intern Med 96: 756–761
29. Konno T, Suzuki H, Katsushima N, Imai A, Tazawa F, Kutzuzawa T, Kitaoka S, Sakamoto M, Yazaki N, Ishida N (1983) Influence of temperature and relative humidity on human rotavirus infection in Japan. J Infect Dis 147: 125–128
30. Kumate J, Isibasi A (1986) Pediatric diarrheal diseases: a global perspective. Pediatr Infect Dis 5: 521–528
31. Lew JF, Valdesuso J, Vesikari T, Kapikian AZ, Jiang X, Estes MK, Green KY (1994) Detection of Norwalk virus or Norwalk-like virus infections in Finnish infants and young children. J Infect Dis 169: 1364–1367
32. Monto AS, Koopman JS (1980) The Tecumseh study. XI. Occurrence of acute enteric illness in the community. Am J Epidemiol 112: 323–333
33. Prasad BV, Burns JW, Marietta E, Estes MK, Chiu W (1990) Localization of VP4 neutralization sites in rotavirus by three-dimensional cryo-electron microscopy. Nature 343: 476–479
34. Rohde JE, Northrop RS (1976) Taking science where the diarrhea is. In: Acute diarrhoea in childhood. Ciba Foundation Symposium 42 (new series). Amsterdam, Excerpta Medica 339–366
35. Snyder JD, Merson MH (1982) The magnitude of the global problem of acute diarrhoeal disease: a review of active surveillance data. Bull WHO 60: 605–613
36. Vesikari T (1994) Bovine rotavirus-based rotavirus vaccines in humans. In: Kapikian AZ (ed) Viral Infections of the Gastrointestinal Tract. Marcel Dekker, New York, pp 419–442
37. Walsh JA, Warren KS (1979) Selective primary health care. An interim strategy for disease control in developing countries. N Engl J Med 301: 967–974
38. World Health Organization (1973) Mortality due to diarrheal diseases in the world. WHO Weekly Epidemiol Rev 48: 409–416

Author's address: A. Z. Kapikian, M. D., Epidemiology Section, Laboratory of Infectious Diseases, National Institute of Allergy and Infectious Diseases, National Institutes of Health, Bethesda, Maryland 20892, U.S.A.

Arch Virol (1996) [Suppl] 12: 21–27

© Springer-Verlag 1996

Rotavirus structure: interactions between the structural proteins

A. L. Shaw[1], R. Rothnagel[1], C. Q.-Y. Zeng[2], J. A. Lawton[3], R. F. Ramig[2], M. K. Estes[2], and B. V. Venkataram Prasad[1]

[1] Verna and Marrs McLean Department of Biochemistry and the W.M. Keck Center for Computational Biology; [2] Division of Molecular Virology [3] Program in Cell and Molecular Biology, Baylor College of Medicine, Houston, Texas, U.S.A.

Summary. Structural studies on rotavirus using electron cryomicroscopy and computer image analysis have permitted visualization of each shell in the triple-layered rotavirus structure. Biochemical results have aided our interpretation of the structural organization of these layers and protein interactions seen in the three-dimensional structure, and have provided a better understanding of the structure-function relationships of the rotavirus structural proteins.

Introduction

Rotaviruses are recognized as the most significant causes of infantile viral gastroenteritis worldwide [7]. In recent years, rotaviruses have been studied by three-dimensional structural analyses using electron cryomicroscopy and computer image processing [11–16, 20, 21]. These studies have provided a detailed structural description of this virus, including the topographical locations of the major structural proteins and their stoichiometric proportions.

Belonging to the family *Reoviridae*, genus *Rotavirus*, rotaviruses are characterized by a segmented double-stranded RNA genome and a non-enveloped multilayered protein capsid [6]. The rotavirus genome consists of 11 segments of double-stranded RNA, each of which encodes one protein. Of the 11 rotaviral proteins, 6 are structural and 5 are nonstructural. The 6 structural proteins are arranged into 3 concentric layers which form the protein capsid. The overall diameter of the mature triple-layered rotavirion is $\sim 1000\,\text{Å}$. The outermost layer is composed of two proteins, VP7 and VP4. VP7, a glycoprotein with an apparent molecular weight of 38 000 (38 K), is the major component of the outer layer. VP7 is present in 780 copies which are arranged on a $T = 13$ left-handed icosahedral lattice. The VP7 layer is perforated with 132 aqueous channels located at all the 5- and 6-coordinated positions of the $T = 13$ icosahedral lattice. The minor component of the outer shell, VP4 is an $\sim 88\,\text{K}$ hemagglutinin protein. VP4 is present in 120 copies and forms spikes that project outward from the VP7 shell. Each spike is composed of a homodimer of VP4 [13, 16], hence, 60 spikes. Treatment of triple-layered particles (TLPs) with chelating agents removes the outer layer and yields double-layered particles (DLPs). The middle protein layer has a diameter of ~ 700 Å and is composed of a $\sim 45K$ protein, VP6. VP6 is the major structural protein in the rotavirus capsid. As is VP7, VP6 is present in 780 copies, assembled on a $T = 13$ lattice. The VP6 layer is also perforated with 132 aqueous channels that lie in register with those in the VP7 layer and continue through to the inner shell. The third, innermost, layer is 520 Å in diameter and is composed of a $\sim 102\,K$ protein, VP2. The VP2 layer houses the RNA genome and several copies each of two large structural proteins, VP1 ($\sim 125\,K$) and VP3 ($\sim 88\,K$).

We are interested in understanding the structural mechanisms underlying each step of rotavirus infection, such as cell entry, transcription, replication, assembly and maturation. Understanding these mechanisms should greatly aid the development of antirotaviral strategies. We report here our current understanding of the various protein-protein interactions in the structure of rotaviruses.

Method

Structural analyses of rotaviruses have been carried out using electron cryomicroscopy and computer image processing techniques. Advances over recent years in the methodology of electron cryomicroscopy and computer image processing have provided us with powerful tools for determining 3-D structures of proteins and macromolecular assemblies.

Specimen preparation for electron cryomicroscopy is well established [5]. A specimen is embedded in a thin layer of vitreous ice by rapidly plunging the specimen grid into liquid ethane near its freezing point. The specimen grid is examined in the electron microscope at near liquid nitrogen temperatures ($-160\,°C$) using low electron dose techniques. The specimen preparation does not require any stain or fixatives and, therefore, prevents dehydration and dessication. Thus, the specimen is preserved in its hydrated state. The image contrast in the absence of stains or metal decoration is caused by the scattering density difference between protein and ice. One of the major advantages of this technique is that it allows structural studies of the specimen not only in its native state, but also under various other physiological conditions. For example, structures of specimens bound to an antibody [13, 19], bound to a receptor [10], or in varied chemical states such as different pH and ionic strengths [17, 18], can be determined.

The electron cryomicrographs are analyzed using computer image processing schemes to extract 3-D structural information [4, 9, 14]. To determine the 3-D structure of an object, it is necessary to combine information content from different views of that object. These different views are sometimes provided by the specimen lying in different orientations as in the case of icosahedral particles, or they may be obtained by tilting the specimen in the microscope. Because of the large depth of focus of conventional electron microscopes, transmission electron micrographs in effect represent two-dimensional (2-D) projections of the specimen. Three dimensional reconstruction procedures are premised on the central-section theorem that states that the 2-D Fourier transform of a projection of a 3-D object is a central section, normal to the direction of the view, in the 3-D Fourier transform of the object. If all of the different orientations of the object can be identified with respect to a common frame of reference, the 3-D Fourier transform of the object can be built from the 2-D Fourier transforms of the different views. Fourier inversion of the resulting 3-D transform yields the 3-D structure of the specimen.

To obtain better understanding of the structural organization of rotaviruses we have studied not only the native viruses [12], but also antibody complexes with the native viruses [13], reassortants [16], and recombinant virus-like particles [15].

Results

The surface representation of the three-dimensional structure of the simian rotavirus (Group A) SA11-4F, reconstructed to 28 Å resolution, is shown in Fig.1A. There are 132 aqueous channels of 3 types; 12 type I channels are located directly on the icosahedral 5-fold axes, 60 type II chanels surround the 5-fold axes, and 60 type III channels surround the icosahedral 3-fold axes. The VP4 spike occupies a portion of the type II channel. The assignment of the spikes to VP4 was determined from our structural analysis of virus complexed with a VP4-specific monoclonal antibody [13].

Our recent studies have shown that the VP4 spikes penetrate the VP7 layer to interact with the VP6 layer. The structure of a reassortant, R-004, was useful in delineating the entire structure of the VP4 spike. R-004 is a reassortant that resulted from a co-infection with SA11-4F and a bovine rotavirus strain, B223. R-004 is genetically identical to SA11-4F with the exception of the VP4 genome segment, which has been exchanged for the bovine strain B223 VP4. The exchange of VP4 results in a reassortant with significantly lowered stability and a different plaque morophology [1, 2]. The structural result of bovine VP4 on a simian background is a spikeless reassortant [16]. A difference map computed between the density maps of SA11-4F and R-004 exposes the entire structure of the VP4 spike, revealing an internal globular domain beneath the VP7 shell. An individual spike isolated from the difference map, shown at three different angles of rotation in Fig.1B, clearly illustrates the dimeric state, the internal globular domain, and other structural features of the spike. The difference map between SA11-4F and R-004, merged with the double-layered particle, reveals the extensive VP4-VP6 interactions (Fig. 1C). Similar results have been obtained by Yeager et al. [21] by difference imaging between native particles and spikeless particles obtained by treating them with alkaline pH (pH 11.5). Thus, the VP4 protein interacts not only with VP7, but also with VP6. From these structural studies using reassortants and native virus, we suggest that (i) VP4 may play an important role in the process of receptor recognition of double-layered progeny during virus budding through the ER membrane, and (ii) assembly of VP4 onto the newly made particles occurs prior to VP7 assembly.

An interesting aspect of the rotavirus infection is trypsin-enhanced infectivity. Trypsin, present at the natural site of rotavirus infection, cleaves VP4 at a conserved arginine residue to produce VP5* ($\sim 60\,$K) VP8* ($\sim 28\,$K). Trypsin cleavage of VP4 is accompanied by a significant increase in rotavirus infectivity which is associated with enhanced cell entry. Preliminary structural analysis with non-trypsinized and trypsinized rotavirus shows that a significant change in the

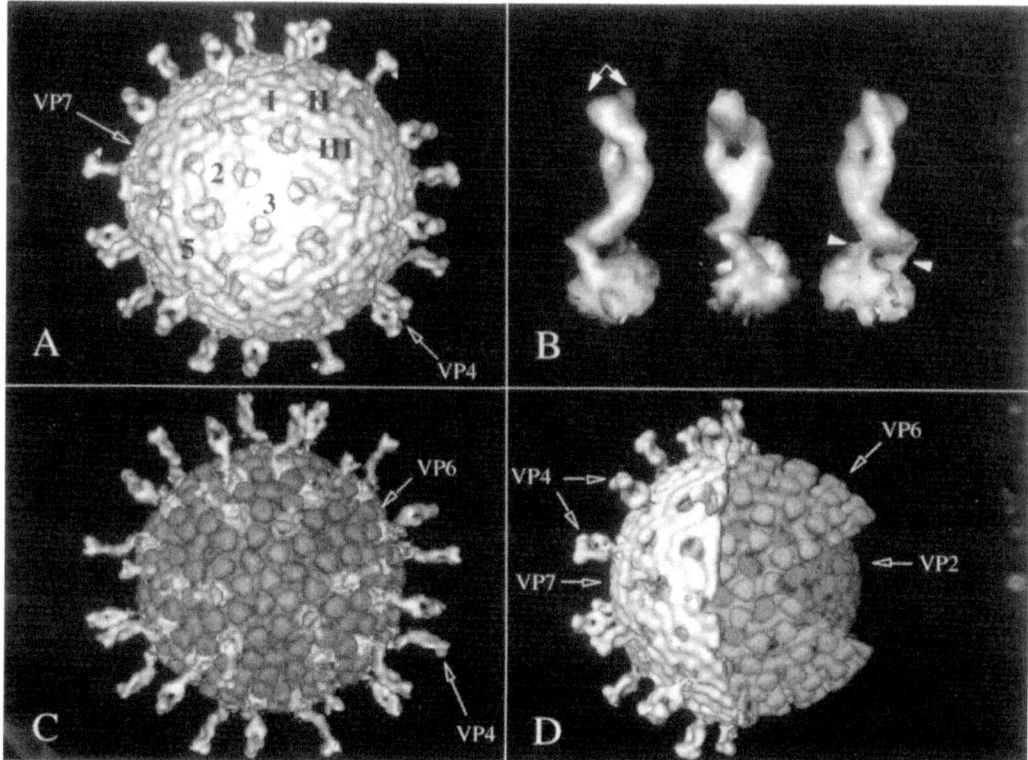

Fig. 1. A) Surface representation of the three-dimensional structure of SA11-4F, computed to a nominal resolution of 28 Å. The outer shell proteins, composed of VP4 and VP7, are indicated by arrows. This layer is composed 780 monomers of VP7 organized on a T = 13 left-handed icosahedral lattice. The strict icosahedral (5-, 3-, 2-fold) axes are indicated. Aqueous channels of the three types, indicated by Roman numerals, penetrate the VP7 layer. Spikes composed of VP4 project outward from the type II channels. **B)** An isolated spike from the difference map of SA11-4F and R-004 rotated at various angles along the vertical illustrates the entire structure of the VP4 spike. The dimeric nature and the distinct asymmetry of the spike is evident at this resolution. At the left, a bilobed structure at the tip of the spike is indicated by the arrows. At the center, a separation in the body of the spike is apparent. At the right, two points of contact between the outer (spike) and inner (globular) portion of the VP4 spike become clear (indicated by arrowheads). **C)** The difference map between SA11-4F and R-004 is shown together with the surface representation of the double-layered particle. A globular domain of VP4 reaches beneath the VP7 layer to interact with VP6. **D)** Cut-away of mature rotavirus to reveal the organization of each of the three protein layers of the rotavirus capsid. The protein composition of each layer is indicated by arrows

structure of the VP4 spike occurs following proteolysis. Presently, we are attempting to resolve the conformational changes following VP4 cleavage that facilitate productive cell entry.

In contrast to the outer layer, the VP6 layer exhibits a bristly surface made of 260 trimers of VP6 located at all the local and strict 3-fold axes of a T = 13 icosahedral lattice. Our structural analysis of the DLPs show that VP6 protein has mainly two domains, a distal globular domain and the lower rod-like domain. The trimers are formed mainly through the interactions between the globular domains. The upper portion of the rod-shaped domain is involved in the interactions between the neighboring VP6 trimers, and the lower portions of the rod-shaped domain interact with VP2. Enclosed within the VP6 layer is the third protein layer (inner layer) made of VP2 [14]. The structural organization of the VP2 layer has been the most difficult to resolve. However, our preliminary studies using baculovirus-expressed VP2 particles [8, 22] have shown that it is a smooth shell with a thickness of about 35 Å. In Fig. 1D, regions of density have been removed at the appropriate radii to reveal the morphology of each of the three capsid layers.

To understand the internal organization in rotavirus, recombinant baculo-virus-expressed virus-like particles [3, 8, 22] have been extremely useful. We have analyzed the 3-D structures of several recombinant virus-like particles made from co-expression of VP2/6, VP1/2/6, VP3/2/6 and VP1/2/3/6 in insect cells to a resolution of 24 Å [15]. Quantitative comparisons of these structures with the native double-layered particles using difference imaging techniques have confirmed that VP2 forms the innermost layer, 35 Å thick and 520 Å in diameter, of the rotavirus structure. The volume of the VP2 layer can accomodate 120 molecules of VP2. It also appears that significant portion of the genomic RNA is icosahedrally ordered into two layers between the radii of 170 Å and 235 Å [23]. Based on these studies it has been proposed that VP1 and VP3 form a complex, positioned below the 5-fold vertices of the VP2 layer, that interacts with the ordered portions of the genomic RNA.

Conclusions

Three-dimensional structural studies on rotaviruses have provided a detailed description of the architectural features of this large (~ 1000 Å in diameter) and complex virus. These structural studies also have provided better insight into some of the biological functions of the virus such as cell entry, neutralization, transcription, and assembly [12, 13, 16, 21, 23]. However, many more questions remain to be answered in greater detail. Further higher resolution structural analysis using electon microscopy techniques and X-ray crystallography, together with the recent advances in the molecular biology of rotaviruses that have resulted in the standardization of cloning and expression of wild-type and mutant genes, and the generation of rotavirus reassortants with unusual phenotypic properties, should permit more detailed understanding of the molecular mechanisms underlying each step of infection with rotaviruses.

Acknowledgements

Our work is supported in part by grants from the NIH (AI 36040 and AI 16687, DK 31044), National Center for Research Resources (RR 02250), NSFBIR-9413229, W.M. Keck Foundation and R. Welch Foundation.

References

1. Chen D, Burns JW, Estes MK, Ramig RF (1989) Phenotypes of rotavirus reassortants depend upon the recipient genetic background. Proc Natl Acad Sci USA 86: 3743–3747
2. Chen D, Estes MK, Ramig RF (1992) Specific interactions between, rotavirus outer capsid proteins VP4 and VP7 determine expression of a cross reactive, neutralizing VP4-specific epitope. J Virol 66: 432–439
3. Crawford SE, Labbe M, Cohen J, Burroughs MH, Zhou Y, Estes MK (1995) Characterization of virus-like particles produced by the expression or rotavirus capsid proteins in insect cells. J Virol 68: 5945–5952
4. Crowther RA (1971) Procedures for three-dimensional reconstruction of spherical viruses by Fourier synthesis from electron micrographs. Phil Trans Roy Soc Lond B 261: 129–228
5. Dubochet J, Adrian M, Chang JJ, Homo JC, Lepault J, McDowall AW, Schultz P (1988) Cryo-electron microscopy of vitrified specimens. Q Rev Biophys 21: 129–228
6. Estes MK (1995) Rotaviruses and their replication. In: Fields BN, Knipe DM, Chanock RM, Hirsch MS, Melnick JL, Monath TP, Roizman B (eds) Virology, vol 2. Raven Press, New York, pp 1625–1655
7. Kapikian AZ, Chanock RM (1995) Rotaviruses. In: Fields BN, Knipe DM, Chanock RM, Hirsch MS, Melnick JL, Monath TP, Roizman B (eds) Virology, vol 2. Raven Press, New York, pp 1657–1708
8. Labbé M, Charpilienne A, Crawford SE, Estes MK, Cohen J (1991) Expression of rotavirusVP2 produces empty corelike particles. J Virol 65: 2946–2952
9. Lawton JA, Prasad BVV (1996) Automated software package for icosahedral virus reconstruction. J Struct Biol 116: 209–215
10. Olson NH, Kolatkar PR, Oliveira MA, Cheng RH, Greve JM, McClelland A, Baker TS, Rossmann MG (1993) Structure of a human rhinovirus complexed with its receptor molecule. Proc Natl Acad Sci USA 90: 507–511
11. Prasad BVV, Wang GJ, Clerx JPM, Chiu W (1987) Cryo-electron microscopy of spherical viruses: an application to rotavirus. Micron and Microscopica Acta 18: 327–331
12. Prasad BVV, Wang GJ, Clerx JPM, Chiu W (1988) Three-dimensional structure of rotavirus. J Mol Biol 199: 269–275
13. Prasad BVV, Burns JW, Marietta E, Estes MK, Chiu W (1990) Localization of CP4 neutralization sites in rotavirus by three-dimensional cryo-electron microscopy. Nature (London) 343: 476–479
14. Prasad BVV, Chiu W (1994) Structure of rotavirus. In: Ramig RF (ed) Rotaviruses. Springer, Berlin, pp 9–29
15. Prasad BVV, Rothnagel R, Zeng CQY, Lawton JA, Chiu W, Estes MK (1995) High resolution difference imaging of baculovirus-expressed subassemblies of rotaviral proteins and the native double-shelled particles (Abstract). Fifth international symposium on double-stranded RNA viruses. March 19–23, Jerba, Tunisia
16. Shaw AL, Rothnagel R, Chen D, Ramig RF, Chiu W, Prasad BVV (1993) Three-dimensional visualization of the rotavirus hemagglutinin structure. Cell 74: 693–701
17. Unwin PNT, Ennis PD (1984) Two configurations of a channel-forming membrane protein. Nature (London) 307: 609–613
18. Wang GJ, Hewlett M, Chiu W (1991) Structural variation of La Crosse virions under different chemical and physical conditions. Virology 184: 455–459
19. Wang G, Porta C, Chen Z, Baker T, Johnson JE (1992) Identification of a Fab interaction footprint site on an icosahedral virus by cryoelectron microscopy and X-ray crystallography. Nature 355: 275–278

20. Yeager M, Dryden KA, Olson NH, Baker TS (1990) Three-dimensional structure of rhesus rotavirus by cryoelectron microscopy and image reconstruction. J Cell Biol 110: 2133–2144
21. Yeager M, Berriman JA, Baker TS, Bellamy AR (1994) Three-dimensional structure of the rotavirus haemagglutinin VP4 by cryo-electron microscopy and difference map analysis. EMBO J 13: 1011–1018
22. Zeng Q, Labbe M, Cohen J, Prasad BVV, Chen D, Ramig RF, Estes MK (1994) Characterization of rotavirus VP2 particles. Virology 201: 55–65
23. Prasad BVV, Rothnagel R, Zeng CQ-Y, Jakana J, Lawton JA, Chiu W, Estes MK (1996) Visualization of ordered genomic RNA and transcriptional complexes in rotavirus. Nature 382: 471–473

Authors' address: B. V. Venkataram Prasad, Verna and Marss McLean Department of Biochemistry and the W.M. Keck Center for Computational Biology, Baylor College of Medicine, Houston, Texas 77030, U.S.A.

Arch Virol (1996) [Suppl] 12: 29–35

Structure and function of rotavirus nonstructural protein NSP3

D. Poncet, C. Aponte, and **J. Cohen**

[1] Laboratoire de Virologie et Immunologie Moléculaires INRA, C.R.J.,
Domaine de Vilvert, Jouy-en-Josas, France

Summary. The genomes of viruses in the family *Reoviridae* consist of segmented double-stranded RNA. There are 10 to 12 segments depending on the genus. The 5′ ends and the 3′ ends of the RNAs present conserved motifs for each virus genus. These conserved motifs have been hypothesized to play a role in genomic segment assortment during virus morphogenesis. Using a set of monoclonal antibodies we have tried to identify rotaviral proteins that bind to RNA during infection in cell culture. This methodology takes advantage of being able to label RNA *in vitro* to high specific activity and also of solid phase processing of RNA-protein complexes. After cross-linking the RNA to protein in infected cells, protein-RNA complexes are precipitated with a specific MAb; then, the RNA in the complex is labeled *in vitro* and the protein or nucleic acid moieties are analyzed by usual protocols. This paper describes results using an anti NSP3 MAb.

In infected cells, we have shown that NSP3 binds to the eleven messenger RNAs, and that a sequence from nucleotides 8 to 15 is protected from digestion with RNAse T1 by NSP3 in the RNA-protein complex. The availability of recombinant protein NSP3 expressed in the baculovirus-insect cell system has allowed the sequence specificity of NSP3 to be studied *in vitro*. The minimal sequence recognized by NSP3 is GACC. The role of NSP3 in rotavirus replication is discussed based on these results and by comparison with other RNA-binding proteins of members of the *Reoviridae* family.

Introduction

The basic feature of viruses in the *Reoviridae* family is the presence of a segmented, double-stranded RNA genome. The mechanism by which genome segment assortment occurs is one of the most puzzling questions about the morphogenesis of these viruses. Conserved motifs present on both ends of the viral mRNAs have been suspected to play a major role in morphogenesis and to be implicated in segment assortment. This hypothesis is primarily based on the fact that reassortment is only observed between viruses of the same group. All the nonstructural proteins of rotavirus, except NSP4 and NSP5, have been shown to bind nucleic acid in non sequence-specific manner [2,6–9]. Using a set of monoclonal antibodies directed against the nonstructural proteins, we have tried to identify viral proteins that, *in vivo*, bind specifically to viral RNA. In this paper, we present the data obtained with an anti-NSP3 antibody.

NSP3 is a nonstructural protein, in regular amounts in rotavirus-infected cells. Sequence analysis of NSP3 [12] suggests this protein consists of two major domains. The first domain is a very conserved basic region that is also predicted to have an alpha-helical structure. A consensus sequence, present in NSP3 between aa 104 and 112, is also present in ssRNA binding proteins of other *Reoviridae* [17]. Recently it has also been shown that group C rotavirus NSP3 binds specifically to double-stranded RNA [11]. The second domain, found in the carboxy-half of NSP3, contains heptapeptide repeats of hydrophobic residues typical of coiled-coil structures. Investigation of distinct forms of NSP3 showed they are all composed of homooligomers [13]. The use of deletion mutants showed that the carboxy terminus of the protein, containing the heptad repeats was responsible for oligomerization [13]. Recent comparisons of several nucleotide sequences of NSP3 from group A rotaviruses have identified a stretch of about 80 nucleotides in the 3' untranslated region that is highly conserved in the NSP3 gene and might play an important role in the regulation expression of the NSP3 gene [16].

Gene 7, that codes for NSP3, is in the group of the early genes (together with genes 5 and 8) whose translation has been reported to be independent of protein synthesis. NSP3 has been proposed to have some role in replication and/or assembly of RNA into viral capsids [5]. The products of these genes are likely to play a regulatory role during the viral growth cycle. The ability of NSP3 to bind RNA also argues for a role of NSP3 in genome replication and the early steps of viral assembly. This report analyzes the conserved motifs present at the ends of the viral RNA segments and summarizes our studies on the structure-function relationships of the nonstructural protein NSP3.

Materials and Methods

Cells and virus

MA104 cells were grown as previously described [10]. Plaque purified stocks of the RF strain of bovine rotavirus (serotype G6), were grown in MA104 cells in the presence of trypsin (0.44 μ/ml Sigma type IX) as previously described [10].

Recombinant baculovirus expressing NSP3 from rotavirus group A (BacNSP3A) was propagated and titrated using a plaque assay in confluent monolayers of *Spodoptera frugiperda* (Sf9) cells in Hink's medium supplemented with 10% fetal bovine serum as described [1]. A baculovirus recombinant (BacNSP3C) containing NSP3 from the group C (Cowden Strain) was constructed using the standard protocol and the transfer vector pBacPak [3].

UV cross-linking of RNA-protein complexes

MA104 cells, infected at a high multiplicity of infection, were irradiated for 13 min. by a germicidal lamp, at a distance of 22 cm, 6 h post infection. Next, a cell lysate was prepared in

RIPA buffer (Tris-Cl 50mM, pH 8.5, 75 mM NaCl, 1% Triton X100, 1% sodium deoxycholate, 20 mM EDTA, 2µg/ml aprotinin). RNA protein complexes were immunoprecipitated using the monoclonal antibody ID3 [1], then treated with RNAse T1, and the RNA moiety was labeled using T4 polynucleotide kinase and $\gamma[^{32}P]$ ATP. The complexes were analyzed by SDS-PAGE or freed of its protein moiety and the resulting RNA was analyzed in a denaturing urea-20% polyacrylamide gel.

Gel retardation assays

Recombinant NSP3 was purified from BacNSP3 A- infected Sf9 cells. Briefly, the cells were collected 2 days post infection and solubilized in 20 mM Hepes, pH 7.5, 100 mM NaCl, 3 mM $MgCl_2$, 0.5% Triton X100, 1 µg/ml leupeptine. After sonication and centrifugation at 100 000 × **g**, the pellet that contained the majority of NSP3 was suspended CAPS 50mM, pH 10.8 buffer containing 5M urea, 10mM DTT, 2% betaine and chromatographed on an ion exchange column (Mono Q, Pharmacia). The peak corresponding to NSP3 was dialyzed against Tris-Cl 50mM, pH 8, 10% glycerol, 5 mM reduced glutathione, 0.5 mM oxidized glutathione. NSP3 from group C rotavirus was prepared using a similar protocol, but without the ion exchange chromatography. RNA probes were either synthesized in an Applied Biosystem 380A DNA synthesizer, using the modified dimethoxytrityl-cyanoethyl RNA phosphoramidite chemistry, or by transcribing the plasmid p71cat with T7RNA polymerase. The plasmid p71cat (14) linearized with Ksp I produced RNA molecules having the exact 5′ and 3′ ends of rotavirus mRNA. Oligoribonucleotides were labeled in vitro with the T4 polynucleotide kinase and $\gamma[^{32}P]$ ATP. Probes and purified protein were incubated for 30 min. on ice in binding buffer (10 mM Hepes pH 7.9, 3 mM $MgCl_2$, 40 mM KCl, 1 mM DTT, 5% glycerol), and then analyzed by electrophoresis on a 8% polyacrylamide gel prepared in 0.5 × TAE buffer. Migration was for 4 hours at 4 °C and autoradiographs were kept for 2–4 hours at − 20 °C.

Results

Sequence analysis

As the complete genome sequences of many viruses in the different genera of the *Reoviridae* have been reported, it has become apparent that a sequence at the 3′ end of each segment is quite specific for each virus genus and group (Fig. 1). The 5′ ends of many groups present a common characteristic and usually contain a short (4 to 10 nucleotide long) AT rich region. In addition, for a given virus, there is more diversity among the 5′ end than in 3′ ends. For all viruses in the *Reoviridae,* strict consensus motifs are not longer than 3 to 7 nucleotides. However, another double-stranded RNA virus, the phage phi6, contains longer conserved motifs for its three mRNAs. The 3′ ends of the mRNAs for all *Reoviridae* sequenced so for have a purine as the last nucleotide. Since reassortment of genomic segments is only possible between viruses of the same group, and impossible between viruses of different groups, it has been hypothesized that these conserved motifs play a major role during morphogenesis. Considering this comparison of the ends of *Reoviridae* mRNAs, one could hypothesize that the 3′ end motif is responsible for restriction of assortment within a group and the 5′ end motif is responsible for the selection of the right set of segments. Of course, secondary structure of the RNA could also be involved in these processes.

```
Orthoreovirus
  Reovirus serotype 3      5'GCUA------------------------UCAUC3'
Orbivirus
  Blue tongue virus        G(AU)--------------------ACUUAC
Rotavirus
  Group A                  GGC(AU)-----------------UGUGACC
      B                    GGC/A(AU)---------------AAAACCC
      C                    GGCU(AU)--------------- UGUGGCU
Cypovirus
  CPV                      AGUAA-------------------GUUAGCC
Phytovirus
  Wound tumor virus        GGUAUU----------------------UGAU
.......................................................................
Bacteriophage phi6         GGG/UAAAAAA-----------GCUCUCUCUCU
```

Fig. 1. Conserved sequences at the 5′ and 3′ ends of different viruses in the *Reoviridae*. The parenthesis indicate the A, U rich regions, from 4 to 10 nucleotides in length. The double-stranded RNA phage phi 6 is included for comparison

Unfortunately to date, limited data are available on the secondary structure of mRNA of *Reoviridae* even though there are several predictions of secondary structure involving the mRNA ends. For group A rotaviruses, the 5′ consensus sequence is 5′ $GGC(A, T)_{6-9}$ and the 3′ consensus sequence is TGTGACC3′ on the cDNA. However genes 2 and 3 sequenced so far have a slightly modified 3′ terminal sequence, respectively TATGACC and CGTGACC.

NSP3 to the 3′ end consensus sequence of viral mRNAs in infected cells

Interactions between viral proteins and RNAs have been studied in rotavirus infected cells. The use of UV cross-linking followed by immunoprecipitation: and labeling with T4 polynucleotide kinase allowed labeling of a band that migrates slightly above the regular NSP3. If the gel analysis is performed in non-reducing conditions, protein bands corresponding to oligomers of NSP3 could also be identified. Analysis of the nucleic acid moiety present in these RNA-protein complexes was performed after extensive treatment with proteinase K. These analyses have shown that nucleic part of the complex is made of a series of discrete bands that range in size from 10 to 50 nucleotides. This labeled material was used as probe with two kinds of targets. The first kind of target consisted of various restriction fragments of a plasmid that contained a full-length rotavirus gene. Only targets containing the 3′ end of the messenger were recoginized by the probe. The second kind of target consisted of the eleven plus strands obtained by *in vitro* synthesis and from denatured genomic RNA. The probe did not hybridize with the plus strand but with the denatured genomic RNA. These hybridization experiments showed that the NSP3-RNA complex contained a part of the plus strand RNA that corresponds to sequences present near the 3′ end of the 11 RNA segments. After treatment of the complex with RNase T1, the size of the 11 RNA between 9 and 23 base pairs and it could be resolved into 8 bands when electrophoresed in denaturing conditions. These RNase T1-

protected fragments were sequenced using the enzymatic method, and their sequences correspond to the 3′ end sequence common to all rotavirus group A genes.

Minimal requirement for RNA recognition by NSP3

Results obtained after UV cross-linking analysis of a mixture of semi-purified NSP3A protein expressed by a recombinant baculovirus and the eleven, mRNAs synthesized *in vitro* were similar to the results obtained in infected cells. Gel retardation assays with a series of synthetic oligoribonucleotides showed that NSP3A binds to RNA in sequence-specific manner. Only the RNAs having at their 3′ end the consensus sequence (AUGUGACC-3′ present on the 3′ ends of all group A rotavirus mRNAs) could complex with recombinant NSP3, as estimated by gel retardation and UV cross-linking studies. Using short oligoribonucleotides, we established that the minimal RNA sequence required for binding of NSP3A is GACC. Modifications of the UGACC oligonucleotide sequence impaired binding of the protein to the RNA. Gel retardation assays also allowed the apparent affinity constant of NSP3 for various oligoribonucleotides to be determined. Furthermore, the recombinant NSP3 protein from rotavirus group C showed specificity for the 3′ end consensus sequence (AUGUGGCU- 3′) of only group C mRNAs.

Discussion

Our studies have clearly demonstrated that NSP3 binds to the 3′ end of rotavirus messenger RNAs *in vivo* during viral infection and *in vitro* in the absence of other viral components. This interaction is sequence specific and for group A rotaviruses, the recognized sequence corresponds to the consensus AUGUGACC. *In vitro,* the minimum required sequence is only the tetraribonucleotide GACC. These short recognized sequences are related to the minimal promoter of the replicase recently described *in vitro* [19]. The corresponding protein of the group C rotavirus recognizes *in vitro* the sequence AUGUGGCU that is also the consensus of the 3′ end of the eleven messengers of this virus group. These very short sequences recognized *in vitro* by NSP3 probably indicate that no secondary structure is implicated in the specificity of the reaction. This situation is fairly unusual, since many of the sequence specific RNA binding proteins recognize a motif in the context of a given secondary structure. However, it cannot be excluded that, *in vivo*, secondary structures of viral mRNAs modulate their interactions with NSP3, and thus balance the frequency of each mRNA independently of their size to get the one to one ratio between each gene in the virus particle.

Among the NSP3 genes sequenced to date, a very conserved basic region (aa 83–150 for group A sequence) having a positive net charge of +10 at pH 7 can be identified. This region also contains the consensus sequence (I/L)XXM(I/L)(S/T)XXG that has been described by Van-Staden *et al.* [18] in ss-RNA binding proteins of orbiviruses (NS2) and reoviruses (sigma NS). The RNA

D. Poncet et al.

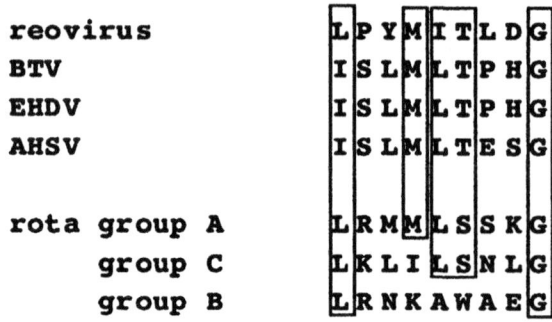

reovirus	L P Y M I T L D G
BTV	I S L M L T P H G
EHDV	I S L M L T P H G
AHSV	I S L M L T E S G
rota group A	L R M M L S S K G
group C	L K L I L S N L G
group B	L R N K A W A E G

Fig. 2. Consensus sequence of ssRNA binding proteins from viruses in the *Reoviridae.* Conserved amino acids that define the consensus motif are boxed

binding activities of NS2 and sigma NS have not been shown to be sequence-specific, but Stamatos and Gomatos [17] have shown that sigma NS prefers reovirus 3′ ends. As illustrated in Fig. 2 this consensus is present in group A as well as in group C rotavirus NSP3, even though these two proteins are only 23% homologous. However, in the rotavirus group B NSP3 that has been recently sequenced [4], this consensus is poorly conserved. It should be noted that rotavirus group C NSP3 has an extra region at the carboxy termini that has been found to also bind specifically to double-stranded RNA and inhibit activation of the interferon-induced protein kinase [11].

The amino acids involved in the interaction of NSP3 with the viral messenger have not been positively identified, but biochemical evidence supports that the amino acids that are cross-linked by UV irradiation to viral RNAs are located in the amino end of the protein (Poncet *et al*, unpubl.). Similarly, construction of chimeric proteins made of NSP3 from group A rotavirus and group C rotavirus has allowed us to show that the region of the protein responsible for sequence specificity binding corresponds to the amino terminus of the protein (Poncet *et al.* unpubl.).

It seems that NSP3 cannot be involved in the selection process of the right set of viral mRNAs since it binds to a sequence common to all virus genes. However, it might act as a condensing agent to bring together the ssRNA templates in preparation for dsRNA synthesis. The exact function of NSP3 remains to be elucidated, but our studies suggest that this protein that discriminates between viral and cellular mRNA, could either protect the viral mRNA from exonuclease degradation or play in viral morphogenesis by transporting viral messenger RNAs from the site of transcription to the site of replication and assembly. This last hypothesis is supported *by* the observation that NSP3 is uniformly distributed in the cytoplasm of infected cells in contrast with many other viral proteins which are localized in large inclusions.

Acknowledgements

The excellent help provided by Delphine Mabon and Annie Charpilienne is gratefully acknowledged. We are also grateful to Michel Bremont who provided the rotavirus group C dsRNA. C.A. is supported in part by a fellowship from CONICIT.

References

1. Aponte C, Mattion NM, Estes MK, Charpilienne A, Cohen J (1993) Expression of two bovine rotavirus non-structural proteins (NSP-2, NSP-3) in the baculovirus system and production of monoclonal antibodies directed against the expressed proteins. Arch Virol 133: 85–95
2. Boyle J and Holmes K (1986) RNA-binding proteins of bovine rotavirus. J Virol 58: 561–568
3. Cohen J, Charpilienne A, Chilmonczyk S, Estes MK (1989) Nucleotide sequence of bovine rotavirus gene 1 and expression of the gene product in baculovirus. Virology 171: 131–140
4. Eiden JJ (1993) Gene-5 of the IDIR agent (Group-B rotavirus) encodes a protein equivalent to NS34 of Group-A rotavirus. Virology 196: 298–302
5. Estes MK, Cohen J (1989) Rotavirus gene structure and function. Microbiol Rev 53: 410–449
6. Gallegos CO, Patton JT (1989) Characterization of rotavirus replication intermediates a model for the assembly of single-shelled particles. Virology 172: 616–627
7. Hua J, Chen X, Patton JT (1994) Deletion mapping of the rotavirus metalloprotein NS53 (NSP1): the conserved cysteine-rich region is essential for virus-specific RNA binding. J Virol 68: 3990–4000
8. Kattoura MD, Chen X, Patton JT (1994) The rotavirus RNA-binding protein NS35 (NSP2) forms 105 multimers and interacts with the viral RNA polymerase. Virology 202: 803–813
9. Kattoura MD, Clapp LL, Patton JT (1992) The rotavirus nonstructural protein, NS35, possesses RNA-binding activity in vitro and in vivo. Virology 191: 698–708
10. L'Haridon R, Scherrer R (1976) In vitro culture of a rotavirus associated with neonatal calf scours. Annales de Recherches Vétérinaires 7: 373–381
11. Langland JO, Pettiford S, Jiang BM, Jacobs BL (1994) Products of the porcine group C rotavirus NSP3 gene bind specifically to double-stranded RNA and inhibit activation of the interferon-induced protein kinase PKR. J Virol 68: 3821–3829
12. Mattion NM, Cohen J, Estes MK (1993) The rotavirus proteins. In: Kapikian A (ed) Virus infection of the gastrointestinal tract. Marcel Dekker, New York, pp 169–249
13. Mattion NM, Cohen J, Aponte C, Estes MK (1992) Characterization of an oligomerization domain and RNA binding properties on rotavirus nonstructural protein NS34. Virology 190: 68–83
14. Poncet D, Aponte C, Cohen J (1993) Rotavirus protein NSP3 (NS34) is bound to the 3' end consensus sequence of viral mRNAs in infected cells. J Virol 67: 3159–3165
15. Poncet D, Laurent S, Cohen J (1994) Four nucleotides are the minimal requirement for RNA recognition by rotavirus by non-structural protein NSP3. EMBO J 13: 4165–4173
16. Rao CD, Das M, Ilango P, Lalwani R, Rao BS, Gowda K (1995) Comparative nucleotide and amino acid sequence analysis of the sequence-specific RNA-binding rotavirus nonstructural protein NSP3. Virology 207: 327–333
17. Stamatos NM, Gomatos PJ (1992) Binding to selected regions of reovirus mRNAs by a nonstructural reovirus protein. Proc Natl Acad Sci USA 79: 3457–3461
18. Van-Staden-V, Huismans-H (1991) A comparison of the genes which encode non-structural protein NS3 of different orbiviruses. J Gen Virol 72: 1073–1079
19. Wentz M, Zeng CQ-Y, Patton JT, Estes MK, Ramig RF (1995) Defining the cis-acting elements for replication of rotavirus RNA. Am Soc Virol meeting Austin TX.

Authors' address: J. Cohen, Laboratoire de Virologie et Immunologie Moleculaires INRA, C.R.J., Domaine de Vilvert, F-78350 Jouy-en-Josas, France.

Arch Virol (1996) [Suppl] 12: 37–51

Genome rearrangements of rotaviruses

U. Desselberger

Clinical Microbiology and Public Health Laboratory, Addenbrooke's Hospital,
Cambridge, UK

Summary. Rotaviruses (and other members of the *Reoviridae* family) undergo rearrangements of their genomes. This review describes evidence of rearranged genomes in rotaviruses. Their structure and functions are reviewed. Possible mechanisms of their emergence are discussed, and the significance of genome rearrangements for viral evolution is considered.

Introduction

Rotaviruses are one out of 9 genera of the *Reoviridae* family infecting a wide variety of vertebrate species including man [36a]. They are the main cause of infantile gastroenteritis in man and also of acute diarrhoea in the young of many mammalian species.

Rotaviruses possess a characteristic double-shelled icosahedral capsid surrounding a core ribonucleoprotein. Both capsids are perforated by numerous channels; the outer capsid carries 5 dozen short spikes [16, 36a].

The genome of rotaviruses consists of 11 segments of double-stranded (ds) RNA molecules ranging in size between 667 and 3 302 base pairs (bps) and yielding a total molecular size of appr. 18 550 bps. Gene-protein assignments have been completed for several strains [16, 36a].

Rotaviruses replicate totally in the cytoplasm of infected cells. The viral core contains the viral RNA-dependent RNA polymerase complex which ensures the formation of capped (non-polyadenylated) mRNA. The transcripts are used for both protein translation and as templates for minus strand RNA synthesis in nascent subviral particles. Those acquire their outer capsid proteins by budding from aggregates termed "viroplasm" through the membrane of the endoplasmic reticulum. Particle release is by cell lysis. Details of replication including involvement of different viral proteins at the different steps are described in [16]. The multi-segmented nature of the viral genome allows reassortment on a large scale in doubly infected cells.

Rotavirus classification is based on serology and gene composition of three major structural proteins:

- VP6, the inner capsid protein coded for by RNA 6, determines group and subgroup specificity. At least seven groups (A-G) and, within group A, at least 4 different subgroups (I, II, I + II, non-I, non-II) have been identified;
- VP7, the major outer capsid protein coded for by RNA 7, 8 or 9 (depending on the virus strain), is a glycoprotein and determines VP7-specific serotype (G type). So far 14 different G types have been identified;
- VP4, the outer capsid protein forming dimeric spikes and coded for by RNA 4, is a protease-sensitive protein and determines VP4-specific serotype (P type). So far over 20 different P types have been differentiated genotypically, but not all of them have been clearly identified as different serotypes.

Designation of strains is illustrated by the following examples (for details see [16]:

- A/hu/Wa G1P1A [8], the human rotavirus strain Wa of group A, G serotype 1 and P serotype 1A [P genotype 8];
- A/bo/UK G6P7 [5], the bovine rotavirus strain UK of group A, G serotype 6 and P serotype 7 [P genotype 5];
- A/eq/L338 G13P[18], the equine rotavirus strain L338 of group A and P genotype [18] (serotype not determined yet).

Rotaviruses occur worldwide. Their epidemiology in man is characterized by cocirculation of group A strains of various G and P type combinations with marked winter peaks in temperate climates, but lack of them in subtropical and tropical regions [for details see ref 28]. There is a combined humoral and cellular immune response to rotavirus infection. The best correlate of protection from infection/reinfection is the presence of virus-specific IgA antibodies in the gut mucosa [28, 36b]. Multivalent vaccines with the potential to protect from severe rotavirus diarrhea are under development [28, 42a].

Discovery of genome rearrangements

Typical profiles of rotavirus RNAs after electrophoretic separation on polyacrylamide gels and silver staining show 4 size classes (I: Segments 1–4; II: Segments 5 and 6; III: Segments 7–9; IV: Segments 10 and 11) [16]. However, these profiles are not always seen.

Pedley et al. [40] investigated group A rotaviruses isolated from chronically infected children with severe combined immunodeficiency (SCID). Serial faecal specimens of such children produced abnormal rotavirus RNA profiles: normal RNA segments were decreased in their relative concentration or even completely lost from the profiles, but additional bands of RNA were seen migrating between RNA segments 1 and 7 (Fig. 1A). Northern blots of atypical profiles often showed multiple hybridizations with segment-specific radiolabelled probes: that of the homologous RNA segment of standard size and several bands of dsRNA which always migrated higher up in the gel (Fig. 1B). This suggested that the additional bands of dsRNA contain segment-specific sequences in the form of covalently bonded concatemers [40]. Internal deletions as seen in defective interfering (DI) particles [23] were not observed [40].

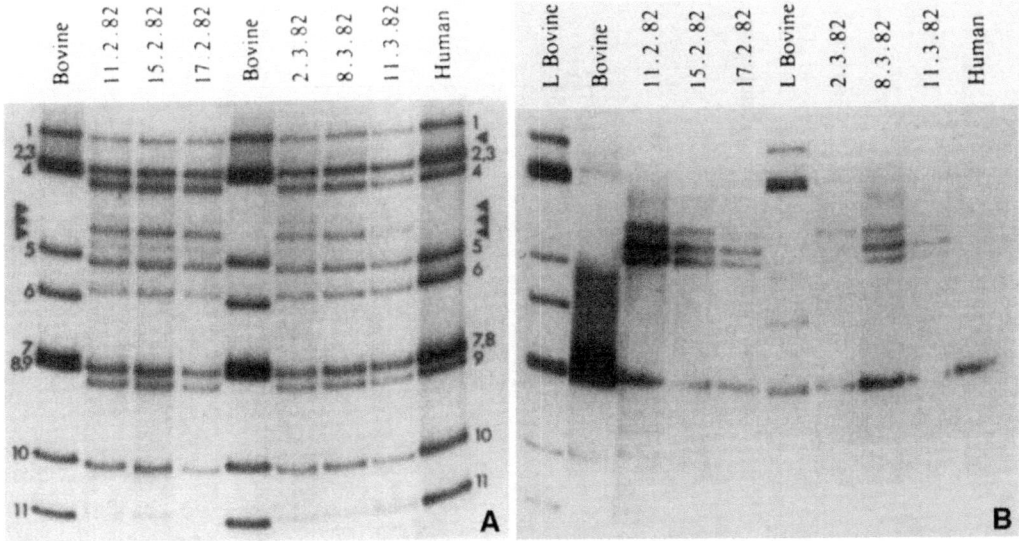

Fig. 1. A. RNA profiles of serial rotavirus specimens obtained from chronically infected patient A.K. (dates indicated on top). All specimens were 3' end-labelled with ^{32}p-pCp [8], separated by PAGE and autoradiographed. Bovine rotavirus and human rotavirus obtained from acute infections served as controls. Order numbers of RNA segments are indicated on both sides, and additional bands are marked by arrowheads. **B.** Hybridization of rotavirus RNA samples of patient A.K. (dates indicated on top), and of human and bovine control RNAs on DPT paper blots to a RNA segment 9-specific radioactive cDNA probe. 3' end-labelled bovine rotavirus RNA (L bovine) and unlabelled bovine and human rotavirus RNAs served as controls. Autoradiogram. From Pedley *et al.* [40]. [With permission of authors and publisher]

Extent of genome rearrangements in rotaviruses

Since their original discovery, genome rearrangements have been described by several independent groups to occur not only in human rotaviruses but also in rotaviruses of a variety of animal species [*Man:* 1, 5, 12, 18, 27, 31, 32, 35; *calves:* 39, 41, 44, 50; *rabbits:* 47, 49; *piglets:* 4, 33; *lambs:* 46]. Whereas the initial observation was in immunodeficient children, the later data from animals and in part also humans were obtained from immunocompetent hosts. (For further details see [9a]).

Sequence data of rearranged genes

Nucleotide sequences of rearranged genes of several group A rotavirus strains of different origin have been obtained, and references and nucleotide sequence accession numbers are summarised in Table 1.

In most cases the genome rearrangement consists of a partial duplication of sequences of the open reading frame (ORF) starting beyond the termination codon and extending then to the 3' end of the normal gene. This is diagrammatically shown in Fig. 2 for rearranged RNA 10 of a human rotavirus isolate [3]; similar changes were also found for rearrangements of other RNAs 10 [31], of

U. Desselberger

Table 1. Sequenced genome rearrangements of rotavirus

RNA Segment[a]	Strain	Origin	Start of reiteration in relation to termination codon	N° of point mutations compared to standard gene	Genbank/EMBL accession number Standard gene	Genbank/EMBL accession number Rearranged gene	Reference
5	brv E	bovine	−596	n/d[b,c]	Z24735	Z12108	[50]
	brv A	bovine	−52	16	L12248	L11575	[25]
6	Lp 14	ovine	23	6	L11596	L11595	[46]
7	H 57	human	0	n/d	n/a[b]	n/a	[35]
10	A 64	human	2	11	D01146	D01145	[3]
	VMRI	human	0	23	n/a	n/a	[31]
11	C7/183	bovine	0	n/a	n/a	n/a	[44]
	C60 X1	pig	6	33	n/a	n/a	[20]
	Alabama	lapine	4	n/a	n/a	n/a	[21]

[a] Genome rearrangements have also been observed in segment 6 of a human strain [40, 46] and segments 7, 8 and 9 of human strains (coding for NSP2 and NSP3) [18, 27, 40] but have so far not been sequenced

[b] Not done; not available

[c] Partial sequence (junction region)

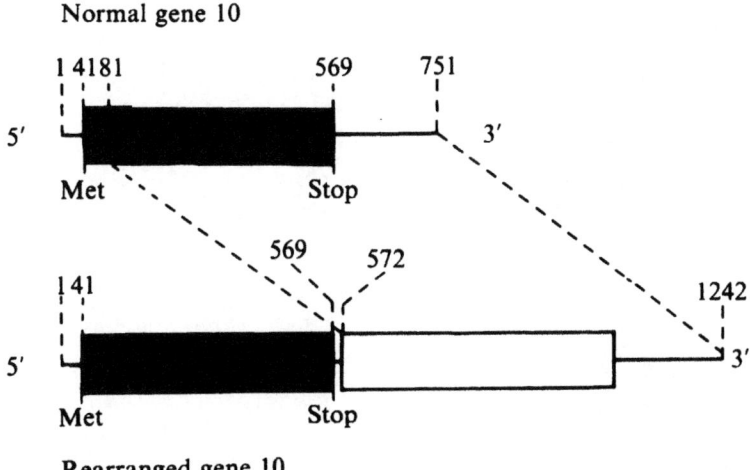

Fig. 2. Diagram of the structures of normal and rearranged genes 10 of a human rotavirus (isolates A28 and A64, respectively). ■, complete ORF; □,duplicated part of ORF of normal gene (untranslated). Solid lines indicate 5' and 3' untranslated regions as well as sequences between normal and duplicated ORFs. From Ballard *et al.* [3]. [With permission of authors and publisher]

RNAs 11 [20, 21, 44], and of one RNA 5 [25]. At various nucleotide positions after the termination codon (0–23; Table 1), the duplication starts reinitiating from various places within the ORF but downstream of the initiation codon, resulting in genes with enormously long 3' untranslated regions (3' UTR; up to 1800–1900 bp; refs. 25,34), in contrast to the relatively short 3' UTRs (17–185 nucleotides) of the standard length genes [11, 16].

In the case of segment 5 of the E variant of bovine rotavirus (brv UK; G6 P7 [5]), the duplication had started before the termination codon, and an extended ORF resulted containing segment 5-specific amino acids as the reiteration had started in-frame [50]. The extended ORF codes for a protein VP5^E of 728 amino acids length which was verified by PAGE of [^{35}S]-methionine labelled proteins of brv E-infected cells [26, 50]. By contrast, rearranged segment 5 of the brv A variant possessed a different structure [25]. Besides the reiteration 52nt before the stop codon (in position 1454), one additonal point mutation in position 808 resulted in a new termination codon (UAG) allowing an ORF of only 258 amino acids (*i.e.* of 31kDa size, slightly more than half the size of the normal product of 491 amino acids, *i.e.* 58 kDa). Thus a gene of 2693 bp resulted of which only 774 (pos. 33–806, *i.e.* 28.7%) coded for a protein!

Matsui *et al.* [31] investigated gene 11 equivalents of rotavirus genomes yielding "short" and "supershort" PAGE profiles and found that "supershort" strain VMRI clearly contained a partial duplication at its 3' end. However, the RNAs 10 of "short" strain DS-1 (G2P1B [4]) and of "supershort" strain 69M (G8P4 [10]) had sequences at their 3' ends which were similar to each other, but not related to any other available rotavirus gene sequence.

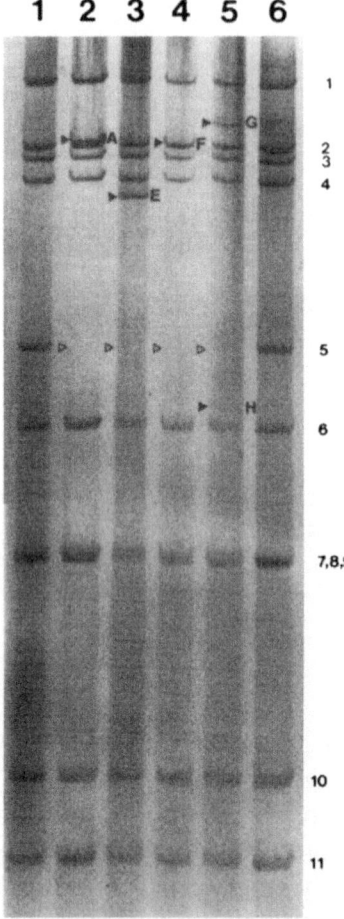

Fig. 3. RNA profiles of plaque-purified bovine rotaviruses (brv) obtained after serial passage at high m.o.i. RNA segment numbers are indicated to the right. Open arrowheads denote missing RNA segments, closed arrowheads additional RNA bands. Tracks 1 and 6: standard brv; track 2: brv A; track 3: brv E; track 4: brv F; track 5: brv G/H (likely to be still a mixture). 2.8 % polyacrylamide-6M urea gel, silver stained. From Hundley *et al.* [26]. [With permission of authors and publisher]

Genome rearrangements generated *in vitro*

After serial passage *in vitro* of bovine rotavirus at high m.o.i., viruses with genome rearrangements (*i.e.* partial duplications), but not genome deletions emerged ([26]; Fig. 3). Bovine rotavirus with a standard genome transformed into brv variants with rearranged RNA segments 5, among others variants brvA and brvE (Fig. 3, tracks 2 and 3; see also section IV). Bovine rotaviruses with rearranged genes are not DI viruses [For details see ref. 9a]. The nvp/pfu ratios were equally low for brv standard and brv A viruses [26].

The *in vitro* generation of viruses with rearranged genomes was reproduced with Chinese lamb rotaviruses (A/ov/Lp14, G10P [15]) by Shen and Bai [45].

Mechanism(s) of genome rearrangements

The sequence data available (see above) allow a formal description of genome rearrangements as partial duplications (concatemer formation). However, it is not clear at which step of the replication cycle the duplication event occurs. It has

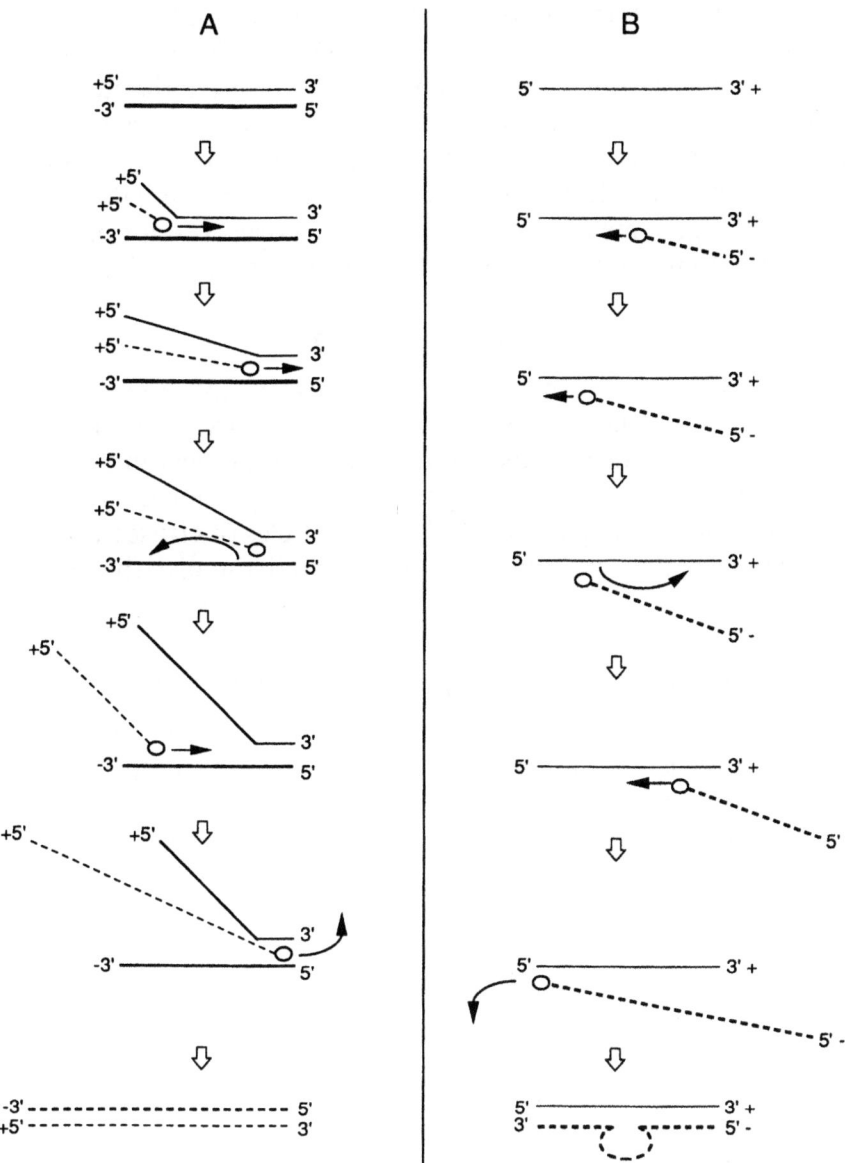

Fig. 4. Diagram of possible mechanisms of the emergence of genome rearrangements. A: During plus strand synthesis; B: during minus strand synthesis (in precore particles). Bold line: minus strand; fine line: plus strand; undulated lines: new synthesised strands; open circle: RNA dependent RNA polymerase; arrows: direction of synthesis

been suggested that the RNA-dependent RNA polymerase of rotaviruses can fall back on its template at various steps of transcription (plus strand synthesis) and reinitiate and retranscribe from that template at different places (Fig. 4). Alternatively, the primary duplication event could occur at the level of replication (negative strand synthesis) (Fig. 4; for further details see ref [9a]).

Most genome rearrangements which were sequenced can be described as intramolecular recombination events, and direct repeats close to the recombina-

tion site were found in some but not other cases of recombination. This is similar
to what was observed in the phi 6 system [36, 37]. In the case of poliovirus
recombination, it was shown that recombination favoured the step of secondary
transcription (from the negative strand of the replicative intermediate, RI) [29].
In the phi 6 system, it was demonstrated very elegantly that recombination also
occurs at the step of negative strand synthesis [37]. Although the mechanism of
recombination in genome rearrangements of rotaviruses is not proven, the data
are consistent with the copy choice model [29, 30; for further details, see ref. 9a].

Biophysical data

Viruses with genome rearrangements had between 450 and 1790 additional
bps packaged amounting to 1.4 to 9.6% of the standard genome size (Fig. 5A).
The viruses differed in density as determined by analytical ultracentrifugation
(Fig. 5B), and the differences in density were directly proportional to the number
of additionally packaged bps (Fig. 5C; [34]. Particles of viruses with up to 10%
additional bps packaged were morphologically indistinguishable from standard
rotavirus [27].

Function of the rearranged genes and their products

Rotaviruses with genome rearrangements which had arisen either naturally (40)
or after serial passage at high m.o.i. in vitro [16] were equally able to reassort
with human rotaviruses carrying a standard genome [2, 6, 9a].

In cases of rearrangements where the normal ORF was extended (brv variant
E, [26, 50]), abrogated (brv variant A; [25]) or mutated (Chinese lamb rotavirus
[46]), functional changes were observed. Bovine rotavirus variants E and
A showed 9- to 60-fold lower yields, respectively, in single-step growth experiments and produced smaller plaques; brv E reaching 40% and brv A only 2% of
the plaques size of standard brv [50].

The analysis of rearrangements of RNA segment 5 were of particular interest
as both extension (brv E) and abrogation (brv A) of the normal ORF were found.
Bovine rotavirus variant A had a truncated VP5 of 258 amino acids length (due
to a termination codon in position 806–808) instead of the authentic size of 491
amino acids, but was viable, non-defective and genetically stable [25, 26]. It was
also found to be associated with the cytoskeleton of the infected cell, like its
normal size counterpart demonstrating that the carboxy-half of VP5 (NSP1,
NS53) is not required for rotavirus replication in vitro [25]. More drastic
reductions of the VP5 gene giving rise to gene products of only 150 or 40–50
amino acids length have been described by Tian et al. [50] and Taniguchi et al.
[48] and are discussed in more detail in ref. [9a].

In passages of the Chinese lamb rotavirus A/ov/Lp14, rearrangement of
RNA 6, the gene coding for the inner capsid protein VP6, was observed in
a similar way [46] as shown by Ballard et al. [3] for segment 10 (Fig. 2), but was
found to be accompanied by a point mutation in nucleotide position 949 (within
the normal ORF) giving rise to a change in amino acid position 309 (from

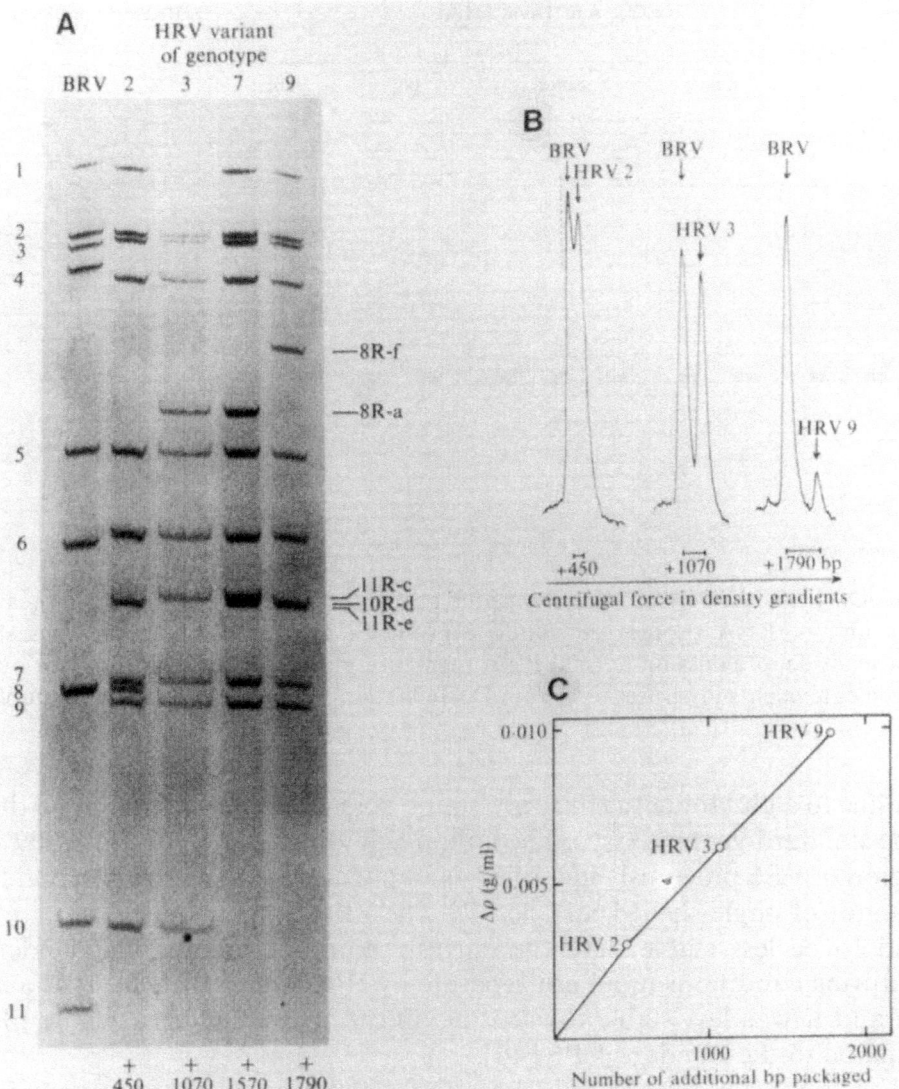

Fig. 5. A. RNA profiles of brv and of human rotavirus variants with rearranged genomes of genotypes 2, 3, 7 and 9 (ref. [27]). Segment numbers (1-11) are indicated to the left, and position and origin of rearranged bands identified to the right (bands a and f being derived from RNA 8, band d from RNA 10 and bands c and e from RNA 11; ref. [27]). The number of additionally packaged bps is indicated at the bottom. 2.8 % polyacrylamide-6M urea gel, silver stained.

B. Scans after analytical equilibrium centrifugation in CsCl of mixtures of single shelled particles containing RNA of standard brv and of human rotaviruses with rearranged genomes of genotypes 2, 3 and 9. The peaks of additionally packaged bps are indicated below scans.

C. Plot of difference in density as determined from data shown in panel B against number of additionally packaged bps. From McIntyre *et al.* [34]. [With permission of authors and publisher]

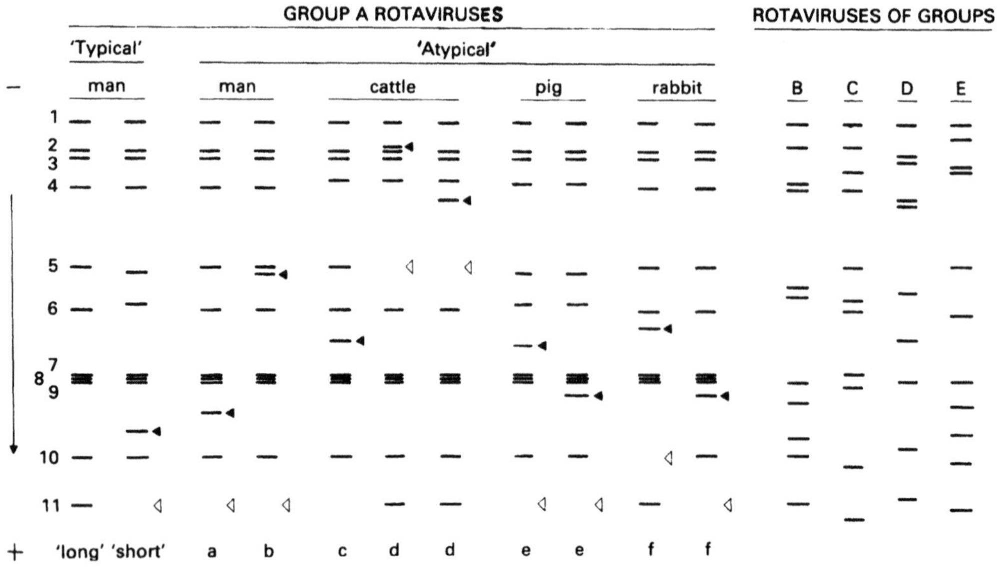

					GROUP A ROTAVIRUSES						ROTAVIRUSES OF GROUPS			

'Typical'   'Atypical'

man | man | cattle | pig | rabbit | B | C | D | E

1 2 3 4 5 6 7 8 9 10 11

+ 'long' 'short' a b c d d e e f f

Fig. 6. Diagram of RNA profiles of various group A rotaviruses with genome rearrangements and of RNA profiles of standard group A and group B-E rotaviruses. Open arrowheads denote missing normal RNA segments, closed arrowheads their various rearranged equivalents in viruses a-f. From Desselberger [9]. [With permission of publisher]

a proline to a glutamine) as the only amino acid difference compared to the VP6 of the standard genome virus. The amino acid difference in position 309 was in a region of VP6 previously identified as important for trimerization and for the formation of single-shelled particles [7]. The VP6 carrying the 309 mutation was found to be less stable than the corresponding standard VP6: Under mild denaturing conditions it did not separate by PAGE as a trimer but as a monomer, and it was less stable towards acidification by almost a whole pH unit compared to the standard VP6 [46].

The analysis of over 500 plaque isolates of a reassortant mixture of human viruses with genome rearrangements and standard bovine rotavirus showed that reassortment was non-random, that there was linkage of occurrence of certain genes, i.e. of RNAs 5, 9 and 11, in reassortants, and that the host cells on which plaque isolates were obtained (MA104 or BSC-1 cells) influenced the frequencies with which certain reassortants were recovered [19].

Genome rearrangements and evolution of rotaviruses

When rotaviruses with genome rearrangements were found in immunocompetent children as a nosocomial infection and circulating over months (Besselaar et al., 1986) and also freely circulating in a variety of elsewise healthy animal hosts (rabbits, calves, pigs), it became clear that the phenomenon was more frequent than originally anticipated. The various forms of rearrangements occurred mainly in genes coding for non-structural proteins (RNAs 5, 8 and

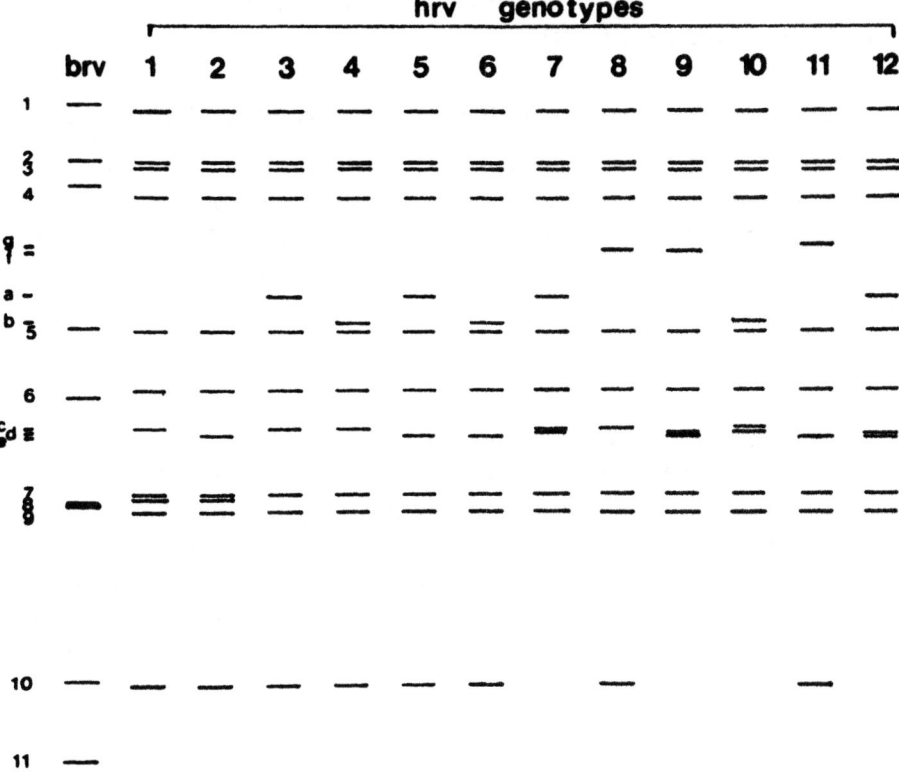

Fig. 7. Diagram of 12 subpopulations (tracks 1–12) of human rotaviruses with various forms of genome rearrangements isolated from a single individual with chronic infection. The brv standard genome is shown for comparison. RNA segments (1–11) are denoted to the left as are rearranged bands (bands c and e derived from RNA 11, band d derived from RNA 10, and bands a, b, f and g from RNA 8). From Hundley *et al.* [27]. [With permission of authors and publisher]

9 (depending on strain), 10 and 11), but were also found in gene 6 [40, 46] and produced RNA profiles of great diversity which were highly atypical for group A rotaviruses (Fig. 6; [9]).

It had further been shown that within a single individual various forms of genome rearrangements (*e.g.* affecting RNA segments 8, 10 and 11) and various combinations thereof in plaque-purified viruses coexisted. At least 12 sub-populations were identified in one isolate (Fig. 7; [27]) and changes in their relative prevalence were observed over time in chronically infected hosts [27, 40]. Thus, multiple rearrangement variants coexisted in a constantly varying (dynamic) equilibrium fulfilling the criteria of the presence of a quasispecies as has been described for the coexistence of various mutants in a number of RNA viruses [13, 22, 24].

In summary, it is proposed that genome rearrangements, besides genetic point mutations [10, 43] and a reassortment continuum [38], are a third

48 U. Desselberger

principle of the evolution of rotaviruses and can contribute to the diversity of
rotaviruses in the field [9, 27, 50].

Genome rearrangements in the other genera of the Reoviridae

Genome rearrangements have also been found involving different RNA segments of several genotypes of bluetongue virus, members of the orbivirus family
[14, 42], and are likely to occur in orthoreoviruses (W. Joklik, pers. commun.).
Thus, this mechanism of genome change seems to be possible for most animal
dsRNA viruses although much less is known than for rotaviruses.

Outlook

Since the original observation of genome rearrangements in rotaviruses 10 years
ago, a lot has been learned about the detailed structure of rearranged genes and
their products, their functions and their significance for the overall diversity of
rotaviruses. There are still gaps in our knowledge about the exact mechanism(s)
by which these genome forms emerge. As a particular form of genetic recombination, genome rearrangements in double-stranded RNA viruses increase their
potential as viral vectors.

References

1. Albert MJ (1985) Detection of human rotaviruses with a "super-short" RNA pattern. Acta Paediatr Scand 74: 975–976
2. Allen AM, Desselberger U (1985) Reassortment of human rotaviruses carrying rear- ranged genomes with bovine rotavirus. J Gen Virol 66: 2703–2714
3. Ballard A, McCrae MA, Desselberger U (1992) Nucleotide sequences of normal and rearranged RNA segments 10 of human rotaviruses. J Gen Virol 73: 633–638
4. Bellinzoni RC, Mattion NM, Burrone O, Gonzalez A, La Torre JL, Scodeller EA (1987) Isolation of group A swine rotaviruses displaying atypical electropherotypes. J Clin Microbiol 25: 952–954
5. Besselaar TG, Rosenblatt A, Kidd AH (1986) Atypical rotavirus from South African neonates. Arch Virol 87: 327–330
6. Biryahwaho B, Hundley F, Desselberger U (1987) Bovine rotavirus with rearranged genome reassorts with human rotavirus. Arch Virol 96: 257–264
7. Clapp LL, Patton JT (1991) Rotavirus morphogenesis: domains in the major inner capsid protein essential for binding to single-shelled particles and for trimerization. Virology 180: 697–708
8. Clarke IN, McCrae MA (1981) A rapid and sensitive method for analysing the genome profiles of field isolates of rotaviruses. J Virol Methods 2: 203–209
9. Desselberger U (1989) Molecular epidemiology of rotaviruses. In: Farthing MJG (ed) Viruses and the Gut. Swan Press, London, pp 55–59
9a. Desselberger U (1996) Genome rearrangements of rotaviruses. Adv Virus Res 46: 69–95
10. Desselberger U, Hung T, Follett EAC (1986) Genome analysis of human rotaviruses by oligonucleotide mapping of isolated RNA segments. Virus Res 4: 357–368
11. Desselberger U, McCrae MA (1994) The rotavirus genome. In: Ramig RF (ed) Rotaviruses. Curr Top Microbiol Immunol 185. Springer, Berlin, pp 31–66

12. Dolan KT, Twist EM, Horton-Slight P, Forrer C, Bell LM, Plotkin SA, Clark HF (1985) Epidemiology of rotavirus electropherotypes determined by a simplified diagnostic technique with RNA analysis. J Clin Microbiol 21: 753–758

13. Domingo E, Martinez-Salas E, Sobrino F, De la Torre JL, Portela A, Ortin J, Lopez-Galindez C, Perez-Brena P, Villanueva M, Najera R, VandePol S, Steinhauer D, dePolo N, Holland J (1985) The quasispecies (extremely heterogeneous) nature of viral RNA genome populations: biological relevance – a review. Gene 40: 1–8

14. Eaton BT, Gould AR (1987) Isolation and characterization of orbivirus genotypic variants. Virus Res 6: 363–382

15. Eiden J, Losonski GA, Johnson J, Yolken R (1985) Rotavirus RNA variation during chronic infection of immunocompromized children. Pediatr Infect Dis 4: 632–637

16. Estes MK (1996) Rotaviruses and their Replication. In: Fields BN, Knipe D, Howley PM, Chanock RM, Melnick JL, Monath TP, Roizman B, Straus SE (eds) Fields Virology, 3rd edn. Lippincott-Raven, Philadelphia, pp 1625–1655

17. Gallegos CO, Patton JT (1989) Characterization of rotavirus intermediates: A model for the assembly of single-shelled particles. Virology 172: 616–627

18. Gault-Frere E, Cassel-Beraud AM, Garbarg-Chenon A (1995) Study of a human group A rotavirus with rearranged RNA segments isolated from an immunodeficient child: Preliminary results. Fifth International Symposium on dsRNA Viruses, Djerba, Abstract P1

19. Graham A, Kudesia G, Allen AM, Desselberger U (1987) Reassortment of human rotavirus possessing genome rearrangements with bovine rotavirus: Evidence for host cell selection. J Gen Virol 68: 115–122

20. Gonzalez SA, Mattion NM, Bellinzoni R, Burrone OR (1989) Structure of rearranged genome segment 11 in two different rotavirus strains generated by a similar mechanism. J Gen Virol 70: 1329–1336

21. Gorziglia M, Nishikawa K, Fukuhara N (1989) Evidence of duplication and deletion in super short segment 11 of rabbit rotavirus Alabama strain. Virology 170: 587–590

22. Holland JJ (1984) Continuum of change in RNA virus genomes. In: Notkins AL, Oldstone MBA (eds) Concepts in Viral Pathogenesis. Springer, New York, pp 137–143

23. Holland JJ, Kennedy SIT, Semler BT, Jones CL, Roux L, Grabau EA (1980) Defective interfering RNA viruses and the host-cell response. In: Fraenkel-Conrat H, Wagner RR (eds) Comprehensive Virology 16. Plenum Press, New York, pp 137–192

24. Holland J, Spindler K, Horodyski F, Grabau E, Nichol S, Vandepol S (1982) Rapid evolution of RNA genomes. Science 215: 1577–1585

25. Hua J, Patton JT (1994) The carboxyl-half of the rotavirus nonstructural protein NS53 (NSP1) is not required for virus replication. Virology 198: 567–576

26. Hundley F, Biryahwaho B, Gow M, Desselberger U (1985) Genome rearrangements of bovine rotavirus after serial passage at high multiplicity of infection. Virology 143: 88–103

27. Hundley F, McIntyre M, Clark B, Beards G, Wood D, Chrystie I, Desselberger U (1987) Heterogeneity of genome rearrangements in rotaviruses isolated from a chronically infected, immunodeficient child. J Virol 61: 3365–3372

28. Kapikian AL, Chanock RM (1996) Rotaviruses In: Fields BN, Knipe D, Howley PM, Chanock RM, Melnick JL, Monath TP, Roizmn B, Straus SE (eds) Fields Virology, 3rd edn. Lippincott-Raven, Philadelphia, pp 1657–1708

29. Kirkegaard K, Baltimore D (1986) The mechanism of RNA recombination in poliovirus. Cell 47: 433–443

30. Lai MMC (1992) RNA recombination in animal and plant viruses. Microbiol Rev 56: 61–79

31. Matsui SM, Mackow ER, Matsuno S, Paul PS and Greenberg HB (1990) Sequence analysis of gene 11 equivalents from "short" and "supershort" strains of rotavirus. J Virol 64: 120–124

32. Matsuno S, Hasegawa A, Mukoyama A, Inouye S (1985) A candidate for a new serotype of human rotavirus. J Virol 54: 623–624

33. Mattion N, Gonzalez SA, Burrone O, Bellinzoni R, La Torre JL, Scodeller EA (1988) Rearrangement of genomic segment 11 in two swine rotavirus strains. J Gen Virol 69: 695–698

34. McIntyre M, Rosenbaum V, Rappold W, Desselberger M, Wood D, Desselberger U (1987) Biophysical characterization of rotavirus particles containing rearranged genomes. J Gen Virol 68: 2961–2966

35. Mendez E, Arias CF, Lopez S (1992) Genomic rearrangement in human rotavirus strain Wa: analysis of rearranged RNA segment 7. Arch Virol 125: 331–338

36. Mindich L, Qiao X, Onodera S, Gottlieb P, Strassman J (1992) Heterologous recombination in the double-stranded RNA bacteriophage phi 6. J Virol 66: 2605–2610

36a. Murphy FA, Fauquet CM, Bishop DHL, Ghabrial SA, Jarvis AW, Martelli GP, Mayo MA, Summers MD (eds) (1995) Virus Taxonomy. Springer Verlag, Wien New York, pp 219–222

36b. Offit PA (1994) Rotaviruses: Immunological determinants of protection against infection and disease. Adv Virus Res 44: 161–232

37. Onodera S, Qiao X, Gottlieb P, Strassman J, Frilander M, Mindich L (1993) RNA structure and heterologous recombination in the double-stranded RNA phage phi 6. J Virol 67: 4914–4922

38. Palese P (1984) Reassortment continuum. In: Notkins AL, Oldstone MBA (eds) Concepts in Viral Pathogenesis. Springer, New York, pp 144–151

39. Paul PS, Young SL, Woode GN, Zheng S, Greenberg HB, Matsui S, Schwartz KJ, Hill HT (1988) Isolation of a bovine rotavirus with a "super-short" RNA electrophoretic pattern from a calf with diarrhoea. J Clin Microbiol 26: 2139–2143

40. Pedley S, Hundley F, Chrystie I, McCrae MA, Desselberger U (1984) The genomes of rotaviruses isolated from chronically infected immunodeficient children. J Gen Virol 65: 1141–1150

41. Pocock DH (1987) Isolation and characterization of two group A rotaviruses with unusual genome profiles. J Gen Virol 68: 653–660

42. Ramig RF, Samal SK, McConnell S (1985) Genome RNAs of virulent and attenuated strains of bluetongue virus serotypes 10, 11, 13, and 17. In: Barger TL, Yochim MM (eds) Bluetongue and Related Orbiviruses. Alan R Liss, New York, pp 389–396

42a. Rennels MB, Glass RI, Dennehy PH, Bernstein DI, Pichichero ME, Zito ET, Mack ME, Davidson BL, Kapikian AZ (1996) Safety and efficacy of high dose rhesus-human reassortant rotavirus vaccines. Report of the national multicenter trial. Pediatrics 97: 7–13

43. Sabara M, Deregt D, Babiuk LA, Misra V (1982) Genetic heterogeneity within individual bovine rotavirus isolates. J Virol 44: 813–822

44. Scott GE, Tarlow D, McCrae MA (1989) Detailed structural analysis of a genome rearrangement in bovine rotavirus. Virus Res 14: 119–128

45. Shen S, Bai ZS (1990) Genome variation and rearrangements of a lamb rotavirus after 96 passages in cell culture at high m.o.i. Chinese J Epidemiol 11: 110–114

46. Shen S, Burke B, Desselberger U (1994) Rearrangements of the VP6 gene of a group A rotavirus in combination with a point mutation affecting trimer stability. J Virol 68: 1682–1688
47. Tanaka TN, Conner ME, Graham DY, Estes MK (1988) Molecular characterization of three rabbit rotaviruses strains. Arch Virol 98: 253–265
48. Taniguchi K, Kojima K, Kobajashi N, Urasawa T and Urasawa S (1994) Properties of a bovine rotavirus variant with gene 5 having a deletion of 500 base parts. Abstract, Twenty-Eighth Joint Working Conference on Viral Diseases, Japan-US Cooperative Medical Science Program, Tokyo, Japan
49. Thouless ME, DiGiacomo RF, Neuman DS (1986) Isolation of two lapine rotaviruses: characterization of their subgroup, serotype and RNA electropherotypes. Arch Virol 89: 161–170
50. Tian Y, Tarlow O, Ballard A, Desselberger U, McCrae MA (1993) Genomic concatermerization/deletion in rotaviruses: A new mechanism for generating rapid genetic change of potential epidemiological importance. J Virol 67: 6625–6632

Author's address: Dr. U. Desselberger, Clinical Microbiology and Public Health Laboratory, Addenbrooke's Hospital, Cambridge, CB2 2QW, UK.

Arch Virol (1996) [Suppl] 12: 53–58

Structure and function of rotavirus NSP1

K. Taniguchi*, K. Kojima, N. Kobayashi, T. Urasawa, and **S. Urasawa**

Department of Hygiene, Sapporo Medical University School of Medicine,
Chuo-ku, Sapporo, Japan

Summary. Studies on the structure and function of the nonstructural proteins (NSP1-NSP5) of rotaviruses are important for dissection of the morphogenesis and replication processes of rotavirus. Above all, NSP1, the product of gene 5, has several interesting features, such as extreme sequence diversity, a highly conserved cysteine-rich region, RNA-binding activity, accumulation on the cytoskeleton, and non-random segregation in reassortment. Recently, comparable NSP1 sequence analysis has been performed on a number of rotavirus strains from various species. Furthermore, characterization of mutants with rearranged NSP1 genes has helped to elucidate the structure-function interaction of NSP1. We isolated and characterized two interesting mutants which have a large deletion including the cysteine-rich region or a nonsense codon at the early portion in the open reading frame (ORF) of the NSP1 gene. In this report, we summarize the structure and function of NSP1.

*

The complete rotavirus virion consists of three concentric protein layers and a genome of 11 segments of double-stranded RNA (dsRNA). The innermost layer (core) is composed of VP1–VP3. Surrounding of the core with VP6 forms double-layered particles. The constituents of the outermost layer are VP4 and VP7, two independent neutralization proteins. In addition to the six structural proteins, five nonstructural proteins (NSP1–NSP5) are encoded by the rotavirus genome composed of 11 segments of dsRNA. The function of most of the nonstructural proteins has not been elucidated well [1,6]. However, accumulation of sequence data and studies using NSP1 expressed in *Escherichia coli* and in insect cells have clarified the structure and function of NSP1 [2–5, 9, 12–14].

The nucleotide sequences for NSP1 have been determined for 20 group A rotavirus strains from various species and for one strain each from group B and group C rotaviruses. Comparative NSP1 sequence analysis has shown this protein generally exhibits host-specificity. This implies that NSP1 interacts with host species-specific cytoplasmic protein (s). In contrast, the fundamental structures of NSP1 of all rotaviruses are quite similar to one another, suggesting that constraints on sequence divergence operates at the level of the overall higher-order structure of NSP1 [22]. Furthermore, there is a correlation between gene 5 sequences and genogroup [4,12,14].

The 5′ and 3′ noncoding regions are predicted to have a similar stem-loop structure. In particular, the 5′-noncoding region seems to contain signals for packaging of RNAs into a replication intermediate (RI) and for regulating mRNA translation [9]. In group A rotaviruses, gene 5 ranges in size from 1564 to 1611, and the total number of amino acids of NSP1 varies from 486 to 495. The overall homology for gene 5 and its protein product is extremely low among strains, in particular among strains from different species. Despite the sequence divergence, the first 150 amino acids of NSP1 exhibit a greater degree of conservation than does the remaining sequence. In particular, the cysteine-rich motif $C-X_2-C_2-X_7-C-X_2-C-X_3-H-X-C-X_2-C-X_5-C$ at amino acids 42–72 is highly conserved in all strains (Fig.1), including the group C Cowden strain, but not in gene 7 of the group B rotavirus (IDIR strain) equivalent to gene 5 of group A rotavirus [5]. This motif may form one or two zinc fingers. The second motif $H-X_2-C/H-X_6-C-X_2-C$ is found at amino acids 314–327 in only a few strains such as UK, RF, and Cat2. NSP1 is a metalloprotein and binds zinc. Since a zinc finger motif has been detected in regulatory and nucleic acid binding proteins, NSP1 might have such a regulatory role in virus replication. However, it has not been demonstrated that this binding activity is due to the putative zinc fingers.

Three basic regions (amino acids 10–39, 79–93, 111–126) are also well conserved (Fig.1). In addition NSP1 is rich in lysine and arginine residues. These features in the sequence imply that NSP1 interacts with nucleic acids. NSP1 indeed has an RNA-binding activity and a specific affinity for all 11 rotavirus mRNAs [8]. NSP1 specifically recognizes an element located near the 5′ ends of viral mRNAs. The RNA binding domain has been mapped within the first 81 amino acids and a cysteine-rich region was found to be essential for this activity [8].

NSP1 is reported to be produced in low levels in infected cells. However, the level of NSP1 production may be dependent on the particular strains, since we detected a high level of NSP1 synthesis in some but not all strains. NSP1 is present in the cytoplasm (cytosol) and shows a punctate and filamentous

```
  1 MATFKDACYH YRKLNKLNGL VLKLGANDAW RPAPIAKYKG WCLDCCQHTD LTYCRGCALF
 61 HVCQWCSQYN RCFLDEEPHL LRMRTFRNQI SRKDIEGLIN MYNTLFPINE RIVDKFISNV
121 KQRRCRNEFL IEWYNHLLLP ITLQALMVEL EGDVYYIFGY FDHMEKENQT PFQFVNMINN
181 YDKLLLDDKN FDRMTNLPVI LQQEYAFRYF SKSRFISKTK KSVNRHDFAN NLMEEMDNPI
241 SLMQVMRNCV NEYMDDKNWN EKCTLIVDMK SYMELMKSSY TEHYSVSQRC KLFTIYKLNI
301 ISKLIKPNYI FSNHGAHALD VHNCKWCQMN NHYKIWDDFR LKKVYNNTMC FIRALMKSNK
361 NVGHRSSQEV VYEYMSNIFI VWKNEKWNKS MQMIFDYLEP VEISGIEYIL LDHELSWEIR
421 GIVMQIMNGE IPRILTFNDV KKIISAIIYD WFDVRYMREM PLIISTTNEL RKMNKRNDLM
481 DEYSYELSDT E
```

Fig. 1. The deduced NSP1 amino acid sequence of a bovine strain A5–13. Cysteine-rich region are boxed, and three basic amino acid sequence regions are underlined

distribution. NSP1 interacts with and accumulates on the cytoskeletal network [10]. The intracellular localization domain was found to reside between amino acids 84 and 176 immediately downstream from the RNA binding domain [9]. NSP1 is detected at a very early stage of virus assembly and is associated with early replicative intermediates (precore RI), which are composed of VP1 and VP3. As precore RI matures into core RI by association with VP2 and then into the double-layered RI, NSP1 is lost from the intermediate. Thus, NSP1 might take viral mRNAs to the cytoskeleton, where with VP1 and VP3 they are assembled into precore RIs.

Reassortment studies *in vivo* and *in vitro* have shown that gene 5 does not segregate randomly but in a non-random fashion [7, 11, 17]. In addition, in the suckling mouse model using reassortants, gene 5 appeared to be involved in the restriction of growth of simian rotavirus strains, and to be correlated with pathogenic phenotype [7, 17].

The structure and function of NSP1 have been studied also by characterizing mutants with rearranged NSP1 genes (Fig. 2). An equine rotavirus, mutant P9△5, has a 308 base pair (bp) deletion in the center of the normal gene 5 and its protein product contains only 150 amino acids [21]. In addition, a bovine rotavirus mutant, brv A, has a gene 5 with a head-to-tail duplication of 1112 bp, but this rearranged gene has a point mutation resulting in the generation of a termination codon near the middle of the ORF. The resultant NSP1 of the

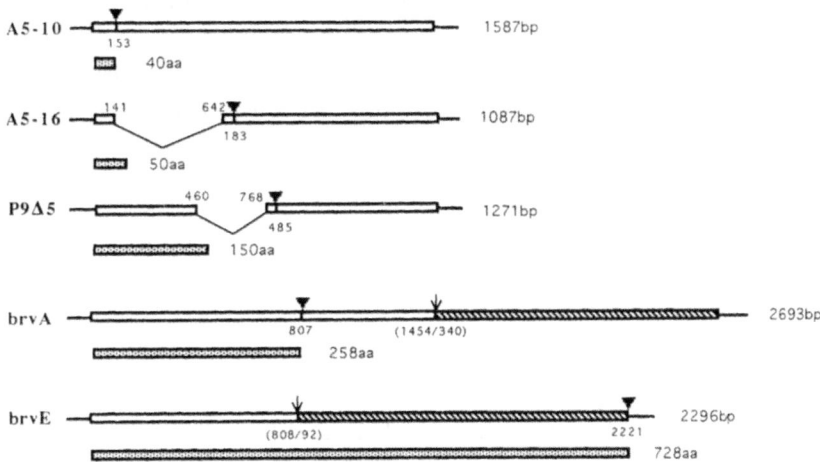

Fig. 2. Structure of NSP1 of nondefective rotavirus mutants with rearranged NSP1 gene ▢: translated NSP1 protein. Truncated NSP1 of A5–10, A5–16, and P9△5 mutants were not detected in the virus-infected cells. ▼ : the location where the first termination codon is found. Figures under the NSP1 gene show the numbers of the first nucleotide of the termination codon. NSP1 genes of A5–16 and P9△5 have a 307-bp (nucleotide no. 461–767) and a 500-bp (nucleotide no. 142–641) deletions, respectively. brv A and brv E show a head-to-tail concatemerization. Arrows indicate the junction points between nucleotide sequence 1454 or 808 of the upstream copy (▢) and nucleotide 340 or 92 of the downstream copy (▨)

brv A mutant produces a truncated NSP1 of 258 amino acids [10]. Because the above two mutants with an NSP1 lacking the carboxy-half are viable, the carboxy-half of NSP1 is not required for rotavirus replication. However, together with the deletion mapping studies [8], the above studies using the rearranged NSP1 gene showed the necessity of having a cysteine-rich region in the amino-terminal half.

We also isolated bovine rotavirus mutants (strain A5) having rearranged NSP1 genes [18] (Fig.2), while studying the antigenic and genomic properties of human, bovine, and porcine rotaviruses in Thailand [15, 16, 19, 20]. Strain A5 has a long RNA pattern and shows subgroup I, G serotype 8 (69M-like VP7), and P serotype 6 (NCDV-like VP4) specificities. From the A5 virus stock, we obtained three clones (A5–10, A5–13, and A5–16) by plaque-purification whose RNA segments 5 showed different migration patterns. Clone A5–16 exhibited an unusual RNA profile; the RNA segment corresponding to segment 5 was lost and an RNA band migrating at the site around RNA segments 7–9 complex appeared.

Sequencing of gene 5 of the three clones was carried out by the dideoxynucleotide chain termination method using single-stranded RNA (ssRNA) transcripts and cloned cDNA. A5–13 gene 5 codes for 491 amino acids. Gene 5 of A5–16 was found to be derived from that of A5–10 and to have a large deletion including the nucleotide sequence encoding the cysteine-rich region. The mechanism for the large deletion is not known. However, it was of note that repeated sequences were found both upstream of the deletion site and at the end of the deletion sequence. When the amino acid sequences of NSP1 of A5–10 and A5–16 were reduced, nonsense codons at the early portions of their gene 5 were found just upstream of the coding region of the cysteine-rich zinc finger motif (Fig. 2). If these termination codons are functional, the A5–10 or A5–16 gene 5 product would produce only a low molecular weight protein; otherwise; a readthrough for A5–10 or a frameshift for A5–16 might occur to produce the normal sized or longer NSP1, respectively. No significant strong protein bands corresponding to the authentic or truncated NSP1 were found in cells infected with A5–10 or A5–16 cloned virus and in *in vitro* translation products of A5–10 or A5–16 gene 5-specific transcripts. Nevertheless, A5–10 and A5–16 were both non-defective and infectious, and the virus titers of these clones were comparable to other animal rotaviruses. The plaque size of A5–13 was extremely large and A5–10 produced medium-sized plaques. In contrast, the plaque size of A5–16, whose gene 5 has a 500 base deletion from A5–10 gene 5, was very small. These results suggest that gene 5 may affect the efficiency of virus replication. Preparation and characterization of reassortants with the gene 5 from A5–10 or A5–16 are required in order to assess the precise association of gene 5 with replication characteristics.

Furthermore, comparable replication in cell culture and plaque formation, though of low efficiency, were found in A-16 whose gene 5 lacks the region encoding the cysteine-rich amino acid sequence. These results imply that the cysteine-rich region is not necessarily required for virus replication, at least *in vitro,* even if the two clones might be exceptional cases. This is in striking contrast to the previous data, which showed the importance of the cysteine-rich region for virus replication [8, 10, 21]. Further studies will be needed to determine whether the gene 5 product itself is actually necessary for virus replication.

References

1. Both GW, Bellamy AR, Mitchell DB (1994) Rotavirus protein structure and function. Curr Top Microbiol Immunol 185: 67–106
2. Bremont M, Charpilienne A, Chabanne O, Cohen J (1987) Nucleotide sequence and expression in *Escherichia coli* of the gene encoding the non-structural protein NCVP2 of bovine rotavirus. Virology 161: 138–144
3. Brottier P, Nandi P, Bremont M, Cohen J (1992) Bovine rotavirus segment 5 protein expressed in the baculovirus system interacts with zinc and RNA. J Gen Virol 73: 1931–1938
4. Dunn SJ, Cross TL, Greenberg HB (1994) Comparison of the rotavirus nonstructural protein NSP1 (NS53) from different species by sequence analysis and Northern blot hybridization. Virology 203: 178–183
5. Eiden JJ (1994) Expression and sequence analysis of gene 7 of the IDIR agent (group B rotavirus): similarity with NS53 of group A rotavirus. Virology 199: 212–218
6. Estes M, Cohen J (1989) Rotavirus gene structure and function. Microbiol Rev 53: 410–449
7. Gombold JL, Ramig RF (1986) Analysis of reassortment of genome segments in mice mixedly infected with rotaviruses SA11 and RRV. J Virol 57: 110–116
8. Hua J, Chen X, Patton JT (1994) Deletion mapping of the rotavirus metalloprotein NS53 (NSP1): the conserved cysteine-rich region is essential for virus-specific RNA binding. J Virol 68: 3990–4000
9. Hua J, Mansell EA, Patton JT (1993) Comparative analysis of the rotavirus NS53 gene: conservation of basic and cysteine-rich regions in the protein and possible stem-loop structures in the RNA. Virology 196: 372–378
10. Hua J, Patton JT (1994) The carboxyl-half of the rotavirus protein NS53 (NSP1) is not required for virus replication. Virology 198: 567–576
11. Kobayashi N, Kojima K, Taniguchi K, Urasawa T, Urasawa S (1994) Genotypic diversity of reassortants between simian rotavirus SA11 and human rotaviruses having different antigenic specificities and RNA patterns. Res Virol 145: 303–311
12. Kojima K, Taniguchi K, Kobayashi N (1996) Species-specific and interspecies relatedness of NSP1 sequences in human, porcine, feline, and equine rotavirus strains. Arch Virol 141: 1–12
13. Mitchell DB, Both GW (1990) Conservation of a potential metal binding motif despite extensive sequence diversity in the rotavirus nonstructural protein NS53. Virology 174: 618–621
14. Palombo EA, Bishop RF (1994) Genetic analysis of NSP1 genes of human rotaviruses isolated from neonates with symptomatic infection. J Gen Virol 75: 3635–3639
15. Pongsuwanna Y, Taniguchi K, Choonthanom M, Chiwakul M Susansook T, Saguan-wongse S, Jayavasu S, Urasawa S (1989) Subgroup and serotype distributions of human, bovine, and porcine rotavirus in Thailand. J Clin Microbiol 27: 1956–1960
16. Pongsuwanna Y, Taniguchi K, Wakasugi F, Sutivijit Y, Chiwakul M, Warachit P, Jayavasu C, Urasawa S (1993) Distinct yearly change of serotype distribution of human rotavirus in Thailand as determined by ELISA and PCR. Epidemiol Infect 111: 407–412

17. Ramig RF, Ward RL (1991) Genomic segment reassortment in rotaviruses and other Reoviridae. Adv Virus Res 39: 163–207
18. Taniguchi K, Kojima K, Urasawa S (1996) Non-defective mutants with NSP1 gene which has a deletion of 500 nucleotides, including a cysteine-rich zinc finger motif-encoding region (nucleotides 156 to 248), or which has a nonsense codon at nucleotides 153 to 155. J Virol 70: 4125–4130
19. Taniguchi K, Urasawa T, Pongsuwanna Y, Choonthanom M, Jayavasu C, Urasawa S (1991) Molecular and antigenic analyses of serotypes 8 and 10 of bovine rotaviruses in Thailand. J Gen Virol 72: 2929–2937
20. Taniguchi K, Urasawa T, Urasawa S (1993) Independent segregation of the VP4 and the VP7 genes in bovine rotaviruses as confirmed by VP4 sequence analysis of G8 and G10 bovine rotavirus strains. J Gen Virol 74: 1215–1221
21. Tian Y, Tarlow O, Ballard A, Desselberger U, McCrae MA (1994) Genomic concatemerization/deletion in rotaviruses: a new mechanism for generating rapid genetic change of potential epidemiological importance. J Virol 67: 6625–6632
22. Xu L, Tian Y, Tarlow O, Harbour D, McCrae MA (1994) Molecular biology of rotaviruses, IX. Conservation and divergence in genome segment 5. J Gen Virol 75: 3413–3421

Authors' address: Dr. K. Taniguchi, Department of Hygiene, Sapporo Medical University School of Medicine, South-1, West-17, Chuo-ku, Sapporo 060, Japan.

Arch Virol (1996) [Suppl] 12: 59–67

Identification of the minimal replicase and the minimal promoter of (−)-strand synthesis, functional in rotavirus RNA replication *in vitro*

M. J. Wentz[1], **C. Q.-Y. Zeng**[1], **J. T. Patton**[2], **M. K. Estes**[1], and **R. F. Ramig**[1,*]

[1]Division of Molecular Virology, Baylor College of Medicine,
One Baylor Plaza, Houston, U.S.A.
[2]Department of Microbiology and Immunology,
University of Miami School of Medicine, Miami, Florida, U.S.A.

Summary. An *in vitro* replication system supporting the initiation and synthesis of complete rotavirus (−)-strands on (+)-strand template RNA (Chen *et al.*, J Virol 68: 7030, 1994) was used to examine several parameters related to rotavirus RNA replication. Coexpression of VP1/2/3 in all possible combinations from baculovirus vectors revealed: [i] Virus-like particles (VLPs) were formed only if VP2 was present, and [ii] VP1/2 and VP1/2/3 VLPs had replicase activity in the *in vitro* system whereas VP2/3 and VP2 VLPs did not. Thus, the minimal replicase is composed of VP1 and VP2 and replicase activity is associated with VP1. *In vitro* replication reactions, using T7 transcripts of porcine rotavirus OSU genome segment 9 as reporter template, were performed to map *cis*-acting elements that regulate replication. Internal deletions and terminal truncations of the reporter RNA localized a replication signal, conferring full template activity, to the 5′-terminal 27 nucleotides (nt 1–27) and the 3′-terminal 26 nucleotides (nt 1037–1062). Further analysis showed that a minimal promoter of (−)-strand synthesis was contained in the 3′-terminal 7 nucleotides (nt 1056–1062); the sequence conserved at the 3′-terminus of all rotavirus genes. Hybrid constructs with this promoter had minimal, but detectable, template activity. This result indicated that upstream sequences between nucleotides 1037–1055 positively regulate the activity of the minimal promoter.

Introduction

Group A rotaviruses are the most significant cause of severe gastroenteritis in young children and animals [4]. While a great deal has been learned about the molecular biology of rotaviruses in the two decades since their discovery, little is known about synthesis of the double-stranded RNA genome.

Replication of rotavirus RNA is an asymmetric process in which synthesis of both plus-and minus-strands occurs in subviral particles. Following viral entry and uncoating, the transcriptase of the double-layered particle synthesizes the 11 viral mRNAs. The free mRNAs are translated and, after an accumulation of viral proteins in the cytoplasm, subviral particles are formed that contain viral structural and nonstructural proteins and the 11 viral mRNAs. In the newly-formed subviral particles, an RNA-dependent RNA polymerase uses the plus-strand mRNAs as templates for the synthesis of minus-strand RNA in a process known as replication. Replication produces the dsRNA genome. Replication and morphogenesis appear to be intimately associated [9].

We recently defined an *in vitro* system capable of replicating rotavirus plus-strand templates to produce *bona fide* rotavirus dsRNA [2]. As templates, this system utilizes either [i] mRNAs synthesized *in vitro* by the rotavirus transcriptase, or [ii] T7 transcripts with native rotavirus 5'-GG...and...CC-3' termini transcribed *in vitro* from rotavirus cDNAs. As replicase, the system utilizes either [i] open cores containing VP1/2/3 derived by stepwise degradation of purified rotavirions, or [ii] virus-like particles made by co-expression of VP1/2/3 from baculovirus vectors in insect cells. In addition, the reaction contains salts and NTPs. The *in vitro* replication system does not require viral nonstructural proteins, and specifically replicates rotavirus templates [2]. The product of the *in vitro* replication system is authentic rotavirus dsRNA as shown by its unit length, the migration of the labeled product strand as ($-$)-strand RNA in strand-separating gels, and the initiation of the ($-$)-strand with a G residue opposite the 3'-terminal C residue of the ($+$)-strand template RNA (Wentz, Patton and Ramig, unpublished data).

Here we report the application of the *in vitro* replication system to: [i] identify particles containing only VP1 and VP2 as the minimal replicase particle, and [ii] identify *cis*-acting elements on a synthetic transcript that are required for replication of the template *in vitro*.

Materials and methods

Cells and virus. MA104 monkey kidney cells were grown and maintained as previously described [12]. *Spodoptera frugiperda* (Sf9) cells were grown and maintained as described [13]. A variant of the group A rotavirus prototype SA11 (simian rotavirus SA11-4F) and porcine rotavirus OSU were from our laboratory stock collection and were cultivated in MA104 cells in the presence of trypsin and purified as described previously (1). Baculovirus expression vectors containing rotavirus genes 1, 2, 3, and 6, were the same as described [2].

Preparation of baculovirus-expressed virus-like particles. VLPs containing VP1/2/3 were prepared by baculovirus coexpression of VP1/2/3/6 followed by dialysis against low ionic strength buffer as described previously [2, 15]. VLPs containing subsets of these proteins were prepared by the same methods after infection of Sf9 cells with the desired subsets of the rotavirus expression vectors.

Preparation of SA11-4F open core particles. Open core particles with replicase activity were prepared from purified SA11-4F as described [2].

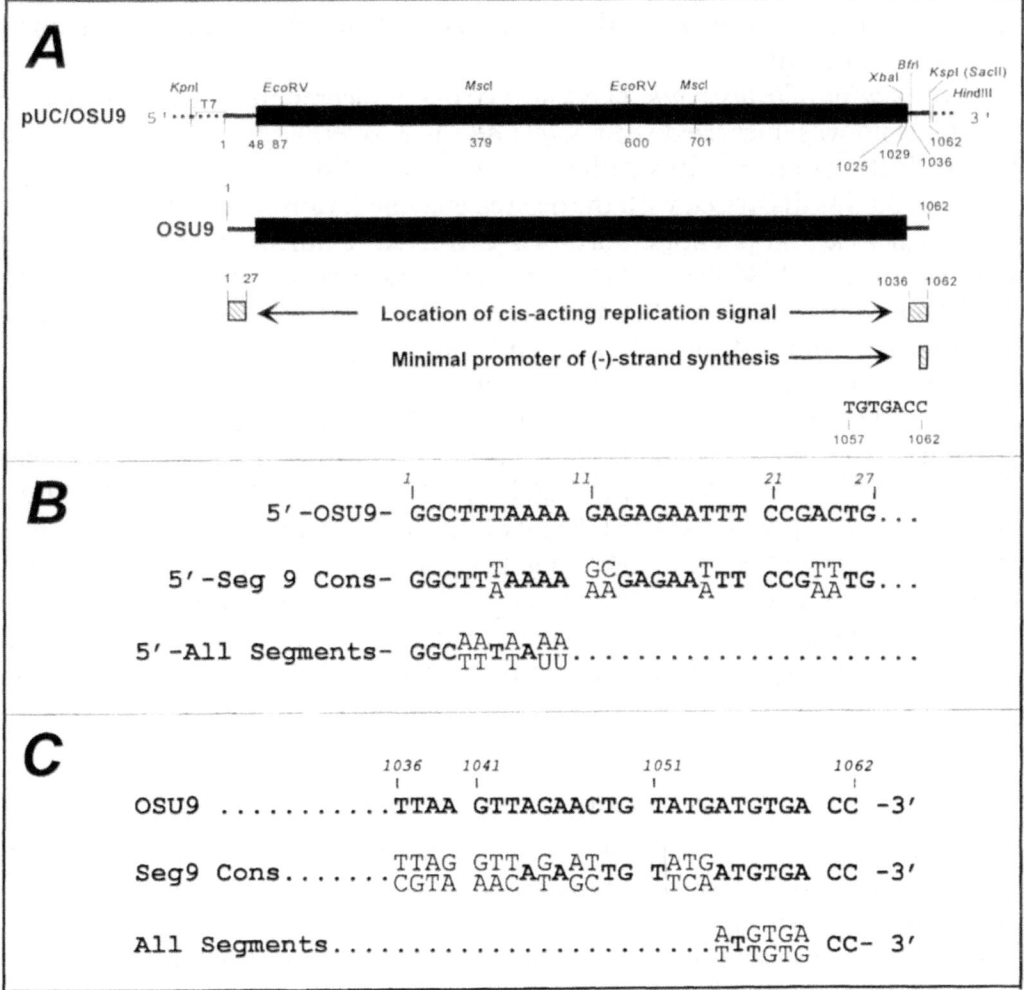

Fig. 1. *Panel A,* Structure of the transcription plasmid (pUC/OSU9) used to make OSU9 transcripts and which served as the backbone for construction of other constructs. The 1062 nucleotide mRNA of OSU segment 9 was PCR cloned behind a T7 promoter so that transcripts initiate with the native 5′-GG. Restriction of the plasmid with *Ksp*I and filling the overhand with T4 DNA polymerase results in native ..CC-3′-termini on the transcripts. Relevant restriction sites are shown. OSU9 shows the structure of the full-length transcript; narrow lines indicate the 5′- and 3′-untranslated regions and the wide line indicates the open reading frame. Cross-hatch boxes below indicate the positions mapped for the *cis*-acting signal conferring full template activity and the location of the minimal promoter of (-)-strand synthesis. *Panel B,* 5′-terminal sequences. *Panel C,* 3′-terminal sequences. Panels B and C show the indicated sequence for OSU9, the consensus sequence derived from segment 9 of virus strains representing all 14 G serotypes, and the consensus sequences at the extreme termini of all rotavirus mRNAs

Construction of transcription vectors. A transcription vector (pUC/OSU9; Figure 1A), for the expression of full-length porcine rotavirus strain OSU segment 9 transcripts with precise 5′ and 3′ termini, was constructed as described [2]. Internal deletions were made by standard techniques using convenient restriction sites or exonuclease III digestions. 3′-terminal truncations were made at convenient restriction sites.

Preparation of transcripts. The wildtype or engineered pUC/OSU9 transcription plasmid was linearized with *KspI* and the 3′-overhang was blunt-ended for the production of transcripts with *bona fide* 3′ ends. For 3′ truncated transcripts, the plasmid was linearized with the desired enzyme. Transcripts were made using T7 polymerase and phenol/chloroform extracted. Unincorporated nucleotides were removed with Sephadex G-50 spin columns. Transcripts were quantitated by Absorbance$_{260}$ and quality was assessed by electrophoresis in 5% polyacrylamide gels buffered with 1X TBE and containing 8 M urea.

In vitro replication of synthetic templates. *In vitro* replication was performed in 20 μl reactions that contained: 100 mM Tris-HCl, pH 7.2; 10 mM MgOAc; 2% polyethylene glycol 8000; 800 U/ml Rnasin; 2 mM dithiothreitol; 1.25 mM each ATP, GTP, CTP; 0.5 mM UTP, 15 μCi α-[^{32}P]-UTP (3000 Ci/mM); 0.75 μg of either VLPs or open cores; and 0.75 μg of the desired transcript. Reactions were run for 4 hours at 37 °C. Following the reaction, the reaction mixture was extracted with phenol-chloroform, ethanol precipitated, sample buffer was added, and the products were resolved in SDS-12% polyacrylamide gels. Dried gels were either exposed to film or product bands were quantitated by β-scan.

Results

VP1/2 particles constitute the minimal replicase in vitro

We previously reported that rotavirus subviral particles containing VP1, VP2, and VP3, derived from either sequential degradation of virions or by coexpression of VP1/2/3/6 from baculovirus vectors, exhibited replicase activity in our *in vitro* replication system. Expression as double-layered particles containing VP6 was necessary to obtain purified, monodisperse particles because expression without VP6 led to extreme aggregation of the particles upon purification [2]. To determine if VP1/2/3 were all required for replicase activity, we expressed VLPs containing all combinations of VP1/2/3, either with or without coexpression of VP6. VLPs containing VP6 were purified and dialyzed against low ionic strength buffer [2, 15]. VLPs without VP6 were purified in the presence of detergent (1% CHAPS or 1% sodium deoxycholate). For each coexpression combination we determined [i] the formation of VLPs, [ii] the protein content of the purified VLPs, [iii] the aggregation state of the purified VLPs, and [iv] the *in vitro* replicase activity associated with the purified VLPs.

The results of these coexpression experiments are summarized in Table 1. The replicase activity of VLPs containing VP1/2/3/6, VP1/2/6, VP1/2/3, and VP1/2, together with the absence of replicase activity in VLPs containing VP2/3/6, VP2/6, VP2/3, and VP2, indicated that VP3 was not a required component for replicase activity. On this basis, we concluded that VP1/2 VLPs constitute the minimal replicase particle *in vitro.* Several other observations were of interest. [i] VLPs were formed only in coexpressions where VP2 was present. [ii] VLPs lacking VP6 could be purified with much less aggregation in the presence of

Table 1. Coexpression of VP1/2/3/6 in various combinations and replicase activity associated with the particles

Proteins Coexpressed[a]	VLPs Formed[b]	Proteins in VLP[c]	VLP Aggregation[d]	Replicase Activity[e]
VP1/2/3/6	+	VP1/2/3/6	−	+ +
VP1/2/6	+	VP1/2/6	−	+ +
VP2/3/6	+	VP2/3/6	−	−
VP2/6	+	VP2/6	−	−
VP1/2/3	+	VP1/2/3	+	+
VP1/2	+	VP1/2	+	+
VP2/3	+	VP2/3	+	−
VP2	+	VP2	+	−
VP1/3	−	ND[f]	ND	ND
VP1	−	ND	ND	ND
VP3	−	ND	ND	ND

[a] Vectors encoding the indicated rotavirus proteins were coexpressed in Sf9 cells as described .
[b] VLP formation assayed by electron microscopy and ability to isolate particles in gradients
[c] Protein content of VLPs verified by Western blot and, for VP3, GTP binding assay
[d] VLP aggregation determined by electron microscopy of purified VLPs
[e] Replicase activity determined in standard assay using full-length OSU9 template
[f] Not determined

nonionic detergent, and these VLPs showed significant replicase activity in the presence of residual detergent remaning after dialysis. [iii] VP1 and VP3 did not form detectable VLPs or oligomeric forms when coexpressed.

Cis-*acting elements required for (−)-strand synthesis map near the 5′-and 3′-termini*

We previously described the construction of a transcription plasmid for the production of transcripts of genome segment 9 of rotavirus OSU [2]. Here, 3'-terminal truncations and additions, internal deletions, and hybrid constructs were made and tested for template activity in the *in vitro* replication system, with the goal of identifying *cis*-acting elements on the (+)-strand templates that regulated (−)-strand synthesis. The transcription plasmids and transcripts shown in Table 2 were made as described and tested in a two template reaction for template activity. Each replication reaction contained equimolar amounts of the test transcript and an internal control transcript known to replicate with equal efficiency to the wildtype OSU9 transcript. The relative activity of the test template is reported relative to the activity of the internal control template. The template activity of wildtype OSU9 template was 100%.

3′-terminal truncations and additions. Truncation of OSU9 by 37 nt (OSU9-3′ *Xba*) or 26 nt (OSU9-3′ *Bfr*) abolished their function as replication templates

M. J. Wentz et al.

Table 2. Transcripts assayed for template activity to identify *cis*-acting elements regulating (−)-strand synthesis

Transcript	Transcript Length[a]	OSU9 Nucleotides Present[b]	Template Activity[c]	Foreign Nucleotides Present
Parental Transcript				
OSU9	1 062	1–1 062	+ + + +	
3′-Terminal Truncations and Additions				
OSU9-3′*Xba*	1 025	1–1 025	−	
OSU9-3′*Bfr*	1 036	1–1 036	−	
OSU9-3′*Bsa*J1	1 061	1–1 061	+ +	
OSU9-3′*Bst*U1	1 063	1–1 062	+ + +	G at nt 1 063
OSU9-3′*Hind*III	1 067	1–1 062	−	GCGGA at nt 1 063–1 067
3′-Internal Deletions				
OSU9-ΔRV	549	1–87 & 600–1 062	+ + + +	
OSU9-ΔMsc	740	1–379 & 701–1 062	+ + + +	
OSU9-ExoIIIΔ108	239	1–107 & 930–1 062	+ + + +	
OSU9-ExoIIIΔ28	142	1–27 & 947–1 062	+ + + +	
OSU9-ExoIIIΔ162	216	1–161 & 1007–1 062	+ + + +	
OSU9-ExoIIIΔ87	378	1–87, 600–854 & 1 025–1 062	+ + + +	
OSU9-ExoIIIΔ115	141	1–115 & 1 036–1 062	+ + + +	
5′-Terminal Deletions				
OSU9-ΔRV	549	1–87 & 600–1 062	+ + + +	
OSU9-ExoIIIΔ38	161	1–37, 92–161 & 1 007–1 062	+ + + +	
OSU9-ExoIIIΔ28	142	1–27 & 947–1 062	+ + + +	
OSU9-Δ3–87	976	1–2 & 87–1 062	+	
Hybrid Construct				
OSU9-L7[d]	540	1 & 1 056–1 062	+	532 nt between 1 & 1 056

[a] Length of tested transcript in nucleotides

[b] Nucleotides of wildtype OSU9 contained within the construct, by nucleotide number

[c] Template activity determined in standard replicase assay containing equimolar amounts of test templa and a control template that replicates with same efficiency as wildtype. Template activity: − = 0% + = 1–25%; + + = 26–75%, + + + = 76–99%; + + + + = 100% as compared to internal control

[d] Contains 532 nt of bacterial sequence between G at residue 1 and indicated sequence at 3′ terminu Requires greater than equimolar test transcript to detect replication

(Table 2). Truncation of OSU9 by a single C residue (OSU9-3′*Bsa*J1) reduced the template activity by approximately 50 % whereas extension of OSU9 by a single G residue (OSU9-3′*Bst*U1) reduced template activity by approximately 10%. Addition of 5 residues at the 3′-terminus of OSU9 (OSU9-3′*Hind*III) abolished function as replication template. We conclude that a required *cis*-acting replication signal lies at or very near the 3′-terminus of the OSU9 reporter template.

3′-terminal internal deletions. A series of overlapping internal deletions, approaching the 3′-terminus of OSU9 from upstream were examined to define the 5′-limit of the *cis*-acting signal mapped near the 3′-terminus of the template.

Overlapping deletions extending to the 5'-side of nt 600 (OSU9-ΔRV) to nt 1 036 (OSU9-ExoIIIΔ115) all had full template activity (Table 2). We conclude that the cis-acting replication signal near the 3'-terminus lies downstream of nt 1 036.

5'-terminal internal deletions. OSU9-ΔRV had full template activity and had only nt 1–87 from the 5'-terminus of OSU9, indicating that any cis-acting signal at the 5'-terminus was upstream of nt 87 (Table 2). Exending the internal deletion in the 5' direction to nt 37 (OSU9-ExoIIIΔ38) or to nt 27 (OSU9-ExoIIIΔ28) yielded templates with full activity. A construct that deleted nt 3–87 (OSU9-Δ3-87) had a reduced template activity of approximately 20 %. We concluded that a cis-acting signal required for full template activity mapped between nt 3–27 near the 5'-terminus of the OSU9 template.

The minimal promoter of (−)-strand synthesis

The cis-acting signal identified at the 3'-terminus of OSU9 contains both 3'-subterminal segment 9-specific sequences (nt 1 036–1 055) and the sequence common to the 3'-terminus of all 11 rotavirus segments (nt 1056–1062; Fig. 1C). To determine if the 3'-terminal 7 nt common to all 11 rotavirus segments functioned as a cis-acting replication signal, we constructed the template OSU9-L7 (Table 2). This template contained 8 nt at the identical positions as they are in OSU9; a G residue at nt 1 and the TGTGACC conserved sequence at the 3'-terminus. Between the rotavirus sequences were 532 nt derived from the pCITE™ vector. OSU9-L7 had 7% template activity as compared to full-length OSU9. We conclude that the 3'-terminal 7 nt of OSU9 comprise the minimal promoter of (−)-strand synthesis.

Discussion

Here, we report three new findings relative to replication of rotavirus RNA in vitro. [i] The minimal composition of particles that have replicase activity in vitro is VP1 and VP2. [ii] Cis-acting signals that confer full template activity on a 1 062 nucleotide (+)-strand reporter template lie at both 5'-and 3'-termini. The 5'-terminal signal lies between nucleotides 1–27 and the 3'-terminal signal lies between nucleotides 1 036–1 062. Templates containing these signals replicate with efficiency equal to the wildtype, full-length template. [iii] A sequence of 7 nucleotides at the extreme 3'-terminus of the (+)-strand template constitutes the minimal promoter of (−)-strand synthesis. This 7 nucleotide signal is identical to the 7 nucleotides conserved at the 3'-terminus of all 11 rotavirus mRNAs (Fig. 1C).

Examination of both single-layered and double-layered VLPs in the in vitro replication system revealed that only VLPs which contain both VP1 and VP2 had replicase activity, regardless of their content of other protein species. On this basis we conclude that VP1/2 particles constitute the minimal replicase particle in the in vitro replication system. We conclusively demonstrated that VP3 is not required in VLPs for replicase activity. VP3 binds GTP specifically and is assumed to be a guanylyltransferase [7, 10] which most likely functions during transcription, rather than replication. Our observation that VP3 is not required for replicase activity is consistent with this notion. VP1 appears from our results to be the replicase, an observation consistent with the presence of motifs common to all RNA polymerases [3, 11] and the ability to cross-link azido-ATP to VP1 of double-layered particles [14]. We found that the presence of VP2 was required for the assembly of VLPs with other proteins, especially VP1 and VP3. This suggests that VP2 acts as a scaffold onto which VP1 and/or VP3 is assembled, since we have been unable to detect particles containing either protein in the absence of VP2. Furthermore, VP2 may impose an active conformation on VP1, as we and others [3] have been unable to detect replicase activity in lysates containing VP1 expressed alone.

Deletion analysis showed that *cis*-acting signals required for efficient template activity of (+)-strand templates mapped in the near-terminal regions at both the 5′-and 3′-termini of the OSU9 reporter template. Specifically, the 5′-terminal signal mapped to nucleotides 1–27 and the 3′-terminal signal to nucleotides 1036–1062. These sites are contained within larger regions identified by Gorziglia and Collins [5]. At each the 5′-and 3′-termini lie conserved sequences that are common to all 11 rotavirus mRNAs [6], and immediately subterminal lie sequences that are conserved within a specific genome segment ([8] Figure 1B, C). The 5′ -and 3′-terminal replication signals identified here contain both the terminal rotavirus-specific sequence and the subterminal segment-specific sequence. In the absence of additional fine-structure mapping of the 5′-and 3′-terminal signals, one cannot yet determine if the signals will be specific (and potentially different) for each mRNA, or if they will be entirely contained within the rotavirus conserved terminal sequence and, thus, be universal for all segments. However, further dissection of the 3′-terminal signal suggests that the fully functional signal may be specific for each mRNA species (below).

A template (OSU9-L7) which contained nucleotide 1 and the 3′-terminal seven nucleotides (TGTGACC; nt 1056–1062) of OSU9, flanking 532 nucleotides of bacterial sequence, was detectably replicated in our *in vitro* system. Although the template activity of this transcript was minimal, it was reproducible and quantifiable as ∼7% of wildtype activity. This result suggests that the 3′-terminal seven nucleotides of OSU9 function as the minimal promoter of (−)-strand synthesis. If this hypothesis is correct, the minimal promoter of (−)-strand synthesis will be common to all 11 rotavirus mRNAs. However, inclusion of template sequences upstream of the 3′-terminal seven nucleotides (*e.g.*, OSU9-△3-87 or OSU9-ExoIII△115) restores significant template activity to the transcript. These results indicate that, if the 3′-terminal seven nucleotides constitute the minimal promoter of (−)-strand synthesis, there are sequences immediately upstream (between nt 1036–1055) that positively regulate the activity of the minimal promoter.

Site-directed mutagenesis studies are currently underway to more precisely define the minimal promoter of (−)-strand synthesis and identify the upstream positive regulatory sequences. Additional studies will also be required to more precisely define the regulatory sequence at the 5′-terminus of the template.

Acknowledgements

We acknowledge support from The National Institutes of Health (grants DK 30144 [MKE]; AI 16687 and AI 36385 [RFR]).

References

1. Chen D, Ramig, RF (1992) Determinants of rotavirus stability and density during CsCl purification and storage. Virology 186: 228–237
2. Chen D, Zeng Q-Y, Wentz M, Gorziglia M, Estes MK, Ramig RF (1994) Template-dependent, *in vitro* replication of rotavirus RNA. J Virol 68: 7030–7039
3. Cohen J, Charpilienne A, Shilmonczyk S, Estes MK (1989) Nucleotide sequence of bovine rotavirus gene 1 and expression of the gene product in baculovirus. Virology 171: 131–140
4. Estes MK (1996) Rotaviruses and their replication. In: Fields BN, Knipe DM, Howley PM (eds) Fields Virology. Lippincott-Raven, Philadelphia, pp 1625–1655
5. Gorziglia M, Collins P (1992) Intracellular amplification and expression of a synthetic analog of rotavirus genomic RNA bearing a foreign marker gene: mapping *cis*-acting nucleotides in the 3'-noncoding region. Proc Natl Acad Sci USA 89: 5784–5788
6. Imai M, Akatani K, Ikagami N, Furuichi Y (1983) Capped and conserved terminal structures in human rotavirus genome double-stranded RNA segments. J Virol 47: 125–136
7. Liu M, Mattion NM, Estes MK (1992) Rotavirus VP3 in insect cells possesses guanylyl-transferase activity. Virology 188: 77–84
8. McCrae MA, McCorquodale JG (1983) Molecular biology of rotaviruses, V. Terminal structure of viral RNA species. Virology 126: 204–212
9. Patton JT (1994) Rotavirus replication. In: Ramig RF (ed) Rotaviruses. Springer, Berlin, pp 107–127 Curr Top Microbiol Immunol vol 185)
10. Pizarro JL, Sandino AM, Pizarro JM, Fernandez J, Spencer E (1991) Characterization of rotavirus guanylyltransferase activity associated with polypeptide VP3. J Gen Virol 72: 352–332
11. Poch O, Saubaget I, Delarue M, Tordo N (1989) Indentification of four conserved motifs among the RNA-dependent polymerase encoding elements. EMBO J 8: 3867–3874
12. Ramig RF (1982) Isolation and genetic characterization of temperature-sensitive mutants of simian rotavirus SA11. Virology 120: 93–105
13. Summers MD, Smith GE (1987) A manual of methods for baculovirus vectors and insect cell culture procedures. Texas Agricultural Experiment Station Bulletin No 1555. Texas A&M University, College Station
14. Valenzuela S, Pizzaro J, Sandino AM, Vasquez M, Fernandez J, Hernandez O, Patton J, Spencer E (1991) Photoaffinity labeling of rotavirus VP1 with 8-azido ATP: identification of the viral RNA polymerase. J Virol 65: 3964–3967
15. Zeng Q-Y, Wentz MJ, Cohen J, Estes MK, Ramig RF (1996) Characterization and replicase activity of double- and single-layered rotavirus-like particles expressed from baculovirus recombinants. J Virol 70 (in press)

Authors' address: Dr. R.F. Ramig, Division of Molecular Virology, Baylor College of Medicine, One Baylor Plaza, Houston, TX 77030, U.S.A.

Arch Virol (1996) [Suppl] 12: 69–77

Rotavirus protein expression is important for virus assembly and pathogenesis

P. Tian[1,2], **J. M. Ball**[1], **C. Q-Y Zeng**[1], and **M. K. Estes**[1]*

[1]Division of Molecular Virology, Baylor College of Medicine, Houston, Texas, U.S.A.
[2]Howard Hughes Medical Institute and Department of Cell Biology,
Baylor College of Medicine, Houston, Texas, U.S.A.

Summary. Rotaviruses have a unique morphogenesis in which particles obtain a transient membrane-envelope as newly made subviral particles bud into the endoplasmic reticulum (ER). This process is mediated by a viral nonstructural glycoprotein, NSP4. We have found that NSP4 has pleiotropic properties that became evident following expression of this protein in eukaryotic cells. NSP4 expressed in insect cells bound double-layered rotavirus particles in a manner similar to receptor-ligand interactions and this interaction is thought to trigger the particle budding process. Expression of NSP4 in insect cells also increases intracellular calcium ($[Ca^{2+}]i$) levels and this effect may explain the toxicity of this protein in eukaryotic cells. Increases in $[Ca^{2+}]i$ levels in insect cells also are observed following exogenous addition to cells of purified NSP4 or of a synthetic peptide of NSP4. Experiments to determine the mechanism by which NSP4 causes an increase in $[Ca^{2+}]i$ showed that Ca^{2+} is released from a subset of the thapsigargin-sensitive store [endoplasmic reticulum (ER)]. However, exogenously added and endogenously expressed NSP4 use different mechanisms to alter the Ca^{2+} permeability of the ER membrane. We hypothesize that NSP4-mediated changes in ER membrane permeability trigger viral budding into the lumen of the ER, and eventually induce cell death and release of virus particles from infected cells. We also propose that release of NSP4 following cell lysis and the concomitant stimulation of a Ca^{2+} signal transduction pathway in neighboring cells contributes to altered ion transport in intestinal epithelium resulting in diarrheal disease.

Introduction

Rotaviruses are non-enveloped, large, complex, triple-layered viruses with a genome of 11 segments of double-standed RNA, and are recognized as the most important cause of severe viral gastroenteritis in humans and animals. Rotaviruses have an unique morphogenesis in which particles obtain a transient membrane envelope that is formed by the budding of newly made subviral particles into the endoplasmic reticulum (ER) [1]. This process is mediated by the rotavirus nonstructural transmembrane ER-specific glycoprotein NSP4 (formerly called NS28) [2–6]. Since the discovery that rotaviruses encode this unique nonstructural glycoprotein [7], we have been interested in understanding the functions of this protein in the replication and assembly process. This paper presents a review of the pleiotropic properties that have been associated with NSP4 and a hypothesis of how NSP4 may function in the viral replication/assembly process and affect rotavirus pathogenesis.

Results and discussion

Early studies indicate NSP4 is a transmembrane ER-specific glycoprotein with multiple functional domains

During rotavirus assembly, subviral particles acquire a transient membrane envelope by budding through the ER membrane. This transient membrane is lost and the mature viral particles acquire the two outer capsid proteins (VP4 and VP7). The mature viral particles then accumulate in the lumen of the ER. These observations stimulated research to investigate this unique morphogenic and budding process. In particular, it was of interest to determine how the two outer capsid proteins are assembled onto particles and whether this process might be a useful model to understand viral-lipid and protein-lipid interactions.

The topography of NSP4 in the ER membrane was predicted to contain a large cytoplasmic domain and three hydrophobic amino terminal domains (Fig. 1). The hydrophobic domains were reported to be responsible for ER

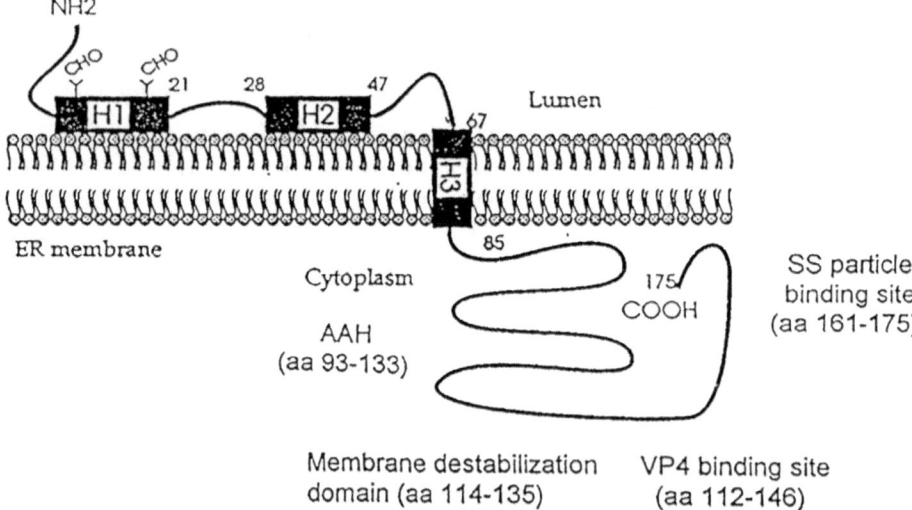

Fig. 1. The predicted or experimentally determined functional domains of NSP4. The two putative glycosylation sites (CHO) are in the first of three predicted hydrophobic domains (H1, H2, H3). The C-terminal cytoplasmic domain, predicted amphipathic alpha helix (AAH), a membrane destabilization domain (MDA), VP4 binding site, and double-layered particle (DLP) binding domains are shown. Modified from Chan *et al.* [8], and Mattion et al.

[25]

retention, since other known ER targeting signals (such as KDEL) are not present in NSP4 [8]. Protease protection and immunocytochemistry studies indicated that the carboxy-terminus of NSP4 was exposed on the outside of microsomes or in the cytoplasm of cells. The cytoplasmic domain was subsequently shown to function as an intracellular receptor which binds VP6 on nascent particles [2]. This binding event was proposed to be the first step in triggering particle budding into the ER which leads to outer capsid formation and subsequent removal of the transient envelope.

Early studies showed that the SA11 (group A simian rotavirus) NSP4 is glycosylated when translated in the presence of microsomes, and that the glycosylation event may be important in the removal of the transient envelope in the ER. Treatment of rotavirus-infected cells with the glycosylation inhibitor, tunicamycin, blocked removal of the envelope membrane from viral particles and the lumen of the ER became filled with enveloped particles [6, 9]. Additionally, the buildup of enveloped particles in the ER occured with a variant strain of SA11 which lacked a glycosylated outer structural glycoprotein VP7 [6, 10]. While it remains possible that a cellular glycoprotein is important in this event, these early results focused studies on NSP4 as a possible key player in the process of rotavirus morphogenesis.

New properties of NSP4 are revealed when expressed in insect cells

Initial studies of the function of NSP4 involved characterization of the properties of the protein expressed in cell-free translation systems, in virus-infected cells, and subsequently in eukaryotic and prokaryotic cells. Studies of NSP4 in virus-infected cells were limited due to the lack of ts mutants which map to NSP4. As an alternative, our laboratory expressed NSP4 in insect cells by constructing baculovirus recombinants [2, 11]. These experiments showed that we could only express NSP4 in first generation baculovirus transfer vectors, in which the construction of the polyhedrin promoter was not yet optimized for a high level of protein expression. Attempts to express NSP4 in higher level expression vectors were uniformly unsuccessful suggesting that this protein might be toxic to insect cells. In spite of these difficulties, microsomes from insect cells expressing NSP4 were shown to specifically bind double-layered rotavirus particles with properties that mimicked receptor-ligand interactions [2]. These results were confirmed by mixing double-layered particles or VP2/6 virus-like particles (VLPs) with purified NSP4 and direct visualization of a specific interaction between NSP4 and subviral particles by electron microscopy (Fig. 2). These studies provided the first evidence that NSP4 acts as an intracellular receptor. Subsequently, these results were confirmed by similar experiments using NSP4 in microsomes from mammalian cells and expressed and purified from bacteria [2, 3, 11a]. Analyses of mutant forms of NSP4 in these different expression systems showed that the cytoplasmic, carboxy-terminus of NSP4 (aa 161–175) serves as a double-layered particle binding domain [3]. Additional studies suggested that NSP4 also contains a binding domain for VP4 between aa 112–146 [3].

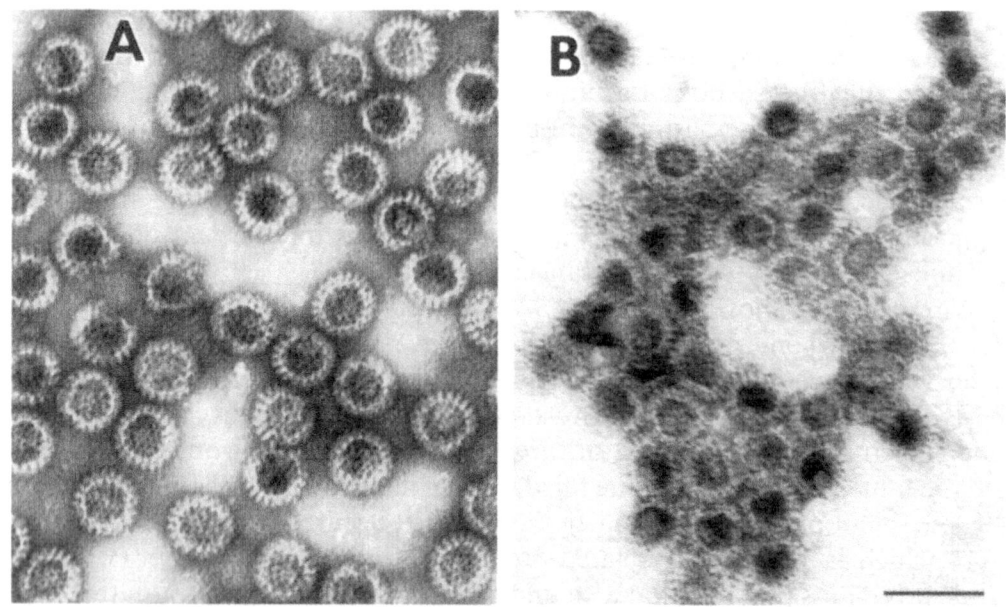

Fig. 2. Electron micrographs of VP2/6 viral like particles (VLPs) (A) or VP2/6 VLPs incubated with purified NSP4 (B). NSP4 was purified from insect cells, mixed with purified VP2/6 VLPs from recombinant baculovirus-infected insect cells, and the mixture was stained with 1% AM. NSP4 causes an aggregation of VP2/6 VLPs. Bar = 100 nm

We attempted to establish mammalian cells that stably express NSP4 with the hope of using these cell lines to identify ts mutants of NSP4. These experiments were uniformly unsuccessful as the cells expressing NSP4 routinely showed unusual morphologies, underwent crisis and died. Those cells that survived were found to have lost the plasmid DNA [11]. These data suggest NSP4 is cytotoxic to mammalian cells.

NSP4 increases basal $[Ca^{2+}]i$ levels by affecting ER membrane permeability

During the course of our experiments to establish a cell line which stably expresses NSP4, a series of independent studies suggested an increasingly important role for calcium in the rotavirus replication process. Treatment of rotavirus particles with chelating agents resulted in the loss of infectivity and the removal of the two outer capsid proteins, VP4 and VP7 [12–14]. Use of the calcium ionophore A23187 to increase the $[Ca^{2+}]i$ during the early stages of replication blocked viral uncoating [15]. Calcium also was found to play a role in the acquisition of specific conformations necessary for the association of structural proteins during virus maturation in the ER, for stabilizing VP7 and for maintaining some neutralization epitopes on VP7 [16]. Thus, the physical and antigenic integrity of rotavirus particles require Ca^{2+}.

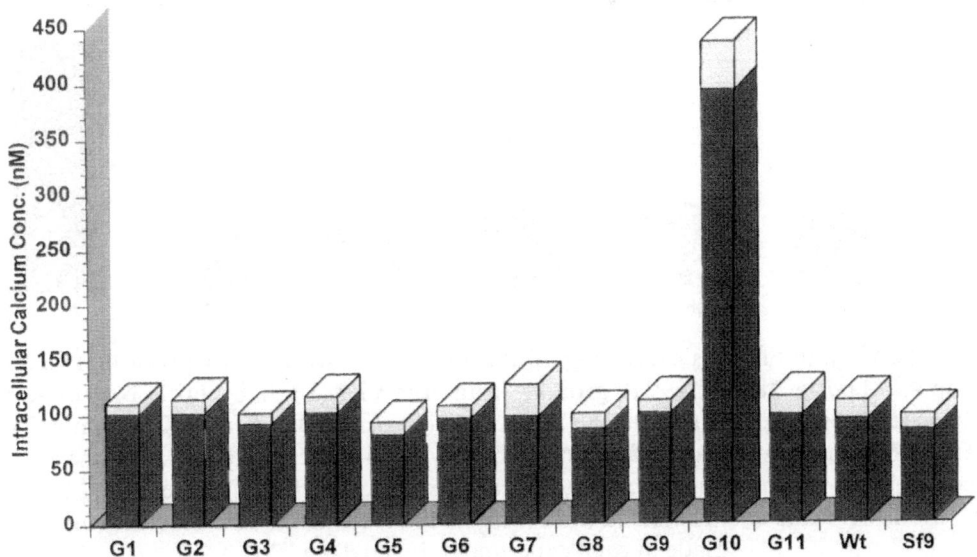

Fig. 3. Intracellular calcium concentration ($[Ca^{2+}]i$) in Sf9 cells infected with wild-type (wt) baculovirus or baculovirus recombinants expressing group A rotavirus genes. G1 to G11 represent baculovirus recombinants expressing rotavirus gene 1 to gene 11. Sf9 cells were infected with each baculovirus at a high multiplicity (MOI of 10). $[Ca^{2+}]i$ was measured between 32 to 36 hours p.i. using the fluorescent indicator fura-2. Each column shows the mean (■) and the standard deviation (□) of three to five independent experiments. N = 3 in G1, G3, G11; N = 4 in G2, G4, G5, G6, G8, G9; and N = 5 in G10, wt and uninfected Sf 9 cells. Three measurements were made in each independent experiment. From Tian *et al.* [11]

Calcium was proposed to have a role in the induction of cytopathic effect in rotavirus-infected cells [17]. We hypothesized that NSP4 might kill infected cells by increasing $[Ca^{2+}]i$ levels, and tested this hypothesis by measuring $[Ca^{2+}]i$ using the fluorescent indicator fura-2 as previously described [11]. Expression of NSP4 in insect cells resulted in a nearly five-fold increase in $[Ca^{2+}]i$ and this effect was not seen in cells expressing any of the other 10 rotavirus proteins or other ER-specific glycoproteins (Fig. 3 and 4). Purified NSP4 and a NSP4 synthetic peptide (residues 114–135, originally made to prepare an antipeptide serum) were added exogenously to insect cells and found to specifically increase $[Ca^{2+}]i$ levels, although the response to the peptide was less than that observed with the protein (2.6-fold increase for the peptide versus 4.5-fold increase for the protein [18]). Possible explanations for this discrepancy include (i) the peptide does not represent the entire functional domain of NSP4, (ii) the peptide does not adopt the correct native conformation, (iii) other remote domains in NSP4 may also play a role in $[Ca^{2+}]i$ change, and (iv) tiny amounts of cellular proteins in the partially purified NSP4 enhance the ability of NSP4 to induce Ca^{2+} release.

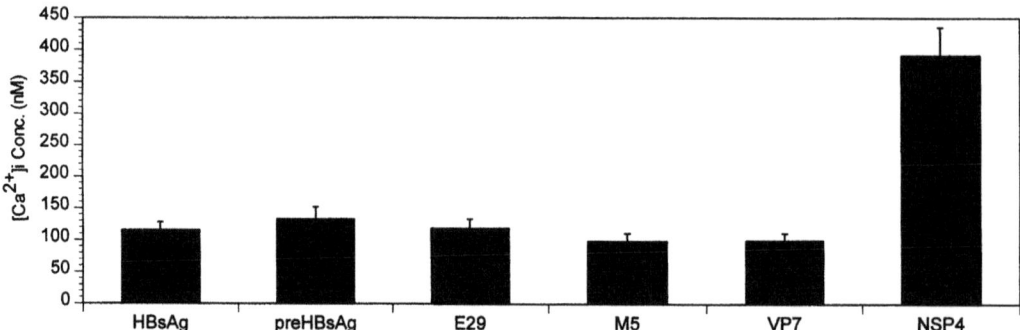

Fig. 4. $[Ca^{2+}]i$ in Sf9 cells expressing viral proteins. The expressed viral proteins include group A rotavirus NSP4 (NSP4), group A rotavirus VP7 (VP7), hepatitis B virus surface antigen (HBsAg), pre hepatitis B virus surface antigen (preHBsAg), a 29 K hepatitis C viral antigen (E29), and a rat brain M5 muscarinic acetylcholine receptor (M5). All these proteins except the M5 receptor are glycoproteins targeted into the ER. The mean and standard deviations were calculated based on three individual experiments. Three measurements were made in each experiment. From Tian [26]

NSP4 increases Ca^{2+} permeability of the ER membrane through two distinct mechanisms

The previous studies revealed that an increase in $[Ca^{2+}]i$ is associated with intracellular expression of NSP4 or exogenous addition of NSP4 or an NSP4 synthetic peptide to the media overlaying insect cell cultures. A series of experiments were performed to determine whether the increases in $[Ca^{2+}]i$ in cells exposed to NSP4 resulted from changes in permeability of the cytoplasmic and/or ER membranes [18]. Influx measurements and analyses of ER permeability in cells treated with a Ca^{2+} ionophore (ionomycin), an inhibitor of the ER Ca^{2+}-ATPase pump (thapsigargin), or a phospholipase C inhibitor indicated that rotavirus NSP4 increased $[Ca^{2+}]i$ levels in insect cells by two mechanisms: (i) compromising the permeability of the ER membrane following the intracellular expression of NSP4, and (ii) activating phospholipase C following the addition of exogenous NSP4 [18]. In both cases, the released Ca^{2+} was from a subset of the thapsigargin-sensitive compartment (ER). The actual mechanism by which NSP4 alters ER permeability is unknown. NSP4 may (i) form a cation channel in the ER membrane, (ii) increase Ca^{2+} leak through the Inositol-1, 4, 5-trisphosphate (InsP3) -receptor, and/or (iii) have action on ER membrane phospholipids leading to membrane disruption. We have proposed that alteration of the Ca^{2+} permeability of the ER membrane during viral expression of NSP4 plays an importent role in virus maturation, and the release of NSP4 following cell lysis and the concomitant stimulation of a Ca^{2+} signal transduction pathway in neighboring cells may contribute to altered ion transport in intestinal epithelium [18].

Table 1. Induction of diarrhea by NSP4 or NSP4 114–135 peptide
in young (6–7 day old) mice

Treatment[a]	Route[b]	Dose (nmol)	% Diarrhea
Purified NSP4	IP	0.1	60
Purified NSP4	IL	0.5	100
Purified VP6	IP	0.5	0
Purified VP6	IL	1.0	0
NSP4 114–135	IP	100.0	69
NSP4 114–35	IL	100.0	71
NSP4 2–22	IP	100.0	0
NSP4 2–22	IL	100.0	0

[a]NSP4 and VP6 were purified as previously described (18). Synthetic peptides were synthesized corresponding to SA11 NSP4 residues 2–22, and 114–135 (24)

[b]IP = intraperitoneal; IL = intraileal (surgical introduction)

Do the pleiotropic properties of NSP4 in insect cells reflect NSP4 function(s) in mammalian cells?

Although the results obtained in insect cells are intriguing, a critical remaining question is, does NSP4 function similarly in mammalian and insect cells? Direct studies in mammalian cells are in progress, but increasing evidence supports data obtained from the expression of heterologous proteins in insect cells. For example, steps in virus assembly have been achieved in insect cells [19, 20], and an increasing number of functional receptors, such as the M5 muscarinic receptor [21], the human B2 bradykinin receptor [22] and the human thrombin receptor [23] have been expressed in insect cells. Futhermore, stimulation of Sf9 cells expressing these receptors with their respective agonists caused a change in $[Ca^{2+}]i$ essentially identical to that observed in mammalian cells. Thus, the Sf9 cells serve as a model system to investigate the mechanisms of action of proteins thought to affect Ca^{2+} homeostasis.

Our results summarized above suggest that NSP4-associated increases in $[Ca^{2+}]i$ might play a role in the physiological consequences of rotavirus infection of intestinal epithelial cells. This hypothesis is supported by the observation that NSP4 and the NSP4 114–135 peptide can induce diarrhea in young mice when administered by the intraperitoneal (IP) or intraileal (IL) route (Table 1; [24]). This observation may be a key to understanding the pathogenesis of rotavirus-induced diarrhea and lead to new approaches to prevent or treat disease.

Acknowledgements

This work was supported in part by Public Health Service Award DK30144 from the National Institutes of Health. We are grateful to Drs. William Schilling and Yanfang Hu for their collaboration and help in the physiology experiments summarized in this review.

References

1. Dubois-Dalcq M, Holmes KV, Rentier B (1984) Assembly of enveloped RNA viruses. Springer, New York
2. Au K-S, Chan W-K, Burns JW, Estes MK (1989) Receptor activity of rotavirus nonstructural glycoprotein NS28. J Virol 63: 4553–4562
3. Au K-S, Mattion NM, Estes MK (1993) A subviral particle binding domain on the rotavirus nonstructural glycoprotein NS28. Virology 194: 665–673
4. Au K-S, Chan W-K, Estes MK (1988) Rotavirus morphogenesis involves an endoplasmic reticulum transmembrane glycoprotein. In: Compans R, Helenius A, Oldston M (eds) Cell Biology of Virus Entry, Replication and Pathogenesis. Alan R Lill, New York
5. Meyer JC, Bergmann CC, Bellamy AR (1989) Interaction of rotavirus cores with the nonstructural glycoprotein NS28. Virology 171: 98–107
6. Petrie BL, Estes MK, Graham DY (1983) Effects of tunicamycin on rotavirus morphogenesis and infectivity. J Virol 46: 270–274
7. Ericson BL, Petrie BL, Graham DY, Mason BB, Estes MK (1983) Rotaviruses code for two types of glycoprotein precursors. J Cell Biochem 22: 151–160
8. Chan W-K, Au K-S, Estes MK (1988) Topography of the simian rotavirus nonstructural glycoprotein (NS28) in the endoplasmic reticulum membrane. Virology 164: 435–442
9. Holmes IH (1993) Rotaviruses. In: Joklik WK (ed) The Reoviridae. Plenum Publishing, New York, pp 359–423
10. Petrie BL (1983) Biologic activity of rotavirus particles lacking glycosylated proteins. In: Compans RW, Bishop DHL (eds) Double-Stranded RNA Viruses. Elsevier, New York, pp 146–156
11. Tian P, Hu Y, Schilling WP, Lindsay DA, Eiden J, Estes MK (1994) The nonstructural glycoprotein of rotavirus affects intracellular calcium levels. J Virol 68: 251–257
11a. Taylor JA, O'Brien JA, Lord VJ, Meyer JC, Bellamy AR (1993) The RER-localized intracellular receptor: A truncated purified soluble form is multivalent and binds virus particles. Virology 194: 807–814
12. Cohen J, Laprote J, Charpilienne A, Scherrer R (1979) Activation of rotavirus RNA polymerase by calcium chelation. Arch Virol 60: 177–186
13. Estes MK, Graham DY, Mason BB (1981) Proteolytic enhancement of rotavirus infectivity: molecular mechanisms. J Virol 39: 879–888
14. Shahrabadi MS, Lee PW (1986) Bovine rotavirus maturation is a calcium-dependent process. Virology 152: 298–307
15. Ludert JE, Michelangeli F, Gil F, Liprandi F, Esparza J (1987) Penetration and uncoating of rotaviruses in cultured cells. Intervirology 27: 95–101
16. Dormitzer PR, Greenberg HB (1992) Calcium chelation induces a conformational change in recombinant herpes simplex virus-1-expressed rotavirus VP7. Virology 189: 828–832
17. Michelangeli F, Ruiz M-C, Del Castillo JR, Ernesto Ludert J, Liprandi F (1991) Effect of rotavirus infection on intracellular calcium homeostasis in cultured cells. Virology 181: 520–527
18. Tian P, Estes MK, Hu Y, Ball JM, Zeng CQ, and Schilling WP (1995) The rotavirus nonstructural glycoprotein NSP4 mobilizes Ca^{2+} from the endoplasmic reticulum. J Virol 69: 5763–5772
19. Crawford SE, Labbe M, Cohen J, Burroughs MH, Zhou Y, Estes MK (1994) Characterization of virus-like particles produced by the expression of rotavirus capsid proteins in insect cells. J Virol 68(9): 5945–5952

20. Belyaev AS, Roy P (1993) Development of baculovirus triple and quadruple expression vectors: Co-expression of three or four bluetongue virus proteins and the synthesis of bluetongue virus-like particles in insect cells. Nucleic Acids Research 21: 1219–1223
21. Hu Y, Rajan L, Schilling WP (1994) Ca^{2+} signaling in Sf9 insect cells and the functional expression of a rat brain M5 muscarinic receptor. Am J Physiol 266 (June): C1736–1743
22. Schilling WP, Chen X, Sinkins WG, Rajan L (1994) Expression of a human bradykinin B2 receptor in Sf9 insect cells using baculovirus expression vector. FASEB J 8: A352 (Abstract)
23. Chen X, Schilling WP (1994) Desensitization of recombinant human thrombin receptor: evidence for active tethered ligand. FASEB J 8:A89 (Abstract)
24. Ball JM, Tian P, Zeng CQ, Morris A, Estes MK (1996) Age-dependent diarrhea is induced by a viral nonstructural glycoprotein. Science 272: 101–104
25. Mattion NM, Cohen J, Estes MK (1994) The Rotavirus Proteins. In: Kapikian A (ed) Viral Infections of the Gastrointestinal Tract, 2nd edn. Marcel Dekker, New York
26. Tian P (1994) Pleiotropic properties of rotavirus nonstructural glycoprotein NSP4 in viral morphogenesis and pathogenesis. Baylor College of Medicine, PhD Thesis, p 119

Authors' address: Prof. M. K. Estes, Ph.D. Division of Molecular Virology, Baylor College of Medicine, Houston, Texas 77030, U.S.A.

Arch Virol (1996) [Suppl] 12: 79–85

A hypothesis about the mechanism of assembly of double-shelled rotavirus particles

H. Suzuki

Department of Public Health, Niigata University School of Medicine, Niigata, Japan

Summary. During double-shelled (ds) particle assembly, subviral particles [possibly single-shelled (ss) particles] acquire the outer capsid protein during their transport across the endoplasmic reticulum (ER) membrane by an exocytosis-like process, probably by a fusion-like mechanism. Fine reticular material is observed around the junction area between virus particles and the ER membrane on the cytoplasmic side of projecting ss particles, suggesting this is the site of assembly of ds particles. It is assumed that the reticular material may correspond to the hetero-oligometric complexes consisting of the non-structural glycoprotein NSP4, the structural proteins VP4 and VP7, and that both VP7 and VP4 may fold onto ss particles as a complex. On the other hand, the budding process simply serves as a vehicle to transport ss particles from the cytoplasm to the ER lumen. Thus, it is assumed that the production of protein complexes may be indispensable for virion assembly, in which NSP4 regulates VP4 folding as an ER chaperone and also the exocytosis-like or fusion-like transport systems through the ER membrane.

Introduction

Rotaviruses have a unique morphogenesis process in which particles obtain a transient membrane envelope that is formed by the budding of newly made subviral particles into the endoplasmic reticulum (ER) [5]. However, the mechanisms by which the envelope is lost and the major outer capsid proteins (VP4 and VP7) are acquired are not yet understood. Recently, we proposed a novel pathway of virus maturation by an exocytosis-like process [27]. In this review, we propose new rotavirus maturation mechanisms based on combining electron microscopic observations with the molecular biology of rotaviruses reported elsewhere.

A. Exocytosis-like process

Kinetic studies have revealed that the VP4 and VP7 components of the outer rotavirus shell appear in ds particles with a lag time of 10–15 minutes, this possibly being the time required for outer capsid assembly [11]. By slowing down the maturation of rotavirus by exposure to a low temperature (4 °C), we succeeded in showing that ds particles are produced through a budding independent process; ss (subviral) particles acquire the outer capsid proteins during their transport across the ER membrane [26, 29]. The localization of VP4 using labeled monoclonal antibody has provided strong supporting evidence that ds particles emerge through the above process, which results in capturing this outer capsid protein as particles mature into virions [29].

B. Budding process

The "envelope" which is formed by the budding process subsequently swells and ruptures, and ss particles are released later during cytolysis [26, 29]. It is concluded that the budding process may simply serve as a vehicle to transport ss particles from the cytoplasm to the ER lumen [26, 29].

C. Mechanisms of the transport of ss particles across the ER membrane for ds particle assembly

1. The first step of virus assembly

As the first step of virus assembly, electron dense granular materials (probably virus precursor) become studded on the ER membrane [26, 29]. Kinetic experiments have indicated that ss particles are the precursors to ds particles [11]. The cytoplasmic domain of the non-structural glycoprotein NSP4 has been proposed to be involved in the morphogenesis of virus particles by functioning as a receptor to bind ss particles prior to their transport across the ER membrane into the ER lumen [2]. Thus, this binding is thought to be the first step which provides the driving force required for particle transfer [2] maturation in as in Semliki Forest virus (SFV) infected cells [23].

2. Interactions between the ER membrane and ss particles

After their binding to the ER membrane via NSP4, ss particles undergo exocytosis-like processes, which are topographically equivalent to opening the ER membrane, probably through a pore of the ER membrane, approximately 60 nm in diameter [26, 29]. Thus, it is assumed that participation between the lipid of the ER membrane and presumably a fusion-like mechanism between the ER membrane and VP6 of ss particles are indispensable for pore widening.

VP6 defines rotavirus antigenic subgroups [8, 12]. The length of VP6, 397 amino acids, is similar in all subgroups, except strain H-2 in which the protein is 399 amino acids [7]. Five regions have been identified as subgroup epitopes [8]. The sequence for amino acid residues 48 and 75 is present in one of the immunodominant sites of VP6, and amino acids 49 and 50 (GlyGly) are strongly conserved among all subgroups. Surprisingly, these amino acids are also strongly conserved in both VP4 of rotaviruses (amino acid 394 to 395) and the E1 glycoprotein of alphaviruses (amino acids 90 to 91) [13, 15] (Fig. 1). The amino

Sequence homologies between SFV E1, rotavirus VP4, and rotavirus VP6

SFV E1 :	^{75}asp	try	gln	cys	lys	val	tyr	thr	gly	val	tyr	pro	phe	met	trp	gly	gly	ala	tyr	cys	phe	cys	asp^{95}
VP4																							
SA11 ;	^{384}ser	–	asn	phe	ser	leu	pro	val	–	gln	trp	–	val	leu	thr	–	–	–	val	ser	leu	his	ser^{401}
WA ;	ser	–	asn	phe	ser	ile	pro	val	–	ala	trp	–	val	–	asn	–	–	–	val	ser	leu	his	phe
UK ;	his	–	ser	phe	ala	leu	pro	val	–	gln	trp	–	val	–	lys	–	–	–	val	thr	leu	his	thr
VP6 ;	^{34}phe	asn	–	ile	ile	–	thr	met	asn	gln	asn	glu	–	gln	thr	–	–	ile	gln	thr	gln	thr	leu^{64}

Fig. 1. The sequence presented corresponds to amino acids 57–97 of the E1 glycoprotein of Semliki Forest virus (SFV) [7], to amino acids 384–401 of the VP4 from a number of rotaviruses [14, 16], and to amino acids 34–56 of the rotavirus VP6 [8]. Identical amino acids are indicated by a hyphen

acid sequence identity between the VP4 of rotaviruses and the E1 of alphaviruses has led to the suggestion that this domain on VP4 may be involved in the entry of rotaviruses into cells [13, 15]. The complete fusion block reported for the SFV Gly91Asp mutant [17], suggests that only key amino acids within the fusion peptide are directly involved in fusion. Thus, it is hypothesized that VP6 might have a new rotavirus putative fusion peptide. Further study is needed to confirm this hypothesis.

Although the fusogenic property of bilayers has been known for some time, it has become clear only recently that two phospholipid bilayers can fuse spontaneously, owing to hydrophobic forces, when the bilayers are brought close together under conditions of membrane tension or high curvature [9, 18]. Thus, with regard to rotaviruses, it is hypothesized that the cytoplasmic domain of NSP4 acts as a receptor-ligand to provide the driving force required for ss particle transfer, and this reaction induces the tension of the ER membrane and close proximity with the putative fusion peptide of VP6 to make an exocytotic fusion pore.

D. Outer shell assembly

In the newly proposed theory, the outer capsid of rotavirus consists of VP4 and VP7 that are added to nascent rotavirus particles during their transport across the ER membrane by the exocytosis-like process [29]. Fine reticular material was observed on the cytoplasmic side of projecting particles at their release from the ER membrane into the ER lumen, suggesting this to be the site of assembly of ds particles, especially around the junction area between virus particles and the ER membrane [29]. Recent data indicate that NSP4, VP4, and VP7 form hetero-oligometric complexes either before or during the maturation process [14]. Thus, it is assumed that the fine reticular material found by electron microscopy may correspond to the molecular complex, and that both VP7 and VP4 may fold directly and simultaneously on ss particles as a complex during transport of ss particles into the ER.

As a last step in this process, pinching off of the above material-encased particles is observed and it is envisaged that the above process may be a virus release step or a last step of ds particle assembly [29]. Based on the kinetics of processing of oligosaccharides attached to membrane-bound vs viral-associated VP7, it is suggested that the former is the precursor for the latter [11]. The cleavage of VP7 between Ala50-Gln51 occurs rapidly after VP7 synthesis [24, 25]. However, the real mechanism of cleavage of VP7 remains unknown. On the basis of these findings and our electron microscopic evidence, it is assumed that the cleavage might account for pinching off of the thread structures, and of retention of the two hydrophobic signal peptide domains, h1 and h2 [25, 30] as remnants of the membrane-bound VP7. This is supported by the reports of others that trimming of carbohydrate side chains may continue after folding, [10, 11] and that the final stages of rotavirus morphogenesis occur in the ER [3, 5].

E. Factors mediating transport of ss particles across the ER membrane

After their binding to the ER membrane via NSP4, ss particles can follow two different pathways, namely budding and exocytosis-like processes. The important question is how transport across the ER membrane is mediated.

First, rotavirus-infected cells treated with tunicamycin (TM, N-linked glycosylation inhibitor) are known to accumulate "enveloped" particles, and ss particles are released into the culture medium [19, 21, 27]. Although TM inhibits glycosylation of both VP7 and NSP4, studies of the maturation of a variant of SA11 (clone 28) which produces a non-glycosylated VP7 have shown that glycosylation of NSP4, but not necessarily VP7, is essential for removal of the envelope [4, 5, 10]. Second, viruses produced in calcium free-medium, and in the presence of the calcium ionophore A23187 were found to be exclusively "enveloped" particles and ss particles [20, 22]. Based on these facts, it is conceivable that glycosylation of NSP4 or/and the presence of calcium are indispensable for virion assembly (Table 1). With regard to the relationship between NSP4 and calcium, it was reported that binding of calcium ions to the cytoplasmic domain of NSP4 might induce conformational changes of NSP4 [1].

Table 1. Relationship between NSP4, calcium ion, and the ER membrane

NSP4	Calcium	ER Membrane
Glycosylation (+)	$Ca^{++}(+)$	open → Exocytosis-like process
	$Ca^{++}(-)$	closed → Budding process
Glycosylation (−)	$Ca^{++}(+)$	closed → Budding process
	$Ca^{++}(-)$	closed → Budding process

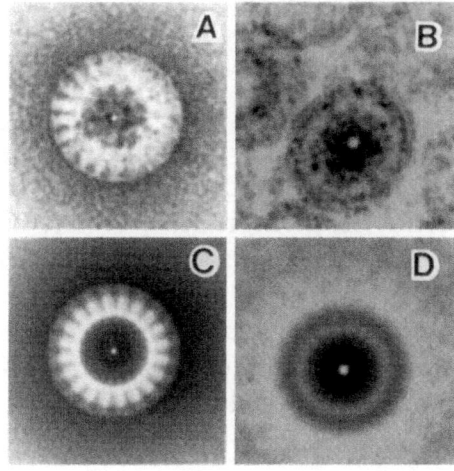

Fig. 2. High power views of KUN human rotavirus strain. Double-shelled (ds) particles by negative staining (A). Twenty pillar-like units of ds particles were enhanced by the photorotation technique [16] (C). "Enveloped" particles by ultra-thin section (B). Twenty pillar-like units of "enveloped" particles were enhanced by the photorotation technique, [16] showing clearly radiating black lines from the core region which correspond to the filamentous substances (D)

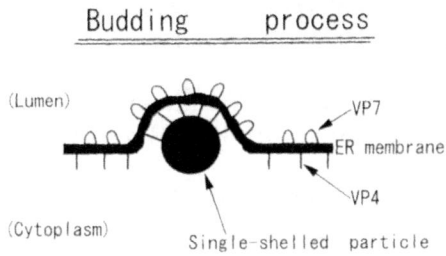

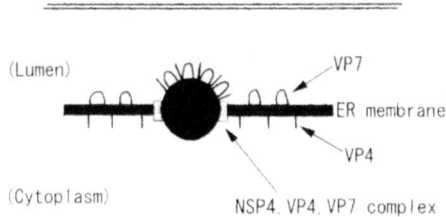

Fig. 3. Schematic illustration of the proposed interconnections among NSP4, VP4, VP7 and a single-shelled particle during the budding and exocytosis-like processes

F. Role of NSP4 in virion assembly

"Envelope" particles which have assembled by the budding process consist of a core, an inner shell, a halo and an "envelope" (Fig. 2). The halo contains filamentous substances [26]. Twenty pillar-like units of inner shell were enhanced by the Markham photorotation technique [16] (Fig. 2), showing that clearly radiating black lines emanate from the core region which correspond to the filamentous substances and also to tube-like channels (Fig. 2C). The localization of VP4 using monoclonal antibody [28] strongly suggested that the filamentous substances in the halo of "enveloped" particles observed in the ultra-thin sections may correspond to VP4.

During the budding process, VP4 may be attached to ss particles prior to particle budding into the ER (Fig. 3). It is hypothesized that this attachment might inhibit the fusion between VP6 and the ER membrane. In contrast, in the exocytosis-like process, after forming NSP4, VP4, and VP7 complexes around the junction area between virus particles and the ER membrane, VP4 may be attached to ss particles with VP7 (Fig. 3). Through this process, V4 would flip-flop and attach onto ss particles. Thus, it is assumed that the production of complexes might be indispensable for virion assembly, in which NSP4 regulates protein folding, especially VP4 folding as an ER chaperone. In conclusion, NSP4 may have two important functions in virion assembly, namely as a receptor to bind ss particles [2] and an ER chaperone.

References

1. Au KS, Chan WK, Estes MK (1989) Rotavirus morphogenesis involves an endoplasmic reticulum transmembrane glycoprotein. In: Compans RW, Helenius A, Oldstone MBA (eds) Cell biology of virus entry, replication, and pathogenesis. Alan R. Liss, New York, pp 257–267
2. Bellamy AR, Both GW (1990) Molecular biology of rotaviruses. In: Maramorosch K, Murphy RA, Shatkin AJ (eds) Advances in virus research. Academic Press, vol 38, pp 1–43
3. Dubois-Dalq M (1984) Assembly of rotaviruses. In: DuboisDalq M, Holmes KV, Rentier B (eds) Assembly of enveloped RNA viruses. Springer, Wien, New York, pp 171–184
4. Estes MK, Graham DY, Ramig RF, Ericson BL (1982) Heterogeneity in the structural glycoprotein (VP7) of simian rotavirus SA11. Virology 122: 8–14
5. Estes MK, Cohen J (1989) Rotavirus gene structure and function. Microbiol Rev 53: 410–449
6. Garoff H, Frischauf AM, Simons K, Lehrach H, Delius H (1980) Nucleotide sequence of cDNA coding for Semliki Forest virus membrane glycoproteins. Nature 288: 236–241
7. Gorziglia M, Hoshino Y, Nishikawa K, Maloy WL, Jones RW, Kapikian AZ, Chanock RM (1988) Comparative sequence analysis of the genomic 6 of four rotaviruses each with a different subgroup specificity. J Gen Virol 69: 1659–1669
8. Greenberg HB, McAuliffe V, Valdesuso J, Wyatt R, Flores J, Kalica A, Hoshino Y, Singh NH (1983) Serological analysis of the subgroup protein of rotavirus, using monoclonal antibodies. Infect and Immunity 39: 91–99
9. Helm CA, Israelachvili JN (1993) Forces between phospholipid bilayers and relationship to membrane fusion. Methods in Enzymology 220: 130–143
10. Kabcenell AK, Atkinson PH (1985) Processing of the rough endoplasmic reticulum membrane glycoproteins of rotavirus SA11. J Cell Biol 101: 1270–1280
11. Kabcenell AK, Poruchynsky MS, Bellamy AR, Greenberg HB, Atkinson PH (1988) Two forms of VP7 are involved in the assembly of SA11 rotavirus in the endoplasmic reticulum. J Virol 62: 2929–2941
12. Kalica AR, Greenberg HB, Wyatt RG, Flores J, Sereno MM, Kapikian AZ, Chanock RM (1981) Genes of human (strain Wa) and bovine (strain UK) rotavirus that code for neutralization and subgroup antigens. Virology 112: 385–390

13. Lopez S, Lopez I, Romero P, Mendez E, Soberon X, Arias CF (1991) Rotavirus YM gene 4: analysis of its deduced amino acids sequence and prediction of the secondary structure of the VP4 protein. J Virol 65: 3738–3745

14. Mass DR, Atkinson PH (1990) Rotavirus protein VP7, NS28, and VP4 from oligomeric structures. J Virol 64: 2632–2641

15. Mackow EAR, Shaw RD, Matsui SM, Dang VPT, Greenberg MH (1988) The rhesus rotavirus gene encoding protein VP3: location of amino acids involved in homologous and heterologous rotavirus neutralization and identification of a putative fusion region. PNAS, U.S.A. 85: 645–649

16. Markham R, Frey S, Hill GJ (1963) Method for the enhancement of image detail and acceleration of structure in electron microscopy. Virology 20: 88–102

17. Mintz PL, Kielian M (1991) Mutagenesis of the putative fusion domain of the Semliki Forest virus spike protein. J Virol 65: 4292–4300

18. Monck J, Fernandez JM (1994) The exocytotic fusion pore and neurotransmitter release. Neuron 12: 707–716

19. Petrie BL, Estes MK, Graham DY (1983) Effect of tunicamycin on rotavirus morphogenesis and infectivity. J Virol 46: 270–274

20. Poruchyosky MS, Maess DR, Atkinson PH (1991) Calcium depletion blocks the maturation of rotavirus by altering the oligomerization of virus-encoded proteins in the ER. J Cell Biol 114: 651–661

21. Sabara M, Babiuk LA, Gilchrist J, Misra V (1982) Effect of tunicamycin on rotavirus assembly and infectivity. J Virol 43: 1082–1090

22. Shahrabadi MS, Lee PWK (1986) Bovine rotavirus maturation is a calcium-dependent process. Virology 152: 298–307

23. Simon K, Garoff H (1980) The budding mechanisms of enveloped animal viruses. J Gen Virol 50: 1–21

24. Sticzaker SC, Whitfeld PL, Christie DL, Bellamy AR, Both GW (1987) Processing of rotavirus glycoprotein VP7: implications for the retention of the protein in the endoplasmic reticulum. J Cell Biol 105: 2897–2903

25. Sticzaker SC, Both GW (1989) The signal peptide of rotavirus glycoprotein VP7 is essential for its retention in the ER as an integral membrane protein. Cell 56: 741–747

26. Suzaki H, Sato T, Konno T, Kitaoka S, Ebina T, Ishida N (1984) Effect of tunicamycin on human rotavirus morphogenesis and infectivity. Arch Virol 81: 363–369

27. Suzuki H, Konno T, Kitaoka S, Sato T, Ebina T, Ishida N (1984) Further observations on the morphogenesis of human rotavirus in MA104 cells. Arch Virol 79: 147–159

28. Suzaki H, Kitaoka S, Konno T, Sato T (1991) The localization and function of the neutralizing protein, VP4, in human rotavirus. Tohoku J Exper Med 163: 73–75

29. Suzuki H, Konno T, Numazaki Y (1993) Electron microscopic evidence for budding process-independent assembly of double-shelled rotavirus particles during passage through endoplasmic reticulum membranes. J Gen Virol 74: 2015–2018

30. Whitfield PL, Tyndall C, Striczaker SC, Bellamy AR, Both GW (1987) Location of signal sequences within the rotavirus SA11 glycoprotein VP7 which direct it to the endoplasmic reticulum. Mol Cell Biol 7: 2491–2479

Authors' address: Dr. H. Suzuki, Department of Public Health, Niigata University School of Medicine, 1-757 Asahimachi-Dori, Niigata 951, Japan.

Arch Virol (1996) [Suppl] 12: 87–91

Development of rotavirus molecular epidemiology: electropherotyping

I. H. Holmes

Department of Microbiology, University of Melbourne, Parkville,
Victoria, Australia

Summary. Early in the era of rotavirology it was realized that the characteristic patterns of bands produced in polyacrylamide gels following electrophoresis of genomic dsRNA were useful for checking the identity of rotavirus isolates. However it was Romilio Espejo who first proposed the use of this technique for epidemiology, although most others did not take the suggestion seriously because the technique was then rather specialized and RNA staining methods were not very sensitive. Using samples collected by Ruth Bishop in Melbourne following the original identification of human rotaviruses, Sue Rodger recorded the "electropherotypes" of all samples available to 1979 and painstakingly compared them, side by side (since minor variations in conditions, especially temperature, alter the relative migration distances of dsRNA bands). These efforts produced the first longitudinal, extensive study of human rotavirus strain variation. Since then, technical improvements have greatly increased the sensitivity of the procedures, and electropherotyping has been recognized as a powerful and economical method for epidemiological studies of rotaviruses.

*

Rotaviruses (family *Reoviridae*) are very frequently associated with acute diarrheal disease in infants and young animals. Virions consist of a core containing the dsRNA genome, and inner and outer capsids exhibiting icosahedral symmetry. In the early days of rotavirus investigations, it was found that analysis of rotaviral dsRNA by polyacrylamide gel electrophoresis (PAGE) produced characteristic migration patterns of the 11 genome segments. These patterns, which were easy to demonstrate and were reproducible for individual samples, were called "electropherotypes". Because at the time (mid 1970s) cultivation and serotyping of rotaviruses was so difficult, it was quickly realized that electropherotyping was very useful for strain identification in the laboratory, although the technique was regarded as rather specialized. Neverthless, in a series of visionary papers, Romilio Espejo [3–6] suggested that it "could become a routine diagnostic procedure". At the time, this appeared very optimistic, but as a result of various improvements of the technique, the prophecy has been adequately fulfilled.

Initially, the methods employed for RNA extraction were somewhat complex and involved either partial purification of the virus, or of the viral RNA, or both [13, 26]. Espejo *et al.* [3] estimated that to produce a visible pattern using ethidium bromide staining of the RNA bands required a stool sample of 1–2 g, containing about 10^{10} rotavirus particles, and many samples were not adequate for this purpose. However, in the early 1980s and using the original methods, it was still possible to carry out the first large scale epidemiological studies based on electropherotyping [1, 16, 24].

These studies provided lasting insights into the variety of rotaviruses in large urban populations and the succession of predominant epidemic strains (*i.e.* distinguishable virus populations; cultivation or isolation is not implied) which could replace each other annually but sometimes persisted through more than one season. Subsequent studies have confirmed these observations, but the nature of the rotaviruses producing the many patterns seen only once or a few times remains to be elucidated. Some may represent human strains imported from other geographic areas, others may be of "animal" origin or may be human-animal reassortants, such as have been identified sporadically in more recent serological surveys. Unfortunately the early studies, with their requirements of large amounts of material, rarely left sufficient amounts of the interesting samples for subsequent study.

In 1982, Herring *et al.* [10] introduced major technical improvements in the methods for extraction of rotaviral RNA and staining of the resultant bands; these still represent the "state of the art". A simplified extraction procedure for application to unprocessed (unpurified) stools samples was devised, involving sodium dodecyl sulphate (SDS) to lyse cells and solubilize proteins, then phenol extraction at acid pH to differentially extract RNA rather than DNA. The most important innovation to enhance sensitivity was the use of silver staining to replace ethidium bromide for visualizing the RNA bands. In addition to the gain in sensitivity, this improved the resolution of closely migrating bands, and thus enhanced the discrimination between similar but not identical electropherotypes. Practical details of a scaled down Herring method for use with Eppendorf tubes have also been published [11].

At first it was assumed that differences in migration of particular segments might indicate differences in the size of the molecules, but in general we now know that this is not so. Instead, it has been found that migration is affected by changes or differences in nucleotide sequence, which are believed to result in local changes in binding (and hence dissociation temperature T_m) between the complementary RNA strands, making the molecules more or less flexible. In practice this means that "electropherotypes" are not absolute but depend on the temperature (and salt concentration) used for electrophoresis. This was noted by Espejo and Puerto [7], and is the reason why it is necessary to coelectrophorese or at least run pairs of samples in the same gel if one wishes to establish whether they have identical or distinguishable electropherotypes.

Electropherotyping provides considerably more information about rotavirus strains than do other diagnostic tests. It is objective and not subject to "false positives", provided that spillover of samples between adjacent lanes in the gel can be excluded. It provided the first insights into the degree of variety among rotaviruses [8]. It demonstrated changes both within and between annual epidemics in the same human populations [15], and showed that rotaviral epidemiology was considerably more complex than that of influenza viruses, for example. The fine distinctions between strains demonstrable by electropherotyping have been particularly useful for elucidating the sources of infection in nosocomial outbreaks [20, 27], and for showing that dual (mixed) infections of infants by two rotaviral strains are not uncommon [20, 29].

Rotavirus serotype cannot be predicted directly from electrophoretic band pattern, although determination of the serotype of one or two of an epidemiological cluster of strains with identical electropherotypes is generally sufficient: all will almost certainly prove to be the same [19]. This can considerably reduce the difficulty and expense of epidemiological surveys in situations for which information on rotavirus serotypes is needed. Unusual band patterns have even led to the discovery of a new serotype in humans [9, 17], but in retrospect this turned out to be coincidental. The "supershort" band pattern was demonstrated to result from a partial duplication of segment 11, which just happened to be associated with the segment encoding VP7 of G8 serotype [21].

Even more markedly aberrant RNA patterns have, however, reliably indicated the existence of "atypical" rotaviruses of distinct serogroups. While the great majority of rotaviruses in humans and most other animal species appear to belong to serogroup A, strains of serogroups B and C can infect humans [1, 12, 23, 25] and those of other groups infect birds [18]. Electropherotyping is certainly the most practical method presently available for detecting these uncommon infections, which are suspected to be mainly zoonoses [2].

Current interest in vaccine development has led to an emphasis on rotavirus serotype determination in community surveys, and thus reagents and methods have been developed for serogrouping and serotyping using enzyme immunoassays. Following the increased availability of sequence information, serotyping by PCR and/or hybridization has become feasible, and genogrouping by RNA-RNA hybridization is very useful for identifying reassortants and relationships between "human" and "animal" rotaviruses, as we will see in the next presentations. Nevertheless, electropherotyping can provide very useful diagnostic and epidemiological information at relatively low cost [14] and with reasonable sensitivity [22, 28]. Even when serotypic information is necessary, electropherotyping can be used to identify subsets of samples for more detailed examination, reducing labor and costs. Thus I believe that electropherotyping still has an important role in studies of rotaviral epidemiology.

References

1. Buitenwerf J, Nuilwijk-van-Alphen M, Schapp GJ (1983) Characterization of rotaviral RNA isolated from children with gastroenteritis in two hospitals in Rotterdam. J Med Virol 12: 71–78
2. Eiden J, Vonderfecht S, Yolken RH (1985) Evidence that a novel rotavirus-like agent of rats can cause gastroenteritis in man. Lancet 2: 8–11
3. Espejo R, Romero P, Calderon E, Gonzalez N (1978) Diagnosis of rotavirus using viral RNA electrophoresis. Bol Med Hosp Infant Mex 35: 323–331
4. Espejo RT, Calderon E, Gonzalez N (1977) Distinct reovirus-like agents associated with acute infantile gastroenteritis. J Clin Microbiol 6: 502–506
5. Espejo RT, Calderon E, Gonzalez N, Salomon A, Martuscelli A, Romero P (1979) Presence of two distinct types of rotavirus in infant and young children hospitalized with acute gastroenteritis in Mexico City, 1977. J Infect Dis 139: 474–477
6. Espejo RT, Munoz O, Serafin F, Romero P (1980) Shift in the prevalent human rotavirus detected by ribonucleic acid segment differences. Infect Immun 27: 293–354
7. Espejo RT, Puerto F (1984) Shifts in the electrophoretic pattern on the RNA genome of rotaviruses under different electrophoretic conditions. J Virol Methods 8: 293–299
8. Estes MK, Graham DY, Dimitrov DH (1984) The molecular epidemiology of rotavirus gastroenteritis. Prog Med Virol 29: 1–22
9. Hasegawa A, Inouye S, Matsuno S, Yamaoka K, Eko R, Suharyono W (1984) Isolation of human rotaviruses with a distinct RNA electrophoretic pattern from Indonesia. Microbiol Immunol 28: 719–722
10. Herring AJ, Inglis NF, Ojeh CK, Snodgrass DR, Menzies JD (1982) Rapid diagnosis of rotavirus infection by direct detection of viral nucleic acid in silver-stained polyacrylamide gels. J Clin Microbiol 16: 473–477
11. Holmes IH (1985) Epidemiology of rotavirus infections based on analysis of genome RNA. In: Tzipori S (ed) Infectious Diarrhoea in the Young. Elsevier, Amsterdam, pp 195–200
12. Hung T, Chen GM, Wang CG, Yao HL, Fang ZY, Chao TX,Chou Zy, Ye W, Chang XJ, Den SS, Liong XQ, Chang WC (1984) Waterborne outbreak of rotavirus diarrhoea in adults in China caused by a novel rotavirus. Lancet 1: 1139–1142
13. Kalica AR, Sereno MM, Wyatt RG, Mebus CA, Chanock RM, Kapikian AZ (1978) Comparison of human and animal rotavirus strains by gel electrophoresis of viral RNA. Virology 87: 247–255
14. Kasempimolporn S, Louisirirotchanakul S, Sinarachatanant P, Wasi C (1988) Polyacrylamide gel electrophoresis and silver staining for detection of rotavirus in stools from diarrheic patients in Thailand. J Clin Microbiol 26: 158–160
15. Konno T, Sato T, Suzuki H, Kitaoka S, Katsushima N, Sakamoto M, Yazaki N, Ishida N (1984) Changing RNA patterns in rotaviruses of human origin: demonstration of a single dominant pattern at the start of an epidemic and various patterns thereafter. J Infect Dis 149: 683–687
16. Lourenco MH, Nicolas JC, Cohen J, Scherrer R, Bricout F (1981) Study of human rotavirus genome by electrophoresis: attempt of classification among strains isolated in France. Ann Virol Inst Pasteur 132E: 161–173
17. Matsuno S, Hasegawa A, Mukoyama A, Inouye S (1985) A candidate for a new serotype of human rotavirus. J Virol 54: 623–624
18. McNulty MS, Todd D, Allan GM, McFerran JB, Greene JA (1984) Epidemiology of rotavirus infection in broiler chickens: recognition of four serogroups. Arch Virol 81: 113–121

19. Nakagomi T, Akatani K, Ikegami N, Katsushima N, Nakagomi O (1988) Occurrence of changes in human rotavirus serotypes with concurrent changes in genomic RNA electropherotypes. J Clin Microbiol 26: 2586–2592
20. Nicolas JC, Pothier P, Cohen J, Lourenco MH, Thompson R, Guimbaud P, Chenon A, Dauvergne M, Bricout F (1984) Survey of human rotavirus propagation as studied by electrophoresis of genomic RNA. J Infect Dis 149: 688–693
21. Nuttall SD, Hum CP, Holmes IH, Dyall-Smith ML (1989) Sequences of VP9 genes from short and supershort rotavirus strains. Virology 171: 453–457
22. Pereira HG, Azeredo RS, Leite JP, Barth OM, Sutmoller F, de-Farias V, Vidal MN (1983) Comparison of polyacrylamide gel electrophoresis (PAGE), immuno-electron microscopy (IEM) and enzyme immunoassay (EIA) for the rapid diagnosis of rotavirus infection in children. Mem Inst Oswaldo Cruz 78: 483–490
23. Pereira HG, Leite JP, Azeredo RS, de-Faria V, Sutmoller F (1983) An atypical rotavirus detected in a child with gastroenteritis in Rio de Janeiro, Brazil. Mem Inst Oswaldo Cruz 78: 245–250
24. Rodger SM, Bishop RF, Birch C, McLean B, Holmes IH (1981) Molecular epidemiology of human rotaviruses in Melbourne, Australia, from 1973 to 1979, as determined by electrophoresis of genome ribonucleic acid. J. Clin Microbiol 13: 272–278
25. Rodger SM, Bishop RF, Holmes IH (1982) Detection of a rotavirus-like agent associated with diarrhea in an infant. J Clin Microbiol 16: 724–726
26. Rodger SM, Holmes IH (1979) Comparison of the genomes of simian, bovine, and human rotaviruses by gel electrophoresis and detection of genomic variation among bovine isolates. J Virol 30: 839–846
27. Rodriguez WJ, Kim HW, Brandt CD, Gradner MK, Parrott RH (1983) Use of electrophoresis of RNA from human rotavirus to establish the identity of strains involved in outbreaks in a tertiary care nursery. J Infect Dis 148: 34–40
28. Shinozaki T, Ushijima H, Tajima T, Kim B, Araki K, Fujii R (1985) Evaluation of four tests for detecting human rotavirus in feces [letter]. Eur J Pediatr 143: 238
29. Spencer EG, Avendano LF, Garcia BI (1983) Analysis of human rotavirus mixed electropherotypes. Infect Immun 39: 569–574

Author's address: Dr. I. H. Holmes, Department of Microbiology, University of Melbourne, Parkville, Victoria 3052, Australia.

Arch Virol (1996) [Suppl] 12: 93–98

Molecular epidemiology of human rotaviruses:
genogrouping by RNA-RNA hybridization

O. Nakagomi and **T. Nakagomi**

Department of Microbiology, Akita University School of Medicine, Akita, Japan

Summary. RNA-RNA hybridization performed under high stringency conditions allows rotavirus isolates to be grouped together based on the overall similarity of their genomic RNA constellation. Classification by this scheme has been termed "genogrouping". Genogrouping has advanced molecular epidemiology of human rotaviruses. Major observations include (i) Interspecies transmission occurs in nature and (ii) Intergenogroup reassortment occurs in nature with or without exchange of serotype-determining genes. Genogrouping is a particularly valuable asset for determining the gene constellation of unusual rotavirus isolates.

Introduction

Group A rotavirus is a species within the genus *Rotavirus*, family *Reoviridae* [1]. The virions contain 11 segments of double-stranded (ds)RNA which can easily be separated by polyacrylamide gel electrophoresis (electropherotyping). Studies based on electropherotyping have contributed to our understanding of the molecular epidemiology of these viruses. Major observations are [15]: (i) Extensive heterogeneity occurs in the electropherotypes of rotavirus isolates. (ii) Despite this variety, two distinct electrophoretic migration patterns, "short" and "long", are readily identified on the basis of the relative mobility of gene segments 10 and 11. (iii) During outbreaks, one strain is predominant at any time. Despite its usefulness in molecular epidemiology, electropherotyping fails to provide information as to how such heterogeneity in electropherotype reflects diversity at the nucleotide sequence level. Sequencing all 11 gene segments of a given strain is not practical. Thus, RNA-RNA hybridization has been used to evaluate the nucleotide sequence similarity of genomic RNA segments among various field isolates. Classification based on the overall genetic relatedness of rotavirus strains has been termed "genogrouping" [13]. The purpose of this paper is to explain the concept of genogroup and to briefly review some of the major findings that genogrouping has thus far achieved. Readers are invited to refer to the following articles for more detailed information [10–13].

The tool: RNA-RNA hybridization in solution

Taking advantage of the endogenous rotavirus transcriptase, which makes full-length positive strands (mRNAs) from the dsRNA genome, genomic RNAs are transcribed *in vitro* in the presence of ^{32}P-GTP to make labeled probes that are colinear with the 11 segments of the genomic RNA. These probes are used for genogrouping by RNA-RNA hybridization as described previously [13]. Genomic RNAs are heat denatured, mixed with the ^{32}P-labeled probes and allowed to hybridize for 16 h at 65 °C. The resulting hybrids are visualized by autoradiography, while genomic dsRNAs are visualized by ethidium bromide staining of the gel.

The hybridization patterns are compared with those of the homologous reaction. Since hybrids are separated on a polyacrylamide gel after hybridization in solution, homologous bands (homoduplex molecules) are identified as bands that co-migrate with the corresponding genomic RNA segments, whereas hybrids consisting of a lower degree of homology can be observed as aberrantly migrating bands often with lesser intensity. The exact identification of the gene segment involved in a hybrid is difficult under these assay conditions. Nevertheless, the number and the relative intensity of the hybrids formed between genomic RNA from a test strain and the probe is indicative of the relative and overall gene homology between two strains. As to hybridization stringency, this assay theoretically allows up to 18% nucleotide sequence mismatch but, in practice, the stringency is much higher (approximately 8%) because the partially homologous RNA-RNA hybrids that are formed at an early time of incubation gradually degrade during incubation at 65 °C [6].

The concept of genogroup

The concept of genogroup is introduced to classify rotaviruses into groups based on their overall genomic RNA constellation, which is revealed by RNA-RNA hybridization performed under high stringency conditions [13]. Initially, Flores *et al.* [2, 3] showed the low sequence relatedness between subgroup I human rotaviruses with short RNA patterns (such as strain DS-1) and subgroup II human rotaviruses with long RNA patterns (such as strain Wa), suggesting that there are at least two distinct "families" of human rotaviruses. This observation was confirmed and extended when Nakagomi and Nakagomi [16] reported AU-1 and similar strains which did not belong to either of the two families but

Fig. 1. Three distinct genogroups among human rotaviruses; *i.e.* the Wa, the DS-1 and the AU-1 genogroup. A panel of genomic RNAs from 24 human rotavirus strains including three representative strains (Wa, KUN, and AU-1) were hybridized with three sets of transcription probes made from Wa, KUN, and AU-1. Notice, for example, that a group of rotaviruses that hybridize very well with the Wa probe do not hybridize with either the KUN (representing the DS-1 genogroup) or the AU-1 probe

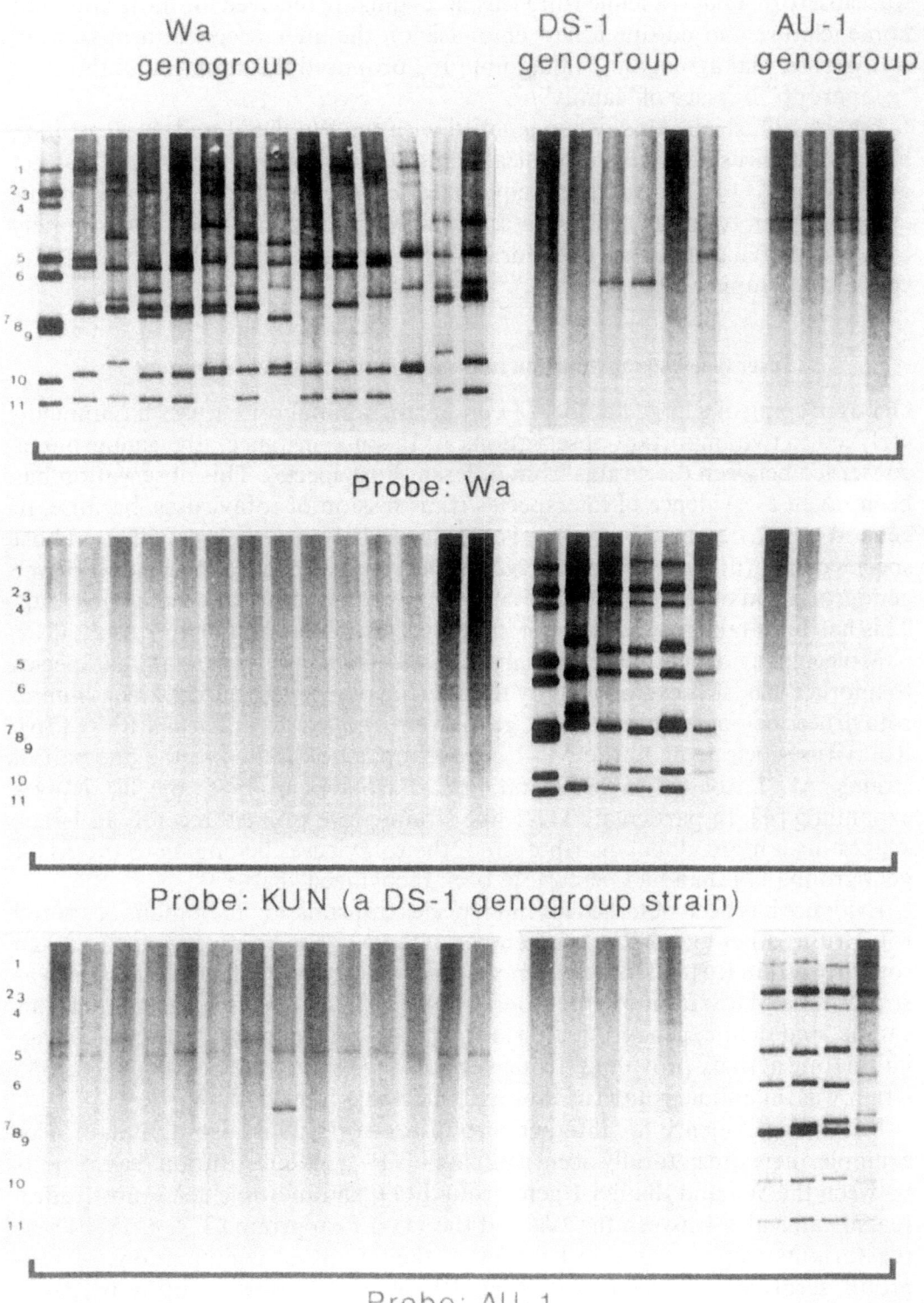

was closely related to a feline rotavirus. Recognizing the need for more coherent nomenclature and putting a new emphasis on the interspecies transmission of rotaviruses, Nakagomi and Nakagomi [16] proposed the adoption of the term "genogroup" in place of "family".

Figure 1 illustrates that, when assayed with the Wa, DS-1 and AU-1 probes, human rotavirus field isolates basically fall into any one of these three distinct genogroups [11]. The relative frequencies for each of the three genogroups among human rotavirus field isolates have been estimated to be approximately 84% for the Wa genogroup, 15% for the DS-1 genogroup, and 1% for the AU-1 genogroup (unpubl. obs.).

Interspecies transmission and intergenogroup reassortment

Genogrouping of a large number of human and animal rotaviruses has immediately led to two important observations: (i) In some instances, the genogroup is conserved between the strains from different host species. This observation has been taken as evidence of interspecies transmission of rotaviruses, because, in general, a high degree of homology is not shared between strains of different host species origin. (ii) Some genomic RNA segments are similar to the genome of one genogroup and other segments are similar to the genome of another genogroup. This has been taken as evidence of intergenogroup reassortment.

Molecular evidence for the transmission of a rotavirus from one animal species to another was, first exemplified by the sharing of a genogroup between human rotaviruses belonging to the AU-1 genogroup and feline rotavirus FRV-1 [16]. Rotaviruses belonging to the AU-1 genogroup is not limited to the one pair of strains, AU-1 isolated in 1982 and FRV-1 isolated in 1985; the list is ever expanding [4]. In particular, AU-1 like strains were isolated recently in Israel, suggesting a more global distribution of the rotaviruses belonging to the AU-1 genogroup [17] than had heretofore been recognized (Table 1).

Evidence has been obtained for interspecies transmission to humans of animal rotaviruses other than FRV-1, a member of the AU-1 genogroup. Israeli human rotavirus strain Ro1845 shares a genogroup with canine and feline strains such as strains K9 and RS-15, and feline strain Cat97 [14]. Interestingly, members of this canine and feline genogroup, termed the K9 genogroup, possess hemagglutinating activity providing strong support of the hypothesis that the Ro1845 strain was an animal rotavirus that had infected a human child [9].

Molecular evidence for intergenogroup reassortants is also abundant. For example, there are naturally occurring single VP7 gene substitution reassortants between the Wa and the DS-1 genogroup [11], and multiple gene substitution reassortants also between the Wa and the DS-1 genogroup [3, 7, 8, 18]. These reassortants were recognized because of their unusual combinations of subgroup, serotype and electropherotype. Rigorous and close scrutiny by genogrouping of human rotavirus field isolates revealed that single nonstructural gene substitution reassortants have repeatedly appeared between the Wa and the DS-1 genogroup [5].

Table 1. A list of human rotavirus strains that were shown by RNA-RNA
hybridization to be members of the AU-1 genogroup ([4, 17])

Strains	G and P types	Place of isolation	Year of isolation
AU-1	G3P3[9]	Akita, Japan	1982
AU1115	G3P3[9]	Yamagata, Japan	1986
AU720	G3P3[9]	Tokyo, Japan	1986
AU228	G3P3[9]	Yamagata, Japan	1987
AU125	G3P3[9]	Akita, Japan	1988
AU242	G3P3[9]	Akita, Japan	1988
AU379	G3P3[9]	Akita, Japan	1989
AU387	G3P3[9]	Akita, Japan	1989
AU785	G3P3[9]	Akita, Japan	1989
AU938	G3P3[9]	Akita, Japan	1989
AU218	G3P3[9]	Akita, Japan	1991
AU229	G3P3[9]	Akita, Japan	1991
Ro5829	G3P3[9]	TelAviv, Israel	1992
Ro5960	G3P3[9]	TelAviv, Israel	1992

Conclusion

Genogrouping by RNA-RNA hybridization under high stringency conditions
has facilitated our understanding of the molecular epidemiology of rotaviruses
circulating in nature. In particular, the concept of genogroup led to the identification of interspecies transmission of rotaviruses from animals to humans and
naturally-occurring genetic reassortment between rotaviruses belonging to different genogroups. Interspecies transmission of genes increases the options for
the rotavirus genome to evolve under natural conditions. Genogrouping will
continue to be a valuable asset for determining the gene constellations of unusual
isolates, at least as an initial step of characterization.

References

1. Murphy FA, Fauquet CM, Bishop DHL, Ghabrial SA, Jarvis AW, Martelli GP, Mayo
   MA, Summers MD (1995) Sixth report of the International Committee on Taxonomy of
   Viruses. Springer, Wien New York, pp 219–222 (Arch Virol [Suppl] 10)
2. Flores J, Perez I, White L, Kalica AR, Marquina R, Wyatt RG, Kapikian AZ, Chanock
   RM (1982) Genetic relatedness among human rotaviruses as determined by RNA
   hybridization. Infect Immun 37: 648–655
3. Flores J, Perez-Schael I, Boeggeman E, Whitel L, Perez M, Purcell R, Hoshino Y,
   Midthun K, Kapikian A, Chanock R (1985) Genetic relatedness among human
   rotaviruses. J Med Virol 17: 135–143
4. Iizuka M, Chiba M, Masamune O, Kaga E, Nakagomi T, Nakagomi O (1994) A highly
   conserved genomic RNA constellation of Japanese isolates of human rotaviruses
   carrying G serotype 3 and P serotype 9. Res Virol 145: 21–24

5. Kaga E, Nakagomi O (1994) Recurrent circulation of single nonstructural gene substitution reassortants among human rotaviruses with a short RNA pattern. Arch Virol 136: 63–71
6. Kaga E, Nakagomi O, Uesugi S (1992) Thermal degradation of RNA-RNA hybrids during hybridization in solution. Mol Cell Probes 6: 261–264
7. Mascarenhas JDP, Linhares AC, Gabbay YB, de Freitas RB, Mendez E, Lopez S, Arias CF (1989) Naturally occurring serotype 2/subgroup II rotavirus reassortants in Northern Brazil. Virus Res 14: 235–240
8. Matsuno S, Mukoyama A, Hasegawa A, Taniguchi K, Inouye S (1988) Characterization of a human rotavirus strain which is possibly a naturally-occurring reassortant virus. Virus Res 10: 167–175
9. Nakagomi O, Mochizuki M, Aboudy Y, Shif I, Silberstein I, Nakagomi T (1992) Hemagglutination by a human rotavirus isolate as evidence for transmission of animal rotaviruses to humans. J Clin Microbiol 30: 1011–1013
10. Nakagomi O, Nakagomi T (1991) Genetic diversity and similarity among mammalian rotaviruses in relation to interspecies transmission of rotavirus. Arch Virol 120: 43–55
11. Nakagomi O, Nakagomi T (1991) Molecular evidence for naturally occurring single VP7 gene substitution reassortant between human rotaviruses belonging to two different genogroups. Arch Virol 119: 67–81
12. Nakagomi O, Nakagomi T (1993) Interspecies transmission of rotaviruses studied from the perspective of genogroup. Microbiol Immunol 37: 337–348
13. Nakagomi O, Nakagomi T, Akatani K, Ikegami N (1989) Identification of rotavirus genogroups by RNA-RNA hybridization. Mol Cell Probes 3: 251–261
14. Nakagomi O, Ohshima A, Aboudy Y, Shift I, Mochizuki M, Nakagomi T, Gotlieb-Stematsky T (1990) Molecular identification by RNA-RNA hybridization of a human rotavirus that is closely related to feline and canine origin. J Clin Microbiol 28: 1198–1203
15. Nakagomi T, Akatani K, Ikegami N, Katsushima N, Nakagomi O (1988) Occurrence of changes in human rotavirus serotypes with concurrent changes in genomic RNA electropherotypes. J Clin Microbiol 26: 2586–2592
16. Nakagomi T, Nakagomi O (1989) RNA-RNA hybridization identifies a human rotavirus that is genetically related to feline rotavirus. J Virol 63: 1431–1434
17. Shif I, Iizuka M, Silberstein I, Mendelson E, Nakagomi O (1994) Rotaviruses belonging to the AU-1 genogroup recovered from Israeli infants with diarrhea. Arch Virol 138: 357–364
18. Ward RL, Nakagomi O, Knowlton DR, McNeal MM, Nakagomi T, Clemens JD, Sack DA, Schiff GM (1990) Evidence of natural reassortants of human rotaviruses belonging to different genogroups. J Virol 64: 3219–3225

Authors' address: Dr. O. Nakagomi, Department of Microbiology, Akita University School of Medicine, Akita 010, Japan.

Arch Virol (1996) [Suppl] 12: 99–111

Classification of rotavirus VP4 and VP7 serotypes

Y. Hoshino and **A. Z. Kapikian**

Epidemiology Section, Laboratory of Infectious Diseases,
National Institute of Allergy and Infectious Diseases,
National Institutes of Health, Bethesda, Maryland, U.S.A.

Summary. Rotaviruses, members of the *Reoviridae* family, are major etiologic agents of acute nonbacterial gastroenteritis of the young in a wide variety of mammalian and avian species, including humans. The need for effective immunoprophylaxis against rotaviral gastroenteritis has stimulated interest in the biochemical, molecular, genetic, and clinical aspects of these agents with the aim of developing safe and effective vaccines. Because neutralizing antibodies appear to play an important role in protection against many viral diseases, rotavirus antigens that induce neutralizing antibodies have played a central role in research and development of a rotavirus vaccine. The VP7 glycoprotein and VP4 spike protein that constitute the outer capsid of a complete rotavirus particle have been shown to be independent neutralization antigens. Since type specificity of the outer capsid proteins of a rotavirus appears to play an important role in protection against disease in experimental animal models, continued efforts have been made for classification and typing of neutralization specificities on the VP7 or VP4 capsid protein. Based on a criterion of > 20-fold differences between the homologous and heterologous reciprocal neutralizing antibody titers, fourteen VP7 (G) serotypes have been established. Studies are underway to characterize and classify the VP4 (P) serotypes among the strains that exhibit the fourteen different G serotypes. Attempts to classify the VP4 serotypes based on the same criterion (*i.e.*, > 20-fold antibody differences) that is applied to classification of VP7 serotypes are in progress. This standard of > 20-fold antibody differences can be applied with hyperimmune serum raised to a reassortant possessing the VP4 encoding gene (and an unrelated VP7 encoding gene). Genotypes can provide leads towards classification but the serotype of a strain should be based on neutralization.

Introduction

Rotaviruses are the single most important known etiologic agents of diarrhea in infants and young children world-wide [2, 24]. The urgent need for effective immunoprophylaxis against rotavirus diarrhea has stimulated interest in the biochemical, molecular, genetic, and clinical aspects of these agents with the aim of developing safe and effective vaccines [10, 24]. Because neutralizing antibodies appear to play an important role in protection against many viral diseases, the two outer capsid rotavirus antigens (VP7 and VP4 [the spike protein]) that induce neutralizing antibodies have played a central role in research and development of a rotavirus vaccine [6, 18, 19, 30, 33]. Classification of rotaviruses into serotypes according to VP7 specificity by neutralization has progressed smoothly by adopting principles applied to other viruses [5, 25]. The cornerstone of this system is that a strain is considered a distinct serotype when a reciprocal >20-fold difference in neutralizing antibody titer is observed between that strain and established serotypes. However, there is considerable confusion regarding the classification of strains according to VP4 specificity because of the lack of specific antisera that recognize VP4 neutralization specificity. As a result, a few strains have been classified by neutralization but most have been assigned numbers according to VP4 genetic relationships. The purpose of this paper is to describe a scheme whereby VP4 classification of rotaviruses can be carried out by neutralization assay using the same standards as employed for VP7.

In 1983, a series of monoclonal antibodies (mAbs) directed against surface proteins (VP7 or VP4) of rhesus monkey rotavirus (RRV) strain MMU18006 were isolated [17]. These studies demonstrated that certain mAbs directed to the major glycoprotein VP7 had neutralizing activities with distinct VP7 serotype or G (VP7 is a glycoprotein) specificity (Table 1). In addition, they showed that certain mAbs directed against the spike protein VP4 also had neutralizing activities with distinct VP4 serotype or P (VP4 is protease sensitive) specificity (Table 2). Thus, these monoclonal antibodies were able to clearly distinguish between VP7 and VP4 neutralization specificities, at least in a one-way assay in which the homotypic titer was at least more than 20-fold greater than the heterotypic titer. Later, naturally occurring "intertypic" rotaviruses, such as

Table 1. Neutralizing activity of monoclonal antibodies directed to rhesus monkey rotavirus (RRV) outer capsid protein VP7 (from [17])

Rotavirus (species of origin)	VP7(G) Serotype	Antibody titer* of indicated monoclonal antibodies	
		954/159/33	952/3/68
RRV (simian)	3	≥ 204 800	≥ 240 800
SA11 (simian)	3	≥ 204 800	12 800
CU–1 (canine)	3	102 400	≥ 240 800
P (human)	3	≥ 16 000	≥ 16 000
Wa (human)	1	< 200	< 200
UK (bovine)	6	< 200	< 200

* Reciprocal of 60% plague reduction neutralization (PRN) antibody titer

Table 2. Neutralizing activity of monoclonal antibodies directed to rhesus monkey
rotavirus (RRV) outer capsid protein VP4 (from [17])

Rotavirus	VP7(G) serotype	Antibody titer* of indicated monoclonal antibodies	
		954/155/25	954/23/4
RRV	3	≥ 204 800	≥ 240 800
SA11	3	< 500	2 000
CU–1	3	< 500	< 500
Wa	1	< 500	< 500
OSU	5	< 500	< 500
UK	6	< 500	< 500

* Reciprocal of 60% PRN antibody titer

Table 3. Characterization by neutralization of naturally-occurring intertypic
porcine rotavirus SB-1A strain (from [21])

Rotavirus	G serotype	Antibody titer* of indicated serum		
		Gottfried	SB–1A	OSU
Gottfried	4	40 960	20 480	< 80
SB–1A	4	40 960	20 480	20 480
OSU	5	< 80	10 240	81 920

* Reciprocal 60% PRN antibody titer of guinea pig hyperimmune antiserum

SB-1A, were found to react with viruses belonging to two distinct G serotypes
(Table 3). This was a result of both VP4 and VP7 outer capsid proteins having
distinct neutralization specificities (Table 3) [21, 22]. This finding was confirmed
by analyzing a single VP7 gene substitution reassortant rotavirus (OSU ×
Gottfried), generated from two antigenically distinct porcine rotavirus strains
(OSU and Gottfried) (Table 4) [20]. Neutralization of Gottfried virus by OSU ×
Gottfried reassortant antiserum is due to Gottfried VP7 antibodies and neutral-
ization of OSU virus by the same antiserum is due to OSU VP4 antiserum.

VP4 has also been shown to induce protection in vivo. For example, in
passive-protection models (i) mice were protected against rotavirus challenge by
homotypic or heterotypic neutralizing mAbs directed to VP4 or VP7 [29, 34],
and (ii) mouse dams hyperimmunized orally with a reassortant containing VP4
and VP7 from two distinct rotavirus serotypes or hyperimmunized parenterally
with baculovirus recombinant-expressed VP4 protected their pups passively
against diarrhea induced by homotypic challenge [27, 35]. In a homologous
system of colostrum-deprived newborn gnotobiotic pigs free of maternal antibodies and a single VP7 gene substitution porcine rotavirus reassortant

Table 4. Characterization by neutralization of a single-gene-substitution reassortant OSU × Gottfried (from [20])

Rotavirus	Antibody titer* of indicated serum		
	Gottfried	OSU × Gottfried	OSU
Gottfried (VP4:2B; VP7:4)	*40 960*	81 920	<80
OSU × Gottfried** (VP4:9; VP7:4)	81 920	≥81 920	20 480
OSU (VP4:9; VP7:5)	<80	40 960	*81 920*

* Reciprocal 60% PRN antibody titer of guinea pig hyperimmune antiserum
** All genes from OSU except VP7 encoding gene which is from Gottfried

Table 5. Serum neutralizing antibody response of gnotobiotic piglets 21 days after oral administration of porcine rotavirus strains (from [20])

No. of Piglets	Infection with rotavirus	Antibody titer* of indicated post-challenge serum		
		OSU	Gottfried	OSU × Gottfried
4	OSU (VP4:9; VP7:5)	960	<20	1 600
4	Gottfried (VP4:2B; VP7:4)	<20	2 720	1 280
4	OSU × Gottfried** (VP4:9; VP7:4)	2 400	2 400	4 480

* Reciprocal mean 60% PRN antibody titer
** All genes from OSU except VP7 encoding gene which is from Gottfried

described above (Table 4), we studied: (i) the antigenicity in the host of the rotavirus outer capsid proteins VP4 and VP7 during a single enteric infection, (ii) the extent of involvement of antibodies induced against VP4 and VP7 in resistance against rotavirus diseases, and (iii) the effectiveness of a single enteric infection by a reassortant rotavirus in providing resistance to both parental viruses [20]. These studies demonstrated that (i) the host immune system responds equally to the two rotavirus outer capsid proteins, VP4 and VP7, after a single oral exposure (Table 5), (ii) antibody to VP4 and antibody to VP7 are each associated with resistance to diarrhea (Table 6), and (iii) a reassortant rotavirus bearing VP4 and VP7 derived from two serotypically distinct viruses induces immunity to both parental viruses (Table 5 and 6). Recently, it was reported that (i) mouse pups born to the dams inoculated intranasally with the SA11 VP7$_{sc}$ expressed at the cell surface by recombinant adenovirus were shown

Table 6. Cross-protection studies in gnotobiotic piglets (from [20])

No. of Piglets	Primary infection with	Challenge with	Duration* of		Onset* of	
			Virus shedding	Diarrhea	Virus shedding	Diarrhea
4	Gottfried (VP4:2B; VP7:4)	OSU (VP4:9; VP7:5)	4.0	4.25	1.5	1.75
4	OSU	Gottfried	3.5	4.5	1.0	1.5
4	OSU × Gottfried** (VP4:9; VP7:4)	OSU	1.3	None	6.5	None
4	OSU × Gottfried	Gottfried	None	None	None	None
2	None	OSU	10.0	5.5	1.0	1.5
2	None	Gottfried	7.0	7.0	1.0	1.0

* Median days following challenge
** All genes from OSU except VP7 encoding gene which is from Gottfried

to be protected passively against diarrhea upon challenge with homologous virus [4], and (ii) mouse dams hyperimmunized intraperitoneally with baculovirus recombinant-expressed RRV VP7 partially protected pups passively against diarrhea induced by RRV virus [13]. With the role of VP4 and VP7 in neutralization and in independent protective immune properties established, a binary system for the classification and nomenclature of the two distinct surface proteins of rotaviruses has become a necessity.

A criterion of >20-fold differences between the homologous and heterologous reciprocal neutralizing antibody titers has been used to establish the serotype of various viruses including polioviruses [5], ECHO viruses [5], coxsackie viruses [5], and rhinoviruses [25]. Based on this criterion, fourteen VP7 (G) serotypes have been established thus far in tests using hyperimmune antiserum to rotavirus (Table 7) [11, 18, 24]. Later, an enzyme-linked immunosorbent assay with anti–VP7 serotype-specific neutralizing or non-neutralizing mAbs, as well as with gene group (allele) typing (genotyping), were found to correlate with neutralization specificity and have been used widely as a proxy method for G typing [1, 3, 7, 8, 16, 18, 37, 40].

Attempts to serotype rotaviruses according to VP4 neutralization specificity have been more difficult than VP7 typing. This has been due to the inability to generate high-titered type-specific hyperimmune antiserum with VP4 specificity. It is of interest that hyperimmune antisera to rotaviruses have a disproportionate neutralizing capability to VP7 when compared to VP4 and because of this, a VP7 typing system was readily established by neutralization [18]. In addition, mAbs directed against the VP4 are, in general, more cross-reactive than those to VP7. Despite such difficulties, however, progress has been made in classification and typing of the VP4 protein (Table 8). With the use of reassortant rotaviruses and hyperimmune antisera to them, bovine and porcine rotaviruses have successfully

Table 7. Serotypic and genotypic classification of group A rotavirus VP7 and subgroup specificities[a]

VP7 (G) serotype[b,c]	Human rotavirus strains		Animal rotavirus strains		
	Strain	Subgroup	Strain	Subgroup	Species
1	Wa, KU, D, M37, RV-4, WI79, K8	II	C60, C86, C95	I	Pig
			T449	I	Cow
2	DS-1, S2, KUN, RV-5, 1076	I	C134	I	Pig
3	P, MO, YO, RV-3, Ito, Nemoto, WI78, McN	II	SA11, SAII-4FM	I	Vervet monkey
			MMU18006	I	Rhesus monkey
	AU-1, AU228, Ro1845, HCR3	I	CU-1, K9, RS-15	I	Dog
	0264	I and II	TAKA, Cat2, Cat97, FRV-1	I	Cat
			H-2, HI-23, HO-5	Not I or II	Horse[d]
			FI-14	I and II	Horse
			CRW-8, C176	I	Pig
			R-2	II	Rabbit
			Ala, C11	I	Rabbit
			EB	I	Mouse
			EW	Not I or II	Mouse
			EHP, EC, EL,	?	Mouse
			LRV1	I	Sheep
4	ST3, ST4, VA70, Hosokawa, Hochi, 57M	II	Gottfried, SB-1A	II	Pig
			BEN-144	?	Pig
			SB-2	I	Pig
5	IAL28	II	OSU, EE, TFR-41, C134	I	Pig
			H-1	I	Horse
6	PA151, PA169	I	NCDV, UK, RF, WC3, Q17, OK, ID	I	Cow
			B641, C486	?	Cow
			LRV2a	I	Sheep

(continued)

Table 7. (continued)

VP7 (G) serotype[b,c]	Human rotavirus strains		Animal rotavirus strains		
	Strain	Subgroup	Strain	Subgroup	Species
7	None	–	Ch2	Not I orII	Chicken
			Ty1[e]	Not I or II	Turkey
			PO-13	I	Pigeon
			993/83	Not I or II	Cow
			678, J2538, A5, Cody	I	Cow
8	69M, B37, HAL1271, HAL1166	I			
	PA171, AU32	II			
9	WI61, F45, 116E	II	ISU-64	I	Pig
	Mc323	I	LRV2c	II	Sheep
10	A64	II	B223, V1005, KK3, 61A, Cr	I	Cow
	Mc35, I321	I	Lp14	?	Sheep
			K923	I	Sheep
11	None	–	YM, A253	I	Pig
12	L26, L27	I	None	–	–
13	None	–	L338	I	Horse
14	None	–	FI23	I	Horse
			FR4, FR8	Not I or II	Horse

[a] From [11] and [19] with additions

[b] VP7 (G) serotype determined by reciprocal cross-neutralization

[c] In addition VP7 genotype has been determined by comparative amino acid sequence anaysis and/or nucleic acid hybridization. Genotyping has been used widely to distinguish among rotavirus strains and, where evaluated, has correlated with serotyping differences by neutralization assay. Thus, genotyping is now used frequently as a proxy method for neutralization for VP7 (G) serotyping

[d] Human rotaviruses with neither subgroup I nor II specificity have also been described but without serotype designation

[e] Recent amino acid sequence analysis of VP7-encoding gene of Ty1 suggests that the virus may not belong to G serotype 7

Table 8. Serotypic and genotypic classification of group A rotavirus VP4[a]

VP4 (P) serotype[b]	VP4 genotype[c]	Human rotavirus strains — Strain	G serotype	Animal rotavirus strains — Strain	G serotype	Species
1 A	8	KU, Wa	1	LRV$_c$[d]	9	Sheep
		P, YO, MO	3			
		VA70, Hochi, Hosokawa	4			
		WI61, F45	9			
B	4	DS-1, S2, RV-5[d]	2	None	—	—
		L26[d]	12			
2 A	6	M37	1	None	—	—
		1076	2			
		McN, RV-3[d]	3			
		ST3	4			
B	6	None	—	Gottfried, BEN-144	4	Pig
3	9	K8	1	FRV-1[d], Cat2[d]	3	Cat
		AU-1[d]	3			
		PA151[d]	6			
4	10	57M[d]	4	None	—	—
		69M	8			
	12	None	—	H-2, FI-14[d]	3	Horse
				FI23[d]	14	Horse
5 A	3	HCR3[d], Ro1845	3	CU-1, K9	3	Dog
				Cat97	3	Cat
B	3	None	—	MMU18006	3	Rhesus Monkey
6	1	None	—	NCDV, C486, J2538	6	Cow
				A5[d], Cody[d]	8	Cow
				SA11-4fM	3	Vervet Monkey
				LRV1[d]	3	Sheep

(continued)

Table 8. (continued)

VP4 (P) serotype[b]	VP4 genotype[c]	Human rotavirus strains		Animal rotavirus strains		
		Strain	G serotype	Strain	G serotype	Species
7	5	None	–	UK, B641, IND[d], OK[d], WC3[d,e]	6	Cow
				61A[d]	10	Cow
8	11	116E[d]	9	B223, B-11, A44[d], KK3[d], Cr[d]	10	Cow
				LRV[d]_a	6	Sheep
9	7	I321[d]	10	C60, C95	1	Pig
		None	–	CRW-8, BEN-307, A131	3	Pig
				BMI-1, SB-1A	4	Pig
				OSU, TFR-41, C134	5	Pig
				YM, A253	11	Pig
10	16	None	–	H-1[d]	5	Horse
11	14	PA169	6	EB, EW, EC, EL	3	Mouse
		HAL1166[d]	8	None	–	–
		Mc35	10			
?	2	None	–	SA11	3	Vervet Monkey
?	13	None	–	MDR-13	3/5	Pig
?	15	None	–	Lp14, K923	10	Sheep
?	17	None	–	993/83	7	Cow
				PO-13	7	Pigeon
?	18	None	–	L338	13	Horse
?	19	None	–	4F	3	Pig
?	20	None	–	EHP	3	Mouse

a From [11] and [19] with additions and slight modifications
b VP4 (P) serotype determined by reciprocal or one-way cross neutralization
c VP4 genotype determined by comparative amino acid sequence analysis and/or nucleic acid hybridization
d Not tested by neutralization; relationship suggested by amino acid sequence and/or nucleic acid hybridization analyses
e Gorziglia and Clark (unpublished data)

been classified into P serotypes using the criterion of a reciprocal > 20-fold antibody difference [20, 28, 31, 38]. More recently, by using a similar approach, equine rotavirus VP4 (P) serotypes have been established based on ≥ 16-fold antibody differences [23]. Classification of human rotavirus VP4 proteins into four serotypes and one subtype has been made by using hyperimmune antisera raised against baculovirus recombinant-expressed VP4 protein [15, 26]. However, a criterion of ≥ 8-fold antibody differences was used to establish a P serotype. A similar approach has been taken to characterize the P serotype of selected porcine and simian rotavirus VP4 protein [27, 32].

A solution to the difficulties in establishing VP4 serotypes by neutralization must rely on the availability of high titered specific antisera with VP4 specificity if the criterion of > 20-fold antibody differences are to be applied. In this regard, such antisera can be produced if single gene substitution reassortants with appropriate VP4 and VP7 components are used as immunogens. Thus, single VP4-gene substitution reassortant rotavirus (Wa × UK or DS-1 × UK) and single VP7-gene substitution reassortant rotavirus (ST3 × DS-1, K8 × DS-1, or 69M × DS-1) were generated (distribution of genes shown in Table 9). Serum of guinea pigs hyperimmunized with each of the reassortants was then utilized in plaque reduction neutralization assays to examine antigenic relationships among the VP4 capsid proteins of various rotavirus strains. As shown in Table 9, three of the five "P" serotypes could be clearly identified in a one-way neutralization assay by the > 20-fold antibody difference criterion, whereas the other two strains were likely serotypically distinct also because the antibody differen-

Table 9. Characterization of VP4 serotyes of selected human rotavirus strains

Rotavirus	G serotype	Reciprocal of 60% PRN antibody titer of hyperimmune antiserum to				
		Wa × UK[a]	DS-1 × UK[a]	ST3 × DS-1[b]	K8 × DS-1[b]	69M × DS-1[b]
Wa	1	*640*	< 80	80	< 80	< 80
K8	1	< 80	< 80	< 80	*2 560*	< 80
DS-1	2	< 80	*640*	ND	ND	ND
DS-1 × EW[c]	3	ND	ND	< 80	< 80	< 80
P	3	320	< 80	< 80	< 80	< 80
ST3	4	< 80	< 80	*2 560*	< 80	< 80
69M	8	< 80	< 80	< 80	< 80	*10 240*
WI-61	9	640	< 80	< 80	< 80	< 80

[a] Immunizing reassortant strain contained the VP4 gene from human rotavirus Wa or DS-1 and ten genes from bovine UK strain

[b] Immunizing reassortant strain contained ten genes including the VP4 gene from ST3, K8 or 69M and only VP7 gene from DS-1

[c] Reassortant strain contained ten genes including the VP4 gene from DS-1 and only the VP7 gene from murine EW strain

ces were at least 16-fold or greater when the starting dilution was 1:80. Characterization of VP4 (P) serotypes of additional human as well as selected animal rotaviruses is underway.

To circumvent the lack of appropriate and readily-available reagents for serotyping and classification of rotavirus VP4, a typing scheme was proposed by Estes and Cohen [10] on the basis of nucleic acid hybridization and sequence analysis of the VP4 gene. This proposal has been largely validated by serological results, and new VP4 genotypes have been added to this sytem [9, 11, 12, 19, 36, 39]. However, a genotype established by nonserological methods is not always identical to a serotype defined by serological methods. For example, human rotavirus 69M strain and equine rotavirus H-2 strain, which are considered to belong to different P genotypes based on the level of VP4 amino acid homology (85.3% vs $\geq$ 89% [suggested standard] [15]), have recently been shown to belong to the same P serotype by neutralization [23, 26].

In conclusion, a rotavirus VP4 (P) serotype classification system should be made on the basis of data obtained by neutralization assay. When the validity of genotyping is confirmed by neutralization methods, the typing of gene 4 alleles [14] could be used as a proxy method for VP4 (P) serotyping in common with the now widely-used typing of VP7-gene alleles [16] for VP7 (G) serotyping.

References

1. Akatani K, Ikegami N (1987) Typing of fecal rotavirus specimens by an enzyme-linked immunosorbent assay using monoclonal antibodies (in Japanese). Clin Virol 15: 61–68
2. Bern C, Glass RI (1994) Impact of diarrheal diseases worldwide. In: Kapikian AZ (ed) Viral Infections of the Gastrointestinal Tract. Marcel Dekker, New York, pp 1–26
3. Birch CJ, Heath RJ, Gust ID (1988) Use of serotype-specific monoclonal antibodies to study epidemiology of rotavirus infection. J Med Virol 24: 45–53
4. Both GW, Lockett LJ, Janardhana V, Edwards SJ, Bellamy AR, Graham FL, Prevac L, Andrew ME (1993) Protective immunity to rotavirus-induced diarrhea is passively transferred to newborn mice from naive dams vaccinated with a single dose of a recombinant adenovirus expressing rotavirus VP7$_{sc}$. Virology 193: 940–949
5. Committee on the Enteroviruses (1957) The Enteroviruses. Am J Pub Health 47: 1556–1566
6. Conner ME, Matson DO, Estes MK (1994) Rotavirus vaccines and vaccination potential. Curr Topics Microbiol Immunol 185: 285–337
7. Coulson BS, Unicomb LE, Pitson GA, Bishop RF (1987) Simple and specific enzyme immunoassay using monoclonal antibodies for serotyping human rotaviruses. J Clin Microbiol 25: 509–515
8. Das BK, Gentsch JR, Cicirello HG, Woods PA, Gupta A, Ramachandran M, Kumar R, Bhan MK, Glass RI (1994) Characterization of rotavirus strains from newborns in New Delhi, India. J Clin Microbiol 32: 1820–1822
9. Dunn SJ, Burns JW, Cross TL, Vo PT, Ward RL, Bremont M, Greenberg HB (1994) Comparison of VP4 and VP7 of five murine rotavirus strains. Virol 203: 250–259
10. Estes MK, Cohen J (1989) Rotavirus gene structure and function. Microbiol Rev 53: 410–449
11. Estes MK (1996) Rotaviruses and their replication. In: Fields BN, Knipe DM, Howley PM (eds) Virology, 3rd edn. Lippincott-Raven, Philadelphia, 1625–1655

12. Fitzgerald TA, Munoz M, Wood AR, Snodgrass DR (1995) Serological and genomic characterization of group A rotavirus from lambs. Arch Virol 140: 1541–1548
13. Fiore L, Dunn SJ, Ridolfi B, Ruggeri FM, Mackow ER, Greenberg HB (1995) Antigenicity, immunogenicity and passive protection induced by immunization of mice with baculovirus-expressed VP7 protein from rhesus rotavirus. J Gen Virol 76: 1981–1988
14. Gentsch JR, Glass RI, Woods P, Gouvea V, Gorziglia M, Flores J, Das BK, Bhan MK (1992) Identification of group A rotavirus gene 4 types by polymerase chain reaction. J Clin Microbiol 30: 1365–1373
15. Gorziglia M, Larralde G, Kapikian AZ, Chanock RM (1990) Antigenic relationships among human rotaviruses as determined by outer capsid protein VP4. Proc Natl Acad Sci USA 87: 7155–7159
16. Gouvea V, Glass RI, Woods P, Taniguchi K, Clark HF, Forrester B, Fang Z-Y (1990) Polymerase chain reaction amplification and typing of rotavirus nucleic acid from stool specimens. J Clin Microbiol 28: 276–282
17. Greenberg HB, Valdesuso J, van Wyke K, Midthun K, Walsh M, McAuliffe V, Wyatt RG, Kalica AR, Flores J, Hoshino Y (1983) Production and preliminary characterization of monoclonal antibodies directed at two surface proteins of rhesus rotavirus. J Virol 47: 267–275
18. Hoshino Y, Kapikian AZ (1994) Rotavirus antigens. Curr Topics Microbiol Immunol 185: 179–227
19. Hoshino Y, Kapikian AZ (1994) Rotavirus vaccine development for the prevention of severe diarrhea in infants and young children. Trends Microbiol 2: 242–249
20. Hoshino Y, Saif LJ, Sereno MM, Chanock RM, Kapikian AZ (1988) Infection immunity of piglets to either VP3 or VP7 outer capsid protein confers resistance to challenge with a virulent rotavirus bearing the corresponding antigen. J Virol 62: 744–748
21. Hoshino Y, Sereno MM, Midthun K, Flores J, Chanock RM, Kapikian AZ (1987) Analysis by plaque reduction neutralization assay of intertypic rotaviruses suggests that gene ressortment occurs in vivo. J Clin Microbiol 25: 290–294
22. Hoshino Y, Sereno MM, Midthun K, Flores J, Kapikian AZ, Chanock RM (1985) Independent segregation of two antigenic specificities (VP3 and VP7) involved in neutralization of rotavirus infectivity. Proc Natl Acad Sci USA 82: 8701–8704
23. Isa P, Snodgrass DR (1994) Serological and genomic characterization of equine rotavirus VP4 proteins identifies three different P serotypes. Virology 201: 364–372
24. Kapikian AZ, Chanock RM (1996) Rotaviruses. In: Fields BN, Knipe DM, Howley PM (eds) Virology, 3rd edn. Lippincott-Raven, Philadelphia, 1657–1708
25. Kapikian AZ, Conant RM, Hamparian VV, Chanock RM, Chapple PJ, Dick EC, Fen JD, Gwaltney JM, Hamre D, Holper JC, Jordan WS, Lennette EH, Melnick JL, Mogabgab WJ, Mufson MA, Phillips CA, Schieble JH, Tyrrell DAJ (1967) Rhinoviruses: A numbering system. Nature 213: 761–763
26. Li B, Hoshino Y, Gorziglia M (1996) Identification of a unique VP4 serotype that is shared by a human rotavirus (69 M strain) and an equine rotavirus (H-2 strain). Arch Virol 141: 155–160
27. Mackow ER, Vo PT, Broome R, Bass D, Greenberg HB (1990) Immunization with baculovirus-expressed VP4 protein passively protects against simian and murine rotavirus challenge. J Virol 64: 1698–1703
28. Matsuda Y, Nakagomi O, Offit PA (1990) Presence of three P types (VP4 serotypes) and two G types (VP7 serotypes) among bovine rotavirus strains. Arch Virol 115: 199–207
29. Matsui SM, Mackow ER, Greenberg HB (1989) Molecular determinant of rotavirus neutralization and protection. Adv Virus Res 36: 181–214

30. Mattion NM, Cohen J, Estes MK (1994) The rotavirus proteins. In: Kapikian AZ (ed) Viral Infections of the Gastrointestinal Tract. Marcel Dekker, New York, pp 169–249

31. Nagesha HS, Holmes IH (1991) VP4 relationships between porcine and other rotavirus serotypes. Arch Virol 116: 107–118

32. Nishikawa K, Fukuhara N, Liprandi F, Green K, Kapikian AZ, Chanock RM, Gorziglia M (1989) VP4 protein of porcine rotavirus strain OSU expressed by a baculovirus recombinant induces neutralizing antibodies. Virology 173: 631–637

33. Offit PA (1994) Immunologic determinants of protection against rotavirus disease. Curr Topics Microbiol Immunol 185: 229–254

34. Offit PA, Shaw RD, Greenberg HB (1986) Passive protection against rotavirus-induced diarrhea by monoclonal antibodies to surface proteins VP3 and VP7. J Virol 58: 700–703

35. Offit PA, Clark HF, Blavat G, Greenberg HB (1986) Reassortant rotaviruses containing structural proteins VP3 and VP7 from different parents induce antibodies protective against each parental serotype. J Virol 60: 491–496

36. Sereno MM, Gorziglia MI (1994) The outer capsid protein VP4 of murine rotavirus Eb represents a tentative new P type. Virology 199: 500–504

37. Shaw RD, Stoner-Ma DL, Estes MK, Greenberg HB (1985) Specific enzyme-linked immunoassay for rotavirus serotypes 1 and 3. J Clin Microbiol 22: 286–291

38. Snodgrass DR, Hoshino Y, Fitzgerald TA, Smith M, Browning GF, Gorziglia M (1992) Identification of four VP4 serological types (P serotypes) of bovine rotavirus using viral reassortants. J Gen Virol 73: 2319–2325

39. Taniguchi K, Urasawa S (1995) Diversity in rotavirus genomes. Seminars Virol 6: 123–131

40. Taniguchi K, Urasawa T, Morita Y, Greenberg HB, Urasawa S (1987) Direct serotyping of human rotavirus in stools by enzyme-linked immunosorbent assay using serotypes 1-, 2-, 3- and 4-specific monoclonal antibodies to VP7. J Infect Dis 155: 1159–1166

Authors' address: Dr. Y. Hoshino, Epidemiology Section, Laboratory of Infectious Diseases, National Institutes of Health, Bethesda, MD 20892, U.S.A.

Arch Virol (1996) [Suppl] 12: 113–118

VP4 and VP7 typing using monoclonal antibodies

B. S. Coulson

Department of Microbiology, The University of Melbourne,
Parkville, Victoria, Australia

Summary. Both rotavirus outer capsid proteins, VP4 and VP7, elicit neutralizing antibodies. Neutralizing mouse monoclonal antibodies (N-MAbs) to VP7 are easily derived and have been used widely and successfully to serotype both stool-derived and culture-adapted rotaviruses by enzyme immunoassay (EIA). Generally, approximately 70% of rotaviruses in stool samples are typable by VP7 EIA, an inexpensive and practical method. Variations in antigenic regions between strains within human rotavirus serotypes 1, 2, 4, and 9 have been recorded. These have been termed monotypes because they are detected with N-MAbs. The molecular basis for monotypes has been determined by mapping mutations selected in N-MAb-resistant antigenic variants, and by sequence analysis of the gene encoding VP7 in newly recognized monotypes. Antigenic regions A, B and C in VP7 are involved. In order to detect all members of a particular VP7 serotype, it is necessary to type with a panel of N-MAbs specific for that serotype.

N-MAbs to VP4 of human rotavirus are difficult to raise and few have proven suitable for VP4 serotyping by EIA. The specificity of the assay for each P type is highest when the VP7 serotype specificity of the capture antiserum is matched to the G type of the rotavirus in the test sample. The VP4 EIA gives similar typing rates to the VP7 typing EIA. N-MAbs directed to VP8*, the smaller subunit of VP4 generated by proteolytic cleavage, are more likely to show serotype specificity. Some N-MAbs that select mutations in the putative fusion region of VP5*, the larger subunit of VP4, show cross-reactivity with extracts of normal, uninfected MA 104 cells and with fetal bovine serum. These N-MAbs also give elevated EIA OD readings with rotavirus-positive, but previously non-reactive fecal samples which have been frozen and thawed repeatedly. Overall, VP8*-reactive N-MAbs appear most suitable for VP4 typing by EIA.

Rotavirus group classification

Rotaviruses are classified serologically into groups, each of which may contain multiple serotypes. Each rotavirus group (A to F) contains viruses that share antigens detectable by immune electron microscopy, EIA and immunofluorescence. Group A, B and C rotaviruses are found in both humans and animals, whereas viruses of other groups have been found in animals only. Group A rotaviruses are the major cause of infantile gastroenteritis in humans and are the major focus of this article.

Rotavirus serotype and monotype classification

Serotypes within group A rotaviruses were originally defined by cross-neutralization, using hyperimmune antisera in plaque reduction or fluorescent focus reduction assays. When it was shown that both outer capsid proteins, VP4 and VP7, elicit neutralizing antibodies [12] and that the genes that encode these proteins (genes 4 and 7, 8 or 9) can segregate independently by virus reassortment, it became necessary to classify rotaviruses by a binary system. Serotypes determined by use of VP7 are termed G (for glycoprotein) types, and those based on VP4 are termed P (for protease-sensitive protein) types [8]. Serotypes defined originally by cross-neutralization assays correspond with particular G types, as hyperimmune antisera predominantly recognize VP7 in neutralization assays. This was confirmed when most neutralizing monoclonal antibodies to VP7 typed virus isolates identically to those typed with hyperimmune serum in fluorescent focus reduction assays [7, 23].

A few neutralizing monoclonal antibodies to VP7 only show reactivity with a subset of strains with a G type. These intraserotypic differences in neutralization epitopes have been termed monotypes [2]. Monotypes have been described within G types 1 and 4 of human rotaviruses [2, 5, 6]. The G4a and G4b monotypes corresponded to the subtypes of G4, which had been defined using cross-adsorbed polyclonal antisera [10]. Monotype-defining monoclonal antibodies selected antigenic variants of rotaviruses with mutations in antigenic regions A, B and C [5]. In order to optimize G typing with monoclonal antibodies, for each G type, a panel of antibodies is raised to strains of different monotypes, which is used to map different antigenic sites and to determine the G serotype.

Definition of serotype based on VP4

Definition of VP4 serotypes has not been so straightforward. Polyclonal antisera recognize the VP4 of the homologous strain only weakly, and interactions between VP4 and VP7 derived from different serotypes in virus reassortants can alter the reactivity of monoclonal antibodies raised to VP4 [1]. The P serotypes of cultivable human rotaviruses have been determined with polyclonal antisera raised against baculovirus recombinant-expressed VP4s [11]. In this system, three distinct P serotypes and one subtype were defined. These corresponded to types of VP4 based on gene 4 sequence variation (termed genotypes). Thus, human rotaviruses of P genotypes 4, 6, 8 and 9 [8] were shown to represent P serotypes 1B, 2A, 1A and 3, respectively.

Antisera to expressed VP4, its subunits VP8* and VP5*, and to VP8*-derived polypeptides showed at most 8-fold difference in titer between P serotypes, showed low titers overall, and have not yet been used for P typing of rotavirus in stools [11, 16, 17].

Fewer N-MAbs have been raised to human rotavirus VP4 than to VP7 [3, 4, 6]. Many of the N-MAbs to VP4 were either cross-reactive between P serotypes (heterotypic), or specific for only a subset of strains (or a single strain) within a P serotype [3, 4, 14, 19, 22]. Heterotypic antibodies usually selected variants with mutations in VP5* (amino acids 305–441) whereas the few P type-specific antibodies studied usually mapped to VP8* [13, 14, 15, 18, 21].

Interestingly, monoclonal antibodies to P1A or P2 rotaviruses are more likely to cross-react with P2 or P1A strains, respectively, than with P1B or P3 strains [3, 19, 21]. Thus, the subtypes P1A and P1B defined using antisera to baculovirus recombinant-expressed VP4 may represent different serotypes, and the P1A and P2 serotypes may be subtypes of the same P type.

Evaluation of monoclonal antibodies to VP4 as reagents for serotyping rotaviruses

Serotyping of cultivable human rotaviruses with monoclonal antibodies to VP4 has been achieved by indirect EIA using the monoclonal antibodies either for detection (DEIA) or for capture of virus [3, 19]. P serotypes determined by DEIA correlated with P serotypes defined with antisera to expressed VP4 using 13 cultivable rotavirus strains. DEIA typing of rotaviruses in stools has been reported [3]. In that case, for optimum results, the polyclonal hyperimmune antiserum used to coat the solid phase needed to be raised to a rotavirus of the same G type as the test strain. Hence, the G type of the rotavirus must be determined prior to P serotyping.

The DEIA was evaluated using 118 stools positive for rotavirus by RNA gel electrophoresis and by enzyme immunoassay using a monoclonal antibody to the group antigen VP6 [3]. These comprised 43% G1, 10% G2, 13% G3 and 10% G4 rotavirus strains, collected from children in Melbourne and Alice Springs hospitalized between 1983 and 1991 with severe gastroenteritis, and 24% G3 strains, collected in 1977 and 1978 from Melbourne neonates with asymptomatic infection. Two of the strains from neonates had been adapted to cell culture and were shown to be P serotype 2 by DEIA, whereas strains from symptomatic older children are almost always P serotype 1A (G1, 3, 4) or P serotype 1B (G2) [20]. As determined by DEIA, the P serotype of virus in these stools was the same as the inferred P type for 98 (83%) of samples. The inferred P type was decided based on the G type, the subgroup, the RNA electropherotype and the species origin (human). Twelve (10%) stools did not react with any P-typing antibody, and eight (7%) were untypable due to cross-reactivity of the P2 antibody (n = 3) or high background readings (n = 5). The DEIA and reverse transcriptase-polymerase chain reaction (RT-PCR) typing with nested primers [9] showed good agreement in P typing virus in nine additional stools (unpubl. results). Both methods typed five strains as P2, one as 1A, and two as not P1A, P1B or P2. One neonatal strain was untypable (G or P) by DEIA, but was P genotype 6 by RT-PCR.

Causes of false-positive and false-negative results obtained by DEIA with monoclonal antibodies to VP4

A group of 39 rotavirus-positive stools collected in Thailand during 1988 from children with diarrhea, each with short RNA electropherotypes also was studied (unpubl. results). These had been stored as stool homogenates, frozen and thawed at least twice after their transport from Thailand. A subset of 13 samples was typed by both DEIA and RT-PCR. Both methods typed four samples (31%) as P1B/P4. However, of five (40%) samples which were P4 by RT-PCR, two were untypable by DEIA due to P2 and P1B antibody reactivity, and three typed as P2 by DEIA. This discrepancy was shown to be due to cross-reactivity of the P2 monoclonal antibody by changing the coating antiserum. Use of antiserum raised to P2 virus ST-3 rather than antiserum raised to P1B virus RV-5, for capture in the DEIA abrogated the P2 reactivity, but not the P1B reactivity.

All of the 39 stools showed high levels of rotavirus VP6 group antigen. However, the level of P1B monoclonal reactivity in each of the 39 stool viruses correlated with the level of G2 monoclonal antibody reactivity, and both were often low. Interestingly, all rotaviruses in stools which had lost their reactivity with the G2 typing monoclonal antibody, showed higher reactivity with the P2 typing monoclonal antibody than with the P1B-reactive monoclonal antibody. Studies with a subset of these stools that were frozen and thawed an additional time and then retested, confirmed that this process reduces G2 and P1B reactivity but increases P2 reactivity. The P2 monoclonal antibody also cross-reacted with normal MA 104 cell antigens present in uninfected cells, fetal bovine serum and with bovine serum albumin (unpubl. results).

These studies with stool-derived virus have shown that neutralizing monoclonal antibodies to VP4 can give spurious results in stools, results which are not predictable from tests with cultivable rotaviruses. The DEIA for P serotype determination has acceptable specificity and sensitivity in the current format, provided that stool specimens are carefully stored and processed, and that electropherotypes and virus subgroups are also determined. Alternative P2-specific monoclonal antibodies, particularly those directed to VP8*, need to be derived and evaluated for P serotyping of human rotaviruses.

Note added in proof

Monotypes within G serotype 2 human rotaviruses now have been described (Coulson BS, Kirkwood CD, Masendycz PJ, Bishop RF, Gerna G (1996) Amino acids involved in distinguishing between monotypes of rotavirus G serotypes 2 and 4. J Gen Virol 77: 239–245).

Acknowledgements

The assistance of Carl Kirkwood, Paul Masendycz and Ruth Bishop with this work is gratefully acknowledged. These studies were supported by grants from the National Health and Medical Research Council of Australia and Royal Children's Hospital Research Foundation.

References

1. Chen D, Estes MK, Ramig RF (1992) Specific interactions between rotavirus outer capsid proteins VP4 and VP7 determine expression of a cross-reactive, neutralizing VP4-specific epitope. J Virol 66: 432–439
2. Coulson BS (1987) Variation in neutralization epitopes of human rotaviruses in relation to genomic RNA polymorphism. Virology 159: 209–216
3. Coulson BS (1993) Typing of human rotavirus VP4 by an enzyme immunoassay using monoclonal antibodies. J Clin Microbiol 31: 1–8
4. Coulson BS, Fowler KJ, Bishop RF, Cotton RGH (1985) Neutralizing monoclonal antibodies to human rotavirus and indications of antigenic drift among strains from neonates. J Virol 54: 14–20
5. Coulson BS, Kirkwood C (1991) Relation of VP7 amino acid sequence to monoclonal antibody neutralization of rotavirus and rotavirus monotypes. J Virol 65: 5968–5974
6. Coulson BS, Tursi JM, McAdam WJ, Bishop RF (1986) Derivation of neutralizing monoclonal antibodies to human rotaviruses and evidence that an immunodominant neutralization site is shared between serotypes 1 and 3. Virology 154: 302–312
7. Coulson BS, Unicomb LE, Pitson GA, Bishop RF (1987) Simple and specific enzyme immunoassay using monoclonal antibodies for serotyping human rotaviruses. J Clin Microbiol 25: 509–515
8. Estes MK (1996) Rotaviruses and their replication. In: Fields BN, Knipe DM, Howley PM, Chanock RM, Melnick JL, Monath TP, Roizman B, Strauss S (eds) Virology. Lippincott-Raven Publishers, Philadelphia, pp 1625–1655
9. Gentsch JR, Glass RI, Woods P, Gouvea V, Gorziglia M, Flores J, Das BK, Bhan MK (1992) Identification of group A rotavirus gene 4 types by polymerase chain reaction. J Clin Microbiol 30: 1365–1373
10. Gerna G. Sarasini A, Matteo A, Parea M, Orsolini P, Battaglia M (1988) Identification of two subtypes of serotype 4 human rotavirus by using VP7-specific neutralizing monoclonal antibodies. J Clin Microbiol. 26: 1388–1392
11. Gorziglia M, Larralde G, Kapikian AZ, Chanock RM (1990) Antigenic relationships among human rotaviruses as determined by outer capsid protein VP4. Proc Natl Acad Sci USA 87: 7155–7159
12. Hoshino Y, Sereno MM, Midthun K, Flores J, Kapikian AZ, Chanock RM (1985) Independent segregation of two antigenic specifications (VP3 and VP7) involved in neutralization of rotavirus infectivity. Proc Natl Acad Sci USA 82: 8701–8704
13. Kirkwood CD, Bishop RF, Coulson BS (1996) Human rotavirus VP4 contains strain-specific and cross-reactive neutralization sites. Arch Virol 141: 587–600
14. Kobayashi N, Taniguchi K, Urasawa S (1990) Identification of operationally overlapping and independent cross-reactive neutralization regions on human rotavirus VP4. J Gen Virol 71: 2615–2623
15. Kobayashi N, Taniguchi K, Urasawa T, Urasawa S (1991) Preparation and characterization of a neutralizing monoclonal antibody directed to VP4 of rotavirus strain K8 which has unique VP4 neutralization epitopes. Arch Virol 121: 153–162

16. Larralde G, Gorziglia M (1992) Distribution of conserved and specific epitopes on the VP8 subunit of rotavirus VP4. J Virol 65: 3213–3218
17. Larralde G, Li B, Kapikian AZ, Gorziglia M (1991) Serotype-specific epitope(s) present on the VP8 subunit of rotavirus VP4 protein. J Virol 65: 3213–3218
18. Padilla-Noriega L, Dunn SJ, Lopez S, Greenberg HB, Arias CF (1995) Identification of two independent neutralization domains on the VP4 trypsin cleavage products VP5* and VP8* of human rotavirus ST-3. Virology 206: 148–154
19. Padilla-Noriega L, Werner-Eckert R, Mackow ER, Gorziglia M, Larralde G, Taniguchi K, Greenberg HB (1993) Serologic analysis of human rotavirus serotypes P1A and P2 by using monoclonal antibodies. J Clin Microbiol 31: 622–628
20. Steele AD, Van Niekerk MC, Mphahlele MJ (1995) Geographic distribution of human rotavirus VP4 genotypes and VP7 serotypes in five South African regions. J Clin Microbiol 33: 1516–1519
21. Taniguchi K, Maloy WL, Nishikawa K, Green KY, Hoshino Y, Urasawa S, Kapikian AZ, Chanock RM, Gorziglia M (1988) Identification of cross-reactive and serotype 2-specific neutralization epitopes on VP3 of human rotavirus. J Virol 62: 2421–2426
22. Taniguchi K, Morita Y, Urasawa S (1987) Cross-reactive neutralization epitopes on VP3 of human rotavirus: analysis with monoclonal antibodies and antigenic variants. J Virol 61: 1726–1730
23. Taniguchi K, Urasawa T, Morita Y, Greenberg HB, Urasawa S (1987) Direct serotyping of human rotavirus in stools by an enzyme-linked immunosorbent assay using serotype 1-, 2-, 3-, 4-specific monoclonal antibodies to VP7. J Infect Dis 155: 1159–1165

Author's address: Dr. B. S. Coulson, Department of Microbiology, The University of Melbourne, Parkville, Victoria 3052, Australia.

Arch Virol (1996) [Suppl] 12: 119–128

© Springer-Verlag 1996

Natural history of human rotavirus infection

R. F. Bishop

Department of Gastroenterology, Royal Children's Hospital,
Flemington Road, Parkville, Victoria, Australia

Summary. Rotavirus infections occur repeatedly in humans from birth to old age. Most are asymptomatic or are associated with mild enteric symptoms. Infection in young children can be accompanied by severe life-threatening diarrhea, most commonly after primary infection. Annual childhood morbidity rates for severe diarrhea are similar worldwide. Mortality rates are low in developed countries but approach 1,000,000 annually in young children in developing countries.

Rotaviruses can be classified into Groups A–E according to antigenic groups on VP6, the major capsid antigen. Only Group A, B and C rotaviruses have been shown to infect humans, and most human rotavirus disease is caused by Group A viruses. These are further classified into G and P types based on identification of antigens on the outer capsid proteins VP7 and VP4 respectively. Most severe infections in young children are caused by serotypes G1–4, and during the last two decades, G1 infections appear to have predominated worldwide. In general the more densely populated countries show the most complex patterns of occurrence of serotypes.

Clinical rotavirus disease can be accompanied by shedding of $>10^{12}$ rotavirus particles/gm feces. The virus is highly infectious and appears to retain infectivity over many months. In temperate climates, disease is most common during the colder months, when it is likely that rapid spread within families and communities occurs. Nosocomial infections are frequent, and rotaviruses can become endemic within obstetric hospital nurseries for the newborn. Few (if any) human rotavirus infections appear to be zoonoses, even though Group A rotaviruses are widespread in the young of all mammalian species. However, infection of humans with reassortant rotavirus strains derived from human-animal sources can occur. The extent to which this contributes to new epidemic strains within particular countries (or worldwide) remains to be determined.

Introduction

Rotaviruses were first identified in humans in 1973 when characteristic particles were observed in the cytoplasm of duodenal epithelial cells obtained from young children admitted to the hospital for treatment of acute diarrhea. Since then, numerous epidemiological studies have confirmed that rotaviruses are the major cause of severe acute diarrhea in children throughout the world [8, 27].

Rotaviruses are now classified as a genus within the family *Reoviridae*. They contain double stranded RNA (dsRNA) that can be separated into 11 distinct bands by gel electrophoresis. The pattern formed by the differing migration of bands from a particular rotavirus is termed an electropherotype. Each of the 11 bands represents a rotavirus gene that codes for a single structural or non-structural viral protein [18]. The complete infectious particle is composed of six structural proteins assembled as a triple layer. The three layers comprise an inner core of VP1, VP2, and VP3 (coded by genes 1, 2, and 3, respectively), an inner capsid of VP6 (coded by gene 6) and an outer capsid of a glycoprotein VP7 (coded by genes 7, 8, or 9 depending on the strain) that is penetrated by 60 spikes of VP4 (coded by gene 4). Rotaviruses have been subdivided into five groups (Groups A, B, C, D and E) based on antigens present on VP6. Group A rotaviruses have been further sudivided on the basis of antigens on VP6 (subgroup I, II), on VP7 (G-types), and on VP4 (P types). Human infections with Groups A, B and C viruses have been recorded. Most severe acute diarrhea in children is caused by rotaviruses of Group A, subgroups I or II, G types 1–4, and P types 1A (8) or 1B (4). This brief review will be confined to observations on the natural history of Group A rotavirus infections in humans and in other mammalian species where relevant. The reader is directed to reviews on rotavirus epidemiology [8, 25–27] and rotavirus genetics and replication [18] for further information. Other comprehensive reviews on particular topics will be cited elsewhere in this text.

Age range of human infection

Rotavirus infection of humans can occur from birth to old age. Rotavirus infection is endemic among newborn babies in obstetric hospital nurseries in many countries, including Australia, UK, USA, Sweden, India and Venezuela [8]. Neonatal infections appear to be nosocomial in origin because they are rarely seen in babies born at home or at village health centres. Results of serological surveys imply that most children have experienced a rotavirus infection by 24 months of age [8, 43]. The sources of infection of young children include siblings and adult family members as well as children and adults within the community [28]. There is ample evidence from serological studies and longitudinal surveillance studies that rotavirus infections pass from member to member of a family, regardless of age. Studies of children (0–3 years old) in day care illustrate the frequency with which young children can be sporadically infected [4, 5]. The broadening of the spectrum of serum neutralizing antibody with age (to include antibody to more than one G type) also implies repeated rotavirus infections [12].

By contrast with the capacity of rotaviruses to cause infection at any age, the clinical consequences of infection appear to be strongly influenced by age. For example, most neonatal infection is asymptomatic or produces only mild symptoms in healthy full term neonates [8]. This cannot be entirely due to the influence of breast feeding, because asymptomatic rotavirus infections have been observed in babies fed artificial milk formulae [16]. The clinical outcome of neonatal infection is probably determined by the balance of a number of factors including the neutralizing antibody spectrum and titer of maternally derived antibody in serum and breast milk; maturity of the infant gut and immune system; presence in the lumen of non-specific rotavirus inhibitory factors; and virulence of rotavirus strains endemic in neonatal nurseries. Breast feeding is associated with a decrease in severity of symptoms and with a decreased incidence of rotavirus infection. A recent case-control study in rural Bangladesh has concluded that the latter effect of breast feeding is mainly attributable to postponement of rotavirus infection, and that the majority of infants have no permanent immunity once breast feeding ceases [13].

The major burden of severe acute disease caused by rotavirus infection falls upon children aged 6–23 months [37, 41]. For example, a longitudinal study conducted in the USA, showed that the development of acute rotavirus diarrhea peaked at 40/100 person years at ages 12–23 months, and decreased to 5/100 person years in adult members of the same families. Children 0–36 months old developed symptoms ranging from severe (requiring hospitalization) in 5%, to moderate (requiring medical attention) in 36% [41]. The resistance to clinical symptoms that is observed in older children and adults is not likely to be due solely to age-altered physiological status of the gut. Severe symptoms have been observed in adults in community outbreaks and in aged people in nursing homes [26, 27]. It is likely that age-related resistance to severe rotavirus is due to active immunity, reinforced by repeated infection throughout life.

Transmission, infectious dose

Rotaviruses are highly infectious when transmitted within the same species, *i.e.*, human–human, calf–calf, *etc*. Replication within the intestinal tract can result in shedding of $\geqslant 10^{10}$ infectious particles (PFU)/ml of feces. Rotaviruses are very durable in the environment and can survive for weeks on surfaces and in potable and recreational water [3]. The infectious dose for the human small intenstine has been calculated as approximately 10 PFU/ml [23]. Rotaviruses show loss of infectivity at pH < 4, so the number that must be ingested to ensure passage of an infectious dose through the stomach is not known. Transmission of rotavirus infection in a susceptible community can occur extremely rapidly. For example, during a multi-island outbreak in the islands of Truk Atoll in the Pacific, rotavirus infections spread from the index patients to involve at least 31% of the population on one island within one week [19]. "Explosive" outbreaks have been observed also in American Indian and Brazilian communities [27, 42]. High nosocomial infection rates have been recorded in most pediatric hospitals and day care centres [4, 40]. Although respiratory spread has been suspected, it appears more likely that infections result from airborne spread after aerosol formation of particles, *e.g.*, from diapers or toilet flushing. Circumstances have occurred where waterborne viruses have been presumed to have caused widespread infection, *e.g.*, after floods in Bangladesh [1].

Seasonality

Climatic factors exert a major influence on the incidence of rotavirus disease (and infections) in areas where there are marked seasonal changes. Rotavirus infections are markedly more common in the cooler months of the year. Peak months may vary from country to country and from year to year [14]. In countries with tropical climates, rotavirus disease is present throughout the year. The differing seasonal patterns in tropical and temperate climates may point to differing patterns of transmission and perhaps to differing reservoirs of infection [14]. In general, tropical countries are also the most populous, with high birth rates together with crowded living conditions and climates optimal for rotavirus survival. The seasonality observed in temperate countries may result more from the occurrence of unfavorable conditions for transmission in summer months (with higher ambient temperatures, low relative humidity and less crowding) rather than from a direct effect of cooler weather favoring transmission. A sentinel study conducted for one season across the USA noted that the peak rotavirus month varied from south-west to north-east, suggesting that areas to the south served as the source of annual epidemics once the hot dry summer weather had ceased [30]. It is also possible that cold weather can have physiological effects that exacerbate the symptoms of rotavirus infection. For example, inducing hypothermia in piglets enhances the severity of symptoms of rotavirus disease [46].

In summary, there is abundant evidence that the occurrence of rotavirus-disease is influenced by seasonal factors, and that in temperate climates the disease is more prevalent in the cooler months. The mechanism underlying this seasonal variation is obscure and probably involves the interplay of many factors including survival of virus in the environment, physiological effects on the host, and degree of crowding of susceptibles. It is fascinating to note that rotavirus disease in infant laboratory mice and in premature newborn babies has also been observed to be seasonal [8, 46], even though both live in controlled environments that appear unaltered year-round.

Epidemiology of differing rotavirus strains

Initial epidemiological studies of electropherotypes of rotaviruses causing severe acute diarrhea revealed worldwide genetic diversity within Group A rotaviruses [8, 27]. The number of co-existent patterns roughly parallels the density of the human population. In any one geographic area, as many as 10 different electropherotypes have been shown to co-exist. Surveys in urban areas in developing countries yield a greater variety of electropherotypes than in urban areas in developed countries. Cities with larger populations can show a greater variety of electropherotypes than those in smaller cities [32]. In any one centre the dominant electropherotype usually changes from year to year and seldom persists for more than two seasons.

Epidemiological surveys based on determination of G serotype indicate that genetic diversity can exist without serotypic diversity, and that many electropherotypes can occur within each serotype. The four major human G types (G1–G4) exist worldwide [8]. G1 types have been dominant worldwide during the past two decades. Epidemics of G2, G3 and G4 types have occurred in a wide variety of geographic settings. Cross-sectional studies have again emphasized that complex patterns of infection (with all four G types co-existing simultaneously) are more common in developing countries than in developed countries. The less common human G types (G5, G8, G9, G10) have usually been found in developing countries [8, 22, 24, 34, 45]. Serological studies imply that these unusual types may be more common in developing countries than is realized [10, 11]. There have been limited epidemiological studies on the occurrence of P types. To date, P types (4) and (8) have been predominant worldwide [47]. It is possible that developing countries will prove to have more complex patterns of occurrence of P types than do developed countries.

The complexity or rotavirus electropherotypes, together with the existence of numerous serotypes (both G types and P types) appears to indicate that successful development of a rotavirus vaccine for use in developing countries could prove difficult [2]. However, numerous longitudinal surveillance studies of children recruited at birth, indicate that severe disease is a consequence of primary infection with any G type, and that reinfection (with homologous or heterologous G types) is usually associated with mild symptoms [6–9, 20, 21]. Severe symptoms have occasionally been observed in reinfections [31] and could prove to be the result of reinfection with a P type differing from that responsible for the primary infection. The current development of techniques to determine P type of rotavirus field strains could clarify this.

Mechanisms contributing to the genomic complexity of rotaviruses

The epidemiology of rotavirus infections worldwide is influenced by the extensive genomic variability of rotavirus strains. This could result from a variety of mechanisms including: sequential point mutations in genes coding for immunologically important rotavirus proteins; reassortment of genes between human rotavirus strains; introduction of genes from animal rotaviruses by reassortment; and genomic rearrangements [2].

Temporal nucleotide sequence variations (coding for deduced amino acid changes) have been described in the VP7 genes of G1 and G4 serotypes of rotaviruses collected from Melbourne children during two, and 16 years, respectively [38, 39]. Serotype G1 strains showed approximately 3% nucleotide sequence diversity between rotavirus strains predominant in 1990 and 1991, but the regions encoding the neutralization epitopes of VP7 were conserved. Serotype G4 strains implicated in four separate epidemics from 1974–1990 showed changes in the sequence coding for region A of VP7 that was associated with

R. F. Bishop

altered binding of serotype G4-specific neutralizing monoclonal antibodies. The sequences coding for the VP8* region of VP4 (containing the proposed serotype-specific neutralization epitopes) were highly conserved both between Melbourne strains, and in relation to standard strains Wa, P and VA70 isolated initially in the USA and Italy. Similar studies are required in other geographical areas in order to determine the degree to which genetic variability contributes to antigenic drift in human rotaviruses.

The recent classification of human rotaviruses into genogroups (Wa-like, DS-1-like, AU-like) using the technique of RNA–RNA hybridization under conditions of high stringency [36] has led to the identification of single and multiple gene substitution reassortants between human rotaviruses of different genogroups. The extension of the technique to include experiments in which genomic RNAs from rotavirus strains from different host species were cross-hybridized with ^{32}P-labelled transcription probes prepared from strains from the same host species has permitted the detection of human-animal rotavirus reassortants. Current knowledge in this areas has been reviewed recently [35].

These techniques have been applied to "unusual" rotaviruses having unexpected combinations of subgroup, G type and P type identified in widely separated geographical areas. Reassortants between human genogroups Wa and DS1 viruses have been reported as the cause of outbreaks in Manipur, India, in 1987–1988 [29], and in a recent outbreak amongst aboriginals in communities in Australia in 1994 (Palombo, Bishop, unpublished). Other reassortants between Wa and DS-1 genogroups have been sporadically identified as occasional causes of severe diarrhea in children in several countries including India, Brazil, Bangladesh and Japan [33, 35, 45], but have not been implicated in outbreaks. The potential for such human inter-genogroup reassortment must frequently exist in any area where mixed infections are not uncommon and where G1–G4 types co-exist.

In addition, there is now convincing evidence that transmission of animal rotaviruses to humans can occur, perhaps as whole virions, but more commonly by a process of genetic reassortment [35]. One of the most common sources of interspecies infections appear to be domestic animals. Gene segments derived from canine or feline strains have been detected in human rotaviruses from infections in Israel and Japan [34]. It is interesting that early experimental attempts at cross-species infection noted the apparent ease with which human rotaviruses could be adapted to infect dogs [44]. The close contact that is common between human infants and domestic animals could potentiate this exchange, and has been identified as a risk factor for rotavirus diarrhea in some settings [17].

Human infection with reassortants possessing bovine (or occasionally porcine or avian) genes has been recorded in Italy, India, Indonesia, Finland, Thailand and the USA [35]. Most strains have appeared as single isolates and have not spread within the communities where they were identified. It is possible that these represent sporadic interspecies transmissions, where the resulting reassortant has either no, or only limited, capacity for secondary transmission and hence does not result in production of an outbreak. An exception to this may have occurred in India where infections with a human-bovine reassortant strain has become endemic in a neonatal nursery [15].

In summary, current data support the belief that interspecies transmission of whole viruses or reassortant viruses occurs, can be responsible for severe disease in some infants, but is unlikely to involve strains that have the capacity to cause significant outbreaks of disease. By contrast, reassortment between human genogroups may occur more commonly than is realized, and may sometimes result in strains able to cause epidemiologically important outbreaks.

Acknowledgments

Valuable assistance in the conduct of epidemiological studies has been received over many years from members of the Department of Gastroenterology, Royal Children's Hospital, in particular Dr Graeme Barnes, Dr Barbara Coulson, and Mrs J. Lund. I am grateful to Mrs Jane Lee for typing this manuscript. The work has been supported by grants from the National Health and Medical Research Council of Australia and the Royal Children's Hospital Research Foundation.

References

1. Ahmed MU, Urasawa S, Taniguchi K, Urasawa T, Kobayashi N, Wakasugi F, Islam AIMM, Sahikh HA (1991) Analysis of human rotavirus strains prevailing in Bangldesh in relation to nationwide floods brought by the 1988 monsoon. J Clin Microbiol 29: 2273–2279
2. Anon (1990) Puzzling diversity of rotaviruses. Lancet 1: 573–574
3. Ansari SA, Springthorpe VS, Sattar SA (1991) Survival and vehicular spread of human rotaviruses: possible relation to seasonality of outbreaks. Reviews of Infectious Diseases 13: 448–461
4. Bartlett AV, Moore M, Gary GW, Starko KM, Erben JJ, Meredith BA (1985) Diarrheal illness among infants and toddlers in day care centres, I. Epidemiology and pathogens. J Pediatr 107: 495–502
5. Bartlett AV, Moore M, Gary GW, Starko KM, Erben JJ, Meredith BA (1985) Diarrheal illness among infants and toddlers in day care centres, II. Comparison with day care homes and households. J Pediatr 107: 503–509
6. Bernstein DI, Sander DS, Smith VE, Schiff GM, Ward RL (1991) Protection from rotavirus reinfection: 2-year prospective study. J Infect Dis 164: 277–283
7. Bhan MK, Lew JF, Sazawal S, Das BK, Gentsch JR, Glass RI (1993) Protection conferred by neonatal rotavirus infection against subsequent rotavirus diarrhea. J Infect Dis 168: 282–287
8. Bishop RF (1994) Natural history of human rotavirus infections. In: Kapikian AZ (ed) Viral infections of the gastrointestinal tract, 2nd edn. Marcel Dekker, New York, pp 131–167
9. Bishop R, Lund J, Cipriani E, Unicomb L, Barnes G (1990) Clinical serological and intestinal immune responses to rotavirus infection of humans. In: de la Maza LM, Peterson EM (eds) Medical Virology 9. Plenum Press, New York, pp 85–109
10. Brussow H, Clark HF, Sidoti J (1991) Prevalence of serum neutralizing antibody to serotype 9 rotavirus W161 in children from South America and Central Europe. J Clin Microbiol 29: 208–211

11. Brussow H, Sidoti J (1991) Antibody to serotype 8 rotavirus in Ecuadorian and German children. Epidemiol Infect 106: 415–420
12. Brussow H, Sidoti J, Barclay D, Sotek J, Dirren H, Freire WB (1990) Prevalence and serotype specificity of rotavirus antibodies in different age groups of Ecuadorian infants. J Infect Dis 162: 615–620
13. Clemens J, Rao M, Ahmed F, Ward R, Huda S, Chakraborty J, Yunus M, Khan MR, Ali M, Kay B, van Loon F, Sack D (1993) Breast-feeding and the risk of life-threatening rotavirus diarrhea: prevention or postponement? Pediatrics 92: 680–685
14. Cook SM, Glass RI, LeBaron CW, Ho M-S (1990) Global seasonality of rotavirus infections. Bulletin WHO 68: 171–177
15. Das M, Dunn SJ, Woode GN, Greenberg HB, Rao CD (1993) Both surface proteins (VP4 and VP7) of an asymptomatic neonatal rotavirus strain (I321) have high levels of sequence identity with the homologous proteins of a serotype 10 bovine rotavirus. Virology 194: 374–379
16. Duffy LC, Riepenhoff-Talty M, Byers TE, La Scolea LJ, Zielezny MA, Dryja DM, Oga PL (1986) Modulation of rotavirus enteritis during breast-feeding. AJDC 140: 1164–1168
17. Engleberg NC, Holburt EN, Barrett TJ, Gary GW, Trujillo MH, Feldman RA, Hughes JM (1982) Epidemiology of diarrhea due to rotavirus on an Indian reservation: Risk factors in the home environment. J Inf Dis 145: 894–898
18. Estes MK (1990) Rotaviruses and their replication. In: Fields BN, Knipe DM, Chanock RM, Hirsch MS, Melnick JL, Monath TP, Roizman B (eds) Fundamental Virology, 2nd edn. Raven Press, New York, pp 619–642
19. Foster SO, Palmer EL, Gary Jr GW, Martin ML, Herrmann KL, Beasley P, Sampson J (1980) Gastroenteritis due to rotavirus in an isolated Pacific Island group: An epidemic of 3, 439 cases. J Inf Dis 141: 32–39
20. Friedman MG, Galil A, Sarov B, Margalith M, Katzir G, Midthun K, Taniguchi K, Urasawa S, Kapikian AZ, Edelman R, Sarov I (1988) Two sequential outbreaks of rotavirus gastroenteritis: Evidence for symptomatic and asymptomatic reinfections. J Inf Dis 158: 814–822
21. Georges-Courbot MC, Monges J, Beraud-Cassel AM, Gouandjika I, Georges AJ (1988) Prospective longitudinal study of rotavirus infections in children from birth to two years of age in central Africa. Ann Inst Pasteur/Virol 139: 421–428
22. Gouvea V, de Castro L, do Carmo Timenetsky M, Greenberg H, Santos N (1994) Rotavirus serotype G5 associated with diarrhea in Brazilian children. J Clin Micro 32: 1408–1409
23. Graham DY, Dufour GR, Estes MK (1987) Minimal infective dose of rotavirus. Arch Virol 92: 261–271
24. Gusmao RH, Mascarenhas JD, Gabbay YB, Linhares AC (1994) Nosocomial transmission of an avian-like rotavirus strain among children in Belém, Brazil. J Diarrheal Dis Res 2: 129–132
25. Haffejee IE (1995) The epidemiology of rotavirus infections: a global perspective. J Ped Gastro Nutr 20: 275–286
26. Hrdy DB (1987) Epidemiology of rotavirus infection in adults. Rev of Infect Dis 9: 461–469
27. Kapikian AZ, Chanock RM (1990) Rotaviruses. In: Fields BN, Knipe DM, Chanock RM, Hirsch MS, Melnick JL, Monath TP, Roizman B (eds) Virology, 2nd edn. Raven Press, New York, pp 1353–1404
28. Koopman JS, Monto AS, Longini (1989) The Tecumseh study XVI: family and community sources of rotavirus infection. Am J of Epidemiol 130: 760–768

29. Krishnan T, Burke B, Shen S, Naik TN, Desselberger U (1994) Molecular epidemiology of human rotaviruses in Manipur: genome analysis of rotaviruses of long eletrophero-type and subgroup I. Arch Virol 134: 279–292

30. LeBaron CW, Lew J, Glass RI Weber JM, Guillermo M, Ruiz-Palacios MD (1990) Annual rotavirus epidemic patterns in North America. JAMA 264: 983–988

31. Linhares AC, Gabbay YB, Mascarenhas JDP, Freitas RB, Flewett TH, Beards GM (1988) Epidemiology of rotavirus subgroups and serotypes in Belem, Brazil: a three-year study. Ann Inst Pasteur/Virol 139: 89–99

32. Masendycz PJ, Unicomb LE, Kirkwood CD, Bishop RF (1994) Rotavirus serotypes causing severe acute diarrhea in young children in six Australian cities, 1989 to 1992. J Clin Microbiol 32: 2315–2317

33. Nakagomi O, Nakagomi T (1991) Molecular evidence for naturally occurring single VP7 gene substitution reassortant between human rotaviruses belonging to two different genogroups. Arch Virol 119: 67–81

34. Nakagomi O, Nakagomi T (1991) Genetic diversity and similarity among mammalian rotaviruses in relation to interspecies transmission of rotavirus. Arch Virol 120: 43–55

35. Nakagomi O, Nakagomi T (1993) Interspecies transmission of rotaviruses studied from the perspective of genogroup. Microbiol Immunol 37: 337–348

36. Nakagomi O, Nakagomi T, Akatani K, Ikegami N (1989) Identification of rotavirus genogroups by RNA–RNA hybridization. Molecular and Cellular Probes 3: 251–261

37. Oyejide CO, Fagbami AH (1988) An epidemiological study of rotavirus diarrhea in a cohort of Nigerian infants: II incidence of diarrhea in the first two years of life. International Journal of Epidemiology 17: 908–912

38. Palombo EA, Bishop RF, Cotton RGH (1993) Intra- and inter-season genetic variability in the VP7 gene of serotype 1 (monotype 1a) rotavirus clinical isolates. Arch Virol 130: 57–69

39. Palombo EA, Bishop RF, Cotton RGH (1993) Sequence conservation within neutraliz-ation epitope regions of VP7 and VP4 proteins of human serotype G4 rotavirus isolates. Arch Virol 133: 323–334

40. Pickering LK, Bartlett AV, Rees RR, Morrow A (1988) Asymptomatic excretion of rotavirus before and after rotavirus diarrhea in children in day care centres. J Pediatr 112: 361–365

41. Rodriguez WJ, Kim HW, Brandt CD, Schwartz RH, Gardner MK, Jeffries B, Parrott RH, Kaslow RA, Smith JI, Kapikian AZ (1987) Longitudinal study of rotavirus infection and gastroenteritis in families served by a pediatric medical practice: clinical and epidemiologic observations. The Pediatr Infect Dis J 6: 170–176

42. Santosham M, Yolken RH, Wyatt RG, Bertrando R, Black RE, Spira WM, Sack RB (1985) Epidemiology of rotavirus diarrhea in a prospectively monitored American Indian population. J Infect Dis 152: 778–783

43. Simhon A, Abed Y, Schoub B, Lasch EE, Morag A (1990) Rotavirus infection and rotavirus serum antibody in a cohort of children from Gaza observed from birth to the age of one year. Int J Epidemiol 19: 160–163

44. Smith M, Tzipori S (1979) Gel electrophoresis of rotavirus RNA derived from six different animal species. Australian J Exp Biol Med Sci 57: 583–585

45. Ward RL, Nakagomi O, Knowlton Dr, McNeal MM, Nakagomi T, Clemens JD, Sack DA, Schiff GM (1990) Evidence for natural reassortants of human rotaviruses belonging to different genogroups. J Virol 64: 3219–3225

46. Woode GN (1982) Rotaviruses in animals. In: Tyrrell DAJ, Kapikian AZ (eds) Virus infections of the gastrointestinal tract. Marcel Dekker, New York, pp 295–314

47. Wu H, Taniguchi K, Wakasugi F, Ukae S, Chiba S, Ohseto M, Hasegawa A, Urasawa T, Urasawa S (1994) Survey on the distribution of the gene 4 alleles of human rotaviruses by polymerase chain reaction. Epidemiol Infect 112: 615–622

Author's address: Dr. R. F. Bishop, Department of Gastroenterology, Royal Children's Hospital, Melbourne, Flemington Rd, Parkville, Victoria 3052, Australia.

Arch Virol (1996) [Suppl] 12: 129–139

Protective immunity against group A rotavirus infection and illness in infants

D. O. Matson

Center for Pediatric Research, Children's Hospital of The King's Daughters
and Eastern Virginia Medical School, Norfolk, Virginia

Summary. Understanding of the protective effect provided by natural rotavirus infections against subsequent rotavirus infections is required for evaluating vaccine development programs. Prior studies of the protective efficacy of natural infections and correlates of natural protection are reviewed and results from several studies presented only in abstract form are summarized to provide a current assessment of knowledge in this area. Six cohort studies have reported rates for the protective efficacy of a natural rotavirus infection against subsequent infection, diarrhea, or severe diarrhea. These efficacy estimates ranged from 0 to 100% and are not directly comparable because of differences in methodology and population monitored. Results from other study designs also have been confusing, until recently. Recent studies have identified immunologic correlates of protection and studies from a cohort of intensely monitored. Mexican children promise to provide a comprehensive assessment of the strength of the protective effect of natural rotavirus infection.

Introduction

Rotaviruses are double-stranded RNA viruses that constitute a genus in the virus family *Reoviridae*. Rotaviruses include six antigenically distinct groups (A to F) that infect a variety of human and non-human hosts. Infections with human group A rotaviruses are common among infants and young children worldwide [1, 6, 19, 20, 23, 34]. Rotaviruses are the leading cause of severe diarrhea among children and are estimated to cause 870 000 deaths annually in developing countries [1, 19, 20]. A worthy goal of research on rotaviruses is to develop a vaccine that prevents severe rotavirus illness and perhaps also rotavirus infection.

Although natural immunity is acquired after early exposure to the virus, descriptions of repeat rotavirus infections and diarrhea during the first few years of life indicate that many children acquire protective immunity only after several exposures [12, 13, 16, 21, 22, 36, 39]. Antigenic variation among circulating strains is high and attempts to identify immunologic correlates of protection have yielded conflicting results. Evaluation of the efficacy of a rotavirus vaccine candidate requires clear knowledge of the protective efficacy provided by natural rotavirus infections against subsequent infection, as well as recognition of immunologic correlates of protection.

Methods

Cohort studies assessing natural protective efficacy and studies of immunologic correlates of protection were identified by literature review. Unpublished results from a large cohort study were abstracted for comparison to published results.

Results

Cohort studies

Six cohort studies have reported rates for the protective efficacy of a natural rotavirus infection against subsequent outcomes (Table 1). The first of these studies, by Bishop and co-workers [3], has provided the model for all subsequent studies. In that study, 106 infants were enrolled at birth during two periods of recruitment. Forty-four of the infants were infected in the first 14 days of life, based upon electron microscopic examination of stool specimens collected daily during that period. Monitoring after the first two weeks of life included testing of fecal specimens when diarrhea occurred, collection of blood specimen at 14 to 20 days of age, then at three-month intervals in the first year of life and six-month intervals thereafter, and collection of acute and convalescent blood specimens for each episode of acute enteritis. Monitoring also included home visits each three months and telephone calls. Infants were examined by a physician at each enteritis episode and the severity of the episode was scored. Rotavirus infections were detected by virus excretion or seroconversion.

Eighty-one (76%) of the 106 infants enrolled in the perinatal period completed the three years of subsequent monitoring. Rotavirus infection was not detected by examination of diarrhea stools or by serology in 37 (45%) of the

Table 1. Published estimates of efficacy of a natural rotavirus infection against various outcomes of reinfection

Study[a]	Rotavirus Infection Outcome	Efficacy Observed
Bishop, 1983 [3]	severe diarrhea	100%
Georges-Courbot, 1988 [14]	diarrhea	87%
Reves, 1989 [35]	diarrhea	0–?%
Bernstein, 1991 [2]	any infection	80%
	diarrhea	70–100%
Bhan, 1993 [4]	diarrhea	46%
Ward, 1994 [43]	diarrhea	93%

[a] First author, year [reference]

children during the 3-year monitoring period. The 44 children with rotavirus infections were equally distributed between the neonatally and nonneonatally infected groups. Of these 44, 12 had two rotavirus infections. In this study, infection of newborn babies with a single, endemic strain did not protect against reinfection but did confer protection (efficacy 100%) against reinfection associated with clinically severe disease. RNA electropherotyping and subsequent antigenic typing indicated that the community strains in this study were of serotypes different, for the most part, from the strain endemic to the hospital.

Georges-Courbot and associates [14] enrolled 223 infants at birth and monitored them for two years. Monitoring included visits to the home and stool collection every other week until age six months then, apparently, a less intense monitoring protocol thereafter. These and additional stool specimens were tested for rotaviruses. One-hundred-eleven (50%) children completed monitoring. Thirty-eight (34%) of the children had a rotavirus infection in the first six months of life. Of these 38 children, 1 (3%) developed rotavirus-associated diarrhea over the next 18 months, compared with 15 (21%) of the 73 children who did not have a rotavirus infection in the first six months of life (efficacy 87%).

In the study by Reves and co-workers [35], 363 rural children in Egypt, born during the two study years, were monitored from birth for diarrhea by twice weekly home visits. Diarrhea stool specimens were tested for rotavirus. Rotavirus-associated diarrhea occurred in 86 (24%) of the children and 12 of these had two rotavirus-associated episodes. The size of the study permitted comparison of age-specific infection rates for first and second episodes of rotavirus diarrhea. Assuming uniformity of risk throughout the study period, and comparing children with no previous rotavirus diarrhea to children with one previous episode of diarrhea, the observed and expected numbers of second episodes of rotavirus-associated diarrhea were equal (age-adjusted rate ratio = 1.01; 95% CL 0.55, 1.86). Despite the size of this study, the confidence interval was quite wide, such that a 50% protective effect could have been missed. An additional limitation to interpreting this study was that the overall rate of diarrhea among children with detected rotavirus-associated episodes was higher than that among children who had no rotavirus-associated diarrhea episodes. This means that the assumption of uniform risk for rotavirus infection probably was not valid. The authors concluded that some partial protection likely occurred because the rotavirus-infected children appeared to have a higher overall rate of exposure to pathogens causing diarrhea.

In the study by Bernstein and co-workers [2], 163 children (81 vaccine, 82 placebo recipients), who participated in a rotavirus vaccine trial that extended through a rotavirus season were monitored through a second rotavirus season. Children were enrolled at 2 to 12 months of age and the group as a whole was about 26 months old at the end of the study. Monitoring included parental instructions to contact study personnel when diarrhea occurred, weekly telephone calls in the first year, and biweekly telephone calls in the second year. Blood was collected for antibody measurement at enrollment, one month after vaccination, in the next summer, and, for the second study year, before and after the rotavirus season. Rotavirus infections were detected by testing of diarrhea stool specimens for rotavirus excretion and/or by serum antibody responses.

Twenty-one children 6 months of age or older at enrollment had serologic evidence of prior infection at enrollment. Because the WC3 vaccine given to the children had no effect on outcome of infection in either year, because high anti-WC3 antibody levels that correlated with protection in the first year did not correlate with protection in the second year, and because antibody responses in vaccine and placebo recipients were similar, vaccine and placebo groups were combined for analysis of the protective effect of natural infection. Of the 60 infants who developed a primary infection in the first year of monitoring, 4 (7%) were infected in the second year compared with 29 (35%) of 82 children not infected in the first year (efficacy 80%). Efficacy rates for protection against symptomatic reinfections ranged from 70 to 100% depending upon which groups were being compared.

A cohort of 238 infants enrolled at birth by Bhan and associates [4] was subsequently monitored until 14 to 23 months of age. Two-hundred-four infants (86%) completed monitoring, which included daily screening for rotavirus infection during the newborn hospital stay and the following four days. Post-neonatal monitoring for diarrhea began at age 3 months and included twice weekly home visits and stool collection. Most (148, 73%) of the infants were infected in the newborn period and all of these were infected (of 71 assessed) with a P genotype 11, G9 rotavirus strain, based upon uniformity of electrophero-types. In the later monitoring period, the incidence of rotavirus infection was 0.23 per year among the neonatally infected children and 0.42 infections per year among the children not infected in the newborn period (efficacy 46%).

The most recently published estimate of the protective effect of natural infection comes from Ward and co-workers [43] who reported the results of monitoring 330 placebo recipients participating in a rotavirus vaccine trial and enrolled at 4 to 20 weeks of age. The dosing period extended for three months, after which monitoring for diarrhea began. That monitoring included biweekly telephone calls and instructions to the parents to report diarrhea episodes. Blood was collected before placebo dosing, a month after the final placebo dose, and before the second rotavirus season. Two-hundred-eighty (85%) of the children completed two years of monitoring. Of the 171 children monitored two years and with complete serologic information, 1 (1%) of 68 infected in the first year had rotavirus diarrhea in the second year compared with 22 (21%) of the 100 children not infected in the first year (efficacy 93%).

Variability among cohort studies

The efficacy estimates from these cohort studies are not directly comparable because of differences in methodology and monitored population. These differences likely influence the rates of detection of rotavirus infection and the potential for measuring protective effects of infection. For example, protective efficacy rates may be lower in regions with high antigenic diversity among circulating strains and a high preponderance of neonatal infection, in the presence of transplacentally acquired antibody, may yield a different protective effect than first infection in older infants. The studies differ in the age of enrollment, use of an operational definition of diarrhea, use of active surveillance for diarrhea, evaluation of the severity of diarrhea episodes, and methods or attempts to detect asymptomatic infections. Study results may have been influenced by unrecognized effects of administration of rotavirus vaccine, high drop-out rates, drop-outs not equally distributed between comparison groups, drop-outs who may have had different clinical characteristics than children who remained in the study, reliance upon serology alone for detection of rotavirus-associated diarrhea, and a lack of or incomplete assessment for, or control of, potential confounding factors. Similar variability of study design has occurred among rotavirus vaccine trials [9]. Despite the variability in studies, the overall trend from these studies is that rotavirus infections result, at least, in subsequent protective immunity.

Correlates of protection

Correlates of natural protection are measures that can be used to predict an individual's or group's susceptibility to infection. As an example, measurement of the titer of serum anti-rotavirus antibody might be one tool for assessing an individual's risk of illness if exposed to the virus. The degree to which a vaccine stimulates responses that achieve the protective level of antibody that correlates with protection is one means of assessing the potential of a vaccine candidate without conducting large efficacy trials.

Epidemiologic correlates of protection, such as age and breast feeding, lack sufficient certainty or permanence to be suitable predictors of protection. Studies attempting to correlate humoral immune factors with protection against rotavirus infection have been conflicting [8, 10, 15, 18, 25, 31, 32, 44, 46]. One factor distinguishing studies where humoral correlates have been found from studies where humoral correlates have not been found is the intensity of monitoring of the study children. In the first study identifying an antibody correlate of protection, by Chiba and associates, [8], children were under continuous surveillance in an orphanage. When outbreaks of infection occurred, accurate information on exposure and infection was assembled. Three studies where exposure history of the children prior to enrollment was unknown failed to find humoral correlates of protection [18, 44, 46]. When adult volunteers, with certain exposure histories, were administered wild-type rotavirus strains, clear evidence of serum antibody correlates of protection were described [15].

We have conducted a series of studies among children attending day care centers. Day care centers provide a setting for epidemiologic studies of rotavirus disease in children where outbreaks of rotavirus gastroenteritis are common and both symptomatic and asymptomatic infections occur after exposure to a single antigenic virus type [30, 33]. In these studies, children were under daily surveillance for diarrhea and at least weekly surveillance for virus excretion. Highlights from these studies [25, 30–32] include the following: 1) A serum anti-rotavirus IgA titer of > 200, IgG titer of > 800, or homotypic G type-specific blocking antibody value of ⩾ 44% (tested at a 1:10 dilution) was associated with protection against infection. 2) The pattern of G type-specific antibody acquisition was influenced by the G type to which the children were exposed. In response to first infections, G type-specific antibody responses were primarily to the infecting type. Repeated infections broadened the spectrum of antibody specificity. 3) The group of children with two rotavirus infections developed levels of G type-specific antibody that correlated with protection against all four prevailing G types, despite exposure to one or two G types. Therefore, a restricted number of rotavirus types may be sufficient to induce broadly protective antibody. 4) The presence of fecal IgA antibody correlated with protection against infection and illness. 5) Each of the antibody correlates demonstrated a gradient of protective effect such that the highest antibody levels correlated with protection against rotavirus infection, lower antibody levels with protection against symptomatic infection, and the lowest antibody levels with no protection. 6) The maintenance of high levels of anti-rotavirus serum antibodies in these children seems to be influenced by continuous exposure to circulating rotaviruses. Increases in serum isotype-specific and G type-specific antibodies occurred during rotavirus seasons; during the interseason period antibody levels tended to decrease.

Our day-care center study results were similar to those observed in the orphanage [8], where children were monitored daily for symptoms, and similar to those observed in animal and volunteer studies [28, 37, 38]. Cellular immunity also has a role in controlling rotavirus infection [11, 29]. However, assays for cellular immunity are too cumbersome at this time to be suitable for large studies, such as vaccine trials. A number of issues remain to be clarified in this area. For example, one puzzling aspect of many of the vaccine trials is that efficacy has been observed when no serum antibody correlates of protection could be demonstrated. In addition, the impact of the diversity of circulating rotavirus antigens [24, 45] on exposure outcome is poorly understood. Current studies have not provided a clear immunologic explanation for the heterotypic protection that longitudinal studies suggest is acquired in the first two to three years of life [5, 17]. Variability of the results of studies of natural protective effects have been complemented by variability in the results of studies of correlates of protection.

A cohort study in Mexico

A cohort study initiated by Ruiz-Palacios and co-workers [39] has combined intensive surveillance for infection with routine antibody determinations to reduce uncertainties of results that might have accrued from a less comprehensive program of child monitoring. In this study, 200 consecutively eligible mother-infant pairs were enrolled, all residing in a single community, at birth and monitored them for two years. Enrollment included a baseline assessment of the conditions of the home and family. Field workers visited each household weekly to interview the mother and to collect specimens. The interviewer solicited data on feeding patterns, occurrence of diarrhea since the previous visit, and the child's stool frequency and consistency the day prior to the visit. The record of the stool pattern for each child provided a baseline for comparison when diarrhea occurred. Stool samples were collected weekly from each child regardless of symoms and an additional stool sample was collected whenever diarrhea occurred. A blood sample was collected in the first week of life and every four months thereafter. Diarrhea severity was evaluated by a physician within 24 hours of notification of the episode and on each subsequent day of illness. Stool specimens were tested for rotavirus by enzyme immunoassay and serum samples were analyzed for the presence of IgA and IgG anti-rotavirus antibodies by enzyme immunoassay using a single lot of rotavirus antigen.

A weakness of analysis of many studies of natural protection is an inability to compare groups of children, such as previously infected or uninfected, for the risk of a particular outcome, such as severe illness, while simultaneously controlling for potential confounding factors observed in the study. The Cox proportional hazards model is one technique for accomplishing such an analysis, but published versions of this method allow analysis of a single failure event per study subject, such as illness. One attribute of the natural history of rotavirus infection is that repeated infections in the same child are common. A statistical method for simultaneously analyzing the risk of subsequent infection among groups of children differing in the number of previous infections was required to analyze our cohort study. The Cox proportional hazards method was generalized to allow for repeat failures (infections) per patient [7].

Highlights of the results from the cohort study [26, 27, 40–42] include the following: 1) Almost exactly half (52%) of the infections were primary and the remainder were repeat. 2) About half (56%) of the infections were detected by rotavirus excretion and about three-fourths (77%) were detected by antibody response. 3) Virtually all children were infected by two years of age and some had as many as five infections in that period 4) First infections generally were the most severe and severity decreased as infection number increased. 5) Protection conferred by infection exhibited a gradient of efficacy against subsequent outcomes, being strongest against moderate-to-severe infection, weaker against mild illness, and weakest against asymptomatic infection. The protective effect against infection indicated the potential for a rotavirus vaccine program to prevent rotavirus infection and not only severe illness. 6) Asymptomatic infection exhibited protection to a degree comparable to that achieved by symptomatic infection. 7) Complete protection against moderate-to-severe infection resulted from the accumulated experience of two infections, whether or not both the infections were symptomatic or asymptomatic. 8) When infections were detected solely by virus excretion, we observed higher efficacy rates than when infections detected by serology were included. 9) Higher serum titers of both anti-rotavirus IgA and IgG were associated with protection against infection and illness. 10) Serum anti-rotavirus IgA and IgG titers continued to rise as the number of infections experienced by a child rose. 11) Breast feeding was a strong protective correlate of protection against infection. 12) The strength of breast feeding's protective effect was comparable to the protective effect observed among children who already had experienced one rotavirus infection.

Conclusions

The purpose of this review was to provide an assessment of the status of studies of the natural protective effect of rotavirus infection against subsequent infection, as well as attempts to identify immunologic correlates of protection.Interpretation of studies of both issues have been complicated by variability among studies. That such variability of study design has a significant effect on the results of individual studies is an important clue to the epidemiology of rotaviruses. Such variability indicates that exposure and host response are diverse and that antigenic diversity likely has an important impact on outcome. The acquisition of heterotypic antibody despite constricted exposure to circulating antigenic types provides an important clue how natural immunity is acquired. In the early newborn period, breast feeding appears to provide a level of protection similar to that observed after one infection has occurred. Once breast feeding stops, children appear to be susceptible to severe, dehydrating illness. Current vaccine candidates provide promise of a vaccination program capable of limiting the risk of this severe illness.

Acknowledgements

Supported by a grant (HD-13021) from the National Institutes of Health, U.S.A. The studies that I participated in were conducted by a number of talented individuals. I thank Drs. Miguel O'Ryan, Raul Velazquez, and Ardythe Morrow in particular for educational conversations.

References

1. Bartlett AV, Bednarz-Prashad AJ, DuPont HL, Pickering LK (1987) Rotavirus gastroenteritis. Ann Rev Med 38: 399–415
2. Bernstein DI, Sander DS, Smith VE, Schiff GM, Ward RL (1991) Protection from rotavirus reinfection: 2-year prospective study. J Infect Dis 164: 277–283
3. Bishop RF, Barnes GL, Cipriani E, Lund JS (1983) Clinical immunity after neonatal rotavirus infection. A prospective longitudinal study in young children. N Engl J Med 309: 72–76
4. Bhan MK, Lew JF, Sazawal S, Das BK, Gentsch JR, Glass RI (1993) Protection conferred by neonatal rotavirus infection against subsequent rotavirus diarrhea. J Infect Dis 168: 282–287
5. Black RE, Brown KH, Becker S, Abdul Alim AAM, Huq I (1982) Longitudinal studies of infectious diseases and physical growth of children in rural Bangladesh, II. Incidence of diarrhea and association with known pathogens. Am J Epidemiol 115: 315–324
6. Black RE, Merson MH, Huq I, Alim A, Yunus MD (1981) Incidence and severity of rotavirus and *Escherichia coli* diarrhoea in rural Bangladesh: Implications for vaccine development Lancet 1: 141–143
7. Carter-Campbell S, Gray RJ (1993) Testing for common hazards in the multiple event failure time setting. Thesis. Harvard School of Public Health

8. Chiba S, Yokoyama T, Nakata S, Morita Y, Urasawa T, Taniguchi K, Urasawa S, Nakao T (1986) Protective effect of naturally acquired homotypic and heterotypic rotavirus antibodies. Lancet 1: 417–421

9. Conner ME, Matson DO, Estes MK (1994) Rotavirus vaccination and vaccine potential. In: Ramig F (ed) Current Topics in Microbiology and Immunology. Springer, New York, 185: 285–337

10. Coulson BS, Grimwood K, Masendycz PJ, Lund JS, Mermelstein N, Bishop RF, Barnes GL (1990) Comparison of rotavirus immunoglobulin A coproconversion with other indices of rotavirus infection in a longitudinal study in childhood. J Clin Microbiol 28: 1367–1374

11. Dharakul T, Rott L, Greenberg HB (1990) Recovery from chronic rotavirus infection in mice with severe combined immunodeficiency: virus clearance mediated by adoptive transfer of immune CD8+ lymphocytes. J Virol 64: 4375–4382

12. Fonteyne J, Zissis G, Lambert JP (1978) Recurrent rotavirus gastroenteritis. Lancet 1: 983

13. Friedman MG, Galil A, Sarov B, et al (1988) Two sequential outbreaks of rotavirus gastroenteritis: Evidence for symptomatic and asymptomatic reinfections. J Infect Dis 158: 814–822

14. Georges-Courbot MC, Monges J, Beraud-Cassel AM, Gouandjika I, Georges AJ (1988) Prospective longitudinal study of rotavirus infections in children from birth to two years of age in Central Africa. Ann Inst Pasteur Virol 139: 421–428

15. Green KY, Kapikian AZ (1992) Identification of VP7 epitopes associated with protection against human rotavirus illness or shedding in volunteers. J Virol 66: 548–553

16. Grinstein S, Gómez JA, Bercovich JA, Biscotti EL (1989) Epidemiology of rotavirus infection and gastroenteritis in prospectively monitored Argentine families with young children. Am J Epidemiol 130: 300–308

17. Gurwith M, Wenman W, Hinde D, Feltham S, Greenberg HB (1981) A prospective study of rotavirus infection in infants and young children. J Infect Dis 144: 218–224

18. Hjelt K, Graubelle PC, Paerregaard A, Nielson OH, Krasilnikoff PA (1987) Protective effect of preexisting rotavirus-specific immunoglobulin against naturally acquired rotavirus infection in children. J Med Virol 21: 39–47

19. Ho MS, Glass RI, Pinsky PF, Anderson LJ (1988) Rotavirus as a cause of diarrheal morbidity and mortality in the United States. J Infect Dis 158: 1112–1116

20. Kapikian AZ, Chanock RM (1990) Rotaviruses. In: Fields BN, Knipe DM, Chanock RM, Hirsch MS, Melnick JL, Monath TP, Roizman B (eds) Virology, 2nd edn. Raven Press, London, 1353–1404

21. Linhares AC, Gabbay YB, Freitas RB, da Rosa ES, Mascarenhas JD, Loureiro EC (1989) Longitudinal study of rotavirus infections among children from Belém, Brazil. Epidem Inf 102: 129–145

22. Mata L, Simhon A, Urrutia JJ, Kronmal RA, Fernandez R, Garcia B (1983) Epidemiology of rotaviruses in a cohort of 45 Guatemalan Mayan Indian children observed from birth to the age of three years. J Infect Dis 148: 452–461

23. Matson DO, Estes MK (1990) Impact of rotavirus infection at a large pediatric hospital. J Infect Dis 162: 598–604

24. Matson DO, Estes MK, Burns JW, Greenberg HB, Taniguchi K, Urasawa S (1990) Serotype variation of human group A rotaviruses in two regions of the USA. J Infect Dis 162: 605–614

25. Matson DO, O'Ryan ML, Herrera I, Pickering LK, Estes MK (1993) Fecal antibody responses to symptomatic and asymptomatic rotavirus infections. J Infect Dis 167: 577–583

26. Matson DO, Velazquez R, Calva JJ, Ruiz-Palacios GM, Estes MK, Pickering LK, Anti-rotavirus IgA antibody status in a cohort of Mexican children monitored from birth to two years of life. IXth International Congress of Virology, Glasgow, Scotland, August 1993
27. Matson DO, Velazquez R, Morrow AL, Shults JN, Ruiz-Palacios GM, Pickering LK (1995) Protective effect of breastfeeding upon first rotavirus infection and illness in a cohort of Mexican children. Society of Pediatric Research
28. Matsui SM, Offit PA, Vo PT, MacKow ER, BenField DA, Shaw RD, Padilla-Noriega L, Greenberg HB (1989) Passive protection against rotavirus-induced diarrhea by monoclonal antibodies to the heterotypic neutralization domain of VP7 and VP8 fragment of VP4. J Clin Microbiol 27: 780–782
29. Offit PA, Dudzik KI (1990) Rotavirus-specific cytotoxic T lymphocytes passively protect against each parental serotype. J Virol 60: 491–496
30. O'Ryan ML, Matson DO, Estes MK, Bartlett AV, Pickering LK (1990) Molecular epidemiology of rotavirus in children attending day care centers in Houston. J Infect Dis 162: 810–816
31. O'Ryan ML, Matson DO, Estes MK, Pickering LK (1994) Anti-rotavirus G type-specific and isotype-specific antibodies in children with natural rotavirus infections. J Infect Dis 169: 504–511
32. O'Ryan ML, Matson DO, Pickering LK, Estes MK (1994) Acquisition of antibody protective against rotavirus infection and illness. Pediatr Infect Dis J 13: 890–895
33. Pickering LK, Bartlett AV, Reves RR, Morrow AL (1988) Asymptomatic excretion of rotavirus before and after rotavirus diarrhea in children in day care centers. J Pediatr 112: 361–365
34. Pickering LK, Evans DJ, Muñoz O, DuPont HL, Coello-Ramirez, P, Vollet J, Conklin RH, Olarte J, Kohl S (1978) Prospective study of enteropathogens in children with diarrhea in Houston and Mexico. J Pediatr 93: 383–388
35. Reves RR, Hossain MM, Midthun K, Kapikian AZ, Naguib T, Zaki AM, DuPont HL (1989) An observational study of naturally acquired immunity to rotaviral diarrhea in a cohort of 363 Egyptian children. Am J Epidemiol 130: 981–988
36. Rodriguez WJ, Kim HW, Brandt CD, Yolken RH, Arrobio JO, Kapikian AZ, Chanock RM, Parrott RH (1978) Sequential enteric illnesses associated with different rotavirus serotypes. Lancet 2: 37
37. Schaller JP, Saif LJ, Cordle CT, Candler E, Winship TR, Smith KL (1992) Prevention of human rotavirus-induced diarrhea in gnotobiotic piglets using bovine antibody. J Infect Dis 165: 623–630
38. Snodgrass DR, Fitzgerald TA, Campbell I, Browning GF, Scott FM, Hoshino Y, Davies RC (1991) Homotypic and heterotypic serological responses to rotavirus neutralization epitopes in immunologically naive and experienced animals. J Clin Microbiol 29: 2668–2672
39. Velázquez FR, Calva JJ, Guerrero ML, Mass D, Glass RI, Pickering LK, Ruiz-Palacios GM (1993) Cohort study of rotavirus serotype patterns in symptomatic and asymptomatic infections in Mexican children. Pediatr Infect Dis J 12: 54–61
40. Velazquez FR, Calva JJ, Matson DO, Guerrero ML, Glass RI, Pickering LK, Ruiz-Palacios GM (1994) Natural protection conferred by rotavirus infection. Implication for vaccine strategies. Mexican Infectious Diseases Society. Monterrey, Nuevo Leon. Enfermedades Infecciosas y Microbiologia 14: 276
41. Velazquez R, Matson DO, Calva JJ, Estes MK, Pickering LK, Ruiz-Palacios GM (1995) Specific serum IgA antibody as a marker of protection against natural rotavirus infection. Infectious Diseases Society of America, San Francisco, California

42. Velazquez FR, Matson DO, Calva JJ, Lourdes Guerrero M, Glass RI, Pickering LK, Ruiz-Palacios GM (1994) Protective effect of a natural rotavirus infection against reinfection and illness. Abstract #22759. Society of Pediatric Research

43. Ward RL, Bernstein DI, for the U.S. Rotavirus Vaccine Efficacy Group (1994) Protection against rotavirus disease after natural rotavirus infection. J Infect Dis 169: 900–904

44. Ward RL, Clemens JD, Knowlton DR, Rao MR, van Loon FP, Huda N, Ahmed F, Schiff GM, Sack DA (1992) Evidence that protection against rotavirus diarrhea after natural infection is not dependent on serotype-specific neutralizing antibody. J Infect Dis 166: 1251–1257

45. Woods PA, Gentsch J, Gouvea V, Mata L, Santosham M, Bai ZS, Urasawa S, Glass RI (1992) Distribution of serotypes of human rotavirus in different populations. J Clin Microbiol 30: 781–785

46. Zheng BJ, Lo SK, Tam JJ, Lo M, Yeung CY, Ng MH (1989) Prospective study of community-acquired rotavirus infection. J Clin Microbiol 27: 2083–2090

Authors' address: Dr. D. O. Matson, Center for Pediatric Research, Children's Hospital of The King's Daughters and Eastern Virginia Medical School, 855 West Brumbleton Avenue Norfolk, Virginia 23510-1001, U.S.A.

Arch Virol (1996) [Suppl] 12: 141–152

Rotavirus immunity in the mouse

M. A. Franco, N. Feng, and **H. B. Greenberg**

Department of Medicine, Microbiology and Immunology,
Stanford University School of Medicine, Stanford,
California, U.S.A.

Summary. Naturally attenuated animal rotaviruses have been tested as anti-rotavirus vaccines with moderate success. The development of improved vaccines will rely on our understanding of the immune mechanism that mediate clearance and protection from rotaviral reinfection. The mouse model of rotavirus infection is a versatile tool for studying these mechanisms: mice have a relative low cost and there is a rapidly increasing number of immunological reagents to study rotavirus immunology. This review covers recent data on the mouse model of rotavirus infection. We show that both effector arms of the immune system (CD8 + T cells and B cells) mediate anti-rotavirus effects *in vivo*.

Introduction

Rotaviruses (family *Reoviridae*, genus *rotavirus*) are the single most important cause of severe dehydrating diarrheal disease in young children in all areas of the world [15]. Group A rotaviruses are responsible for the deaths of 800,000 infants per year worldwide and approximately 1 billion dollars per year in health costs in the US. [17, 18]. The development of a rotavirus vaccine is a priority of both the NIH and the WHO. Vaccines currently being tested use a Jennerian or modified Jennerian strategy and have met with variable or moderately successful results [1, 19]. The rational design of vaccines with improved efficacy will clearly need a better understanding of the factors that regulate the resolution of primary rotavirus infection and the induction of immunity to reinfection. To understand these factors we, and others, have focused our attention on a murine model of rotavirus infection. Findings from this model system will be reviewed below [3, 34].

The mouse model

Although mice are susceptible to infection with some heterologous rotaviruses (rotavirus derived from species other than mice) [29] these are always semipermissive infections and heterologous rotaviruses do not efficiently spread from mouse to mouse. Because of this, recent studies of the murine model have concentrated on studying infection with murine rotaviruses. At least six distinct strains of group A murine rotaviruses have been described [6, 33]. The genes encoding VP4 and VP7 of most of these strains have been sequenced and the serotypic diversity of most of the murine strains have been analyzed [6, 31, 33]. Two distinct P type sequences have been identified among murine strains while the G serotype of all strains identified to date is most closely related to G serotype 3 [6, 31, 33]. Several murine rotaviruses have been adapted to replication in cell culture, although in general yields tend to be low and plaque morphology poor [3, 16].

Mice become maturationally resistant to rotavirus diarrheal disease at approximately 15 days of age [3, 9, 35]. Because of this, studies that evaluate the active immune mechanisms against rotavirus induced diarrheal disease are very difficult to perform [36]. Studies of the protective mechanisms against rotavirus induced diarrheal disease are thus limited to passive treatment of suckling mice, either via immunization of the dam or passive transfer of cells, serum or milk to their pups [20, 24, 27]. Both homologous and heterologous rotaviruses can be used to challenge the passively treated pups. While this passive immunity model has provided much useful information, its direct relevance to humans, in which active viral immunization is probably necessary in most circumstances, remains controversial.

Recently, an active model of rotavirus immunity to infection using adult mice has been described [3, 34]. Although adult mice do not develop diarrhea following rotavirus infection, they are just as susceptible to infection with some murine rotaviruses as are sucklings. Adult mice shed murine rotaviruses in feces for only slightly shorter duration than do sucklings, they shed it in comparable quantities, and appear to mount immune responses that are very similar, both qualitatively and quantitatively, to those of suckling mice ([3], S. Ishida Unpubl. data). Of note, adult mice develop increased fluid accumulation in their small bowel during primary infection and they develop complete and long lasting immunity to rechallenge after primary infection with murine rotaviruses [3]. Hence, the active immunization adult mouse model employing murine rotaviruses (and the acute diarrheal disease model in suckling mice) appears to offer many similarities to infection in humans and can be used to characterize in detail several aspects of rotaviral immunity.

Primary immune responses in mice

a) Cellular

The mouse model has been used to study the cytotoxic and T helper cell immune response to primary rotavirus infection [2, 14, 25, 26, 27]. In general, these studies have examined the nature and specificity of cellular responses to heterologous, as opposed to homologous, infection. Although it has not yet been systematically examined, it is our impression that cytotoxic responses are more readily detected in mice following heterologous rather than homologous infection. Whether this paradoxical observation is an artifact of the cytotoxicity assay systems employed or a fundamental difference in immune response to hetero-logous *vs.* homologous viruses remains to be determined.

Cytotoxic T lymphocytes (CTLs) generated after heterologous infection in mice are cross-reactive, MHC restricted, and detectable among intraepithelial as well as systemic lymphocytes [25, 26]. The major target of CTLs appears to be located on VP7 but other proteins may also serve as targets [13, 23]. An immunodominant K^b restricted CTL epitope has been mapped to the signal sequence of VP7 [14]. VP6 has been identified as a target of rotavirus-specific T helper cells [2].

b) Humoral immune response

Both local and systemic responses are induced after oral infection, although the relative titers of these responses may differ significantly in different circumstan-ces (see section IV below) [10]. Immune responses to VP2, VP4, VP6, VP7, and NSP 2 have been detected in serum after primary infection of both suckling and adult mice and they appear comparable in these groups (S. Ishida unpubl. data). The time course of the local and systemic humoral response after primary infection has recently been examined ([3], S. Ishida unpubl. data).

Elispot analysis of B cell responses to heterologous infection has been carried out and has demonstrated a very substantial and long lived response, most pronounced in the lamina propria [32]. The long term duration of detectable local antibody following primary infection is consistent with the enduring immunity to reinfection in the mouse following a single exposure to virus [3]. Mice appear to develop more complete and long lasting immunity to reinfection (Fig. 1) than do humans (data not shown).

Immune effector mechanisms involved in resolution of primary infection and prevention of reinfection

The resolution of primary rotavirus infection is dependent on immune mechan-isms since both SCID and RAG-2 knockout mice (both mice lack T and B cells) become chronically infected after exposure to rotavirus [30, 12]. In contrast to these studies, nude mice (that lack thymus dependent T cells) were reported to resolve rotavirus infection in the absence of a rotavirus specific antibody response, suggesting that neither T nor B cells were necessary for clearance of primary rotavirus infection [8]. Because of the inconsistency between the SCID-RAG-2 knockout mice data with the nude mice data we decided to re-examine this latter model. We found that after infection with our virulent strains of murine rotavirus, nude mice on a Balb/c background shed viral antigen for two to three more days than did their heterozygous littermates but resolve primary infection completely (M. F. Franco unpubl. results). This result suggests that T cells do participate in the timely clearance of primary rotavirus infection, but are not neccesary for this function. The mechanism governing the T cell independent resolution of primary infection observed in the nude mice is currently under study.

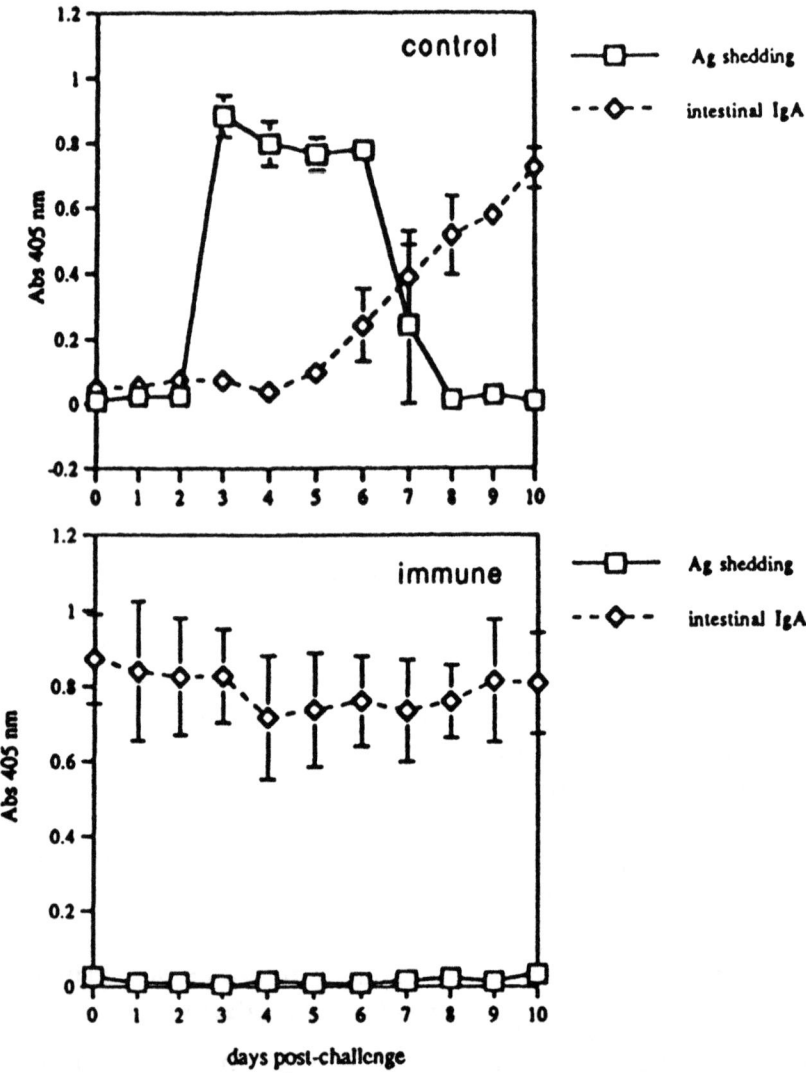

Fig. 1. Evaluation of active protection following challenge with murine rotavirus. Naive (nonimmune) and immune adult Balb/c mice were orally challenged with wild type EC virus 6 weeks after initial exposure of the immune group. Stool samples, collected daily from all animals, were suspended in diluent containing protease inhibitor and stored at $-70\,°C$. Stool suspensions were quantitated for levels of rotavirus antigen (open squares) and secretory IgA (open diamonds) by ELISA. The absorbance profiles at 405 nm are shown for the nonimmune (top) and immune (bottom) groups of mice ($+/-$ SD, n = 4). Reproduced from reference [3] with permission

Early studies indicated that passively transferred immune CD8+ T cells were capable of clearing chronic infection in SCID mice and preventing primary rotavirus illness in suckling mice (Fig. 2) [5, 27]. To evaluate the role of actively induced CD8+ T cells in rotavirus immunity, we compared rotavirus infection in β_2 microglobulin deficient mice and congenic C57BL/6 control mice [12].

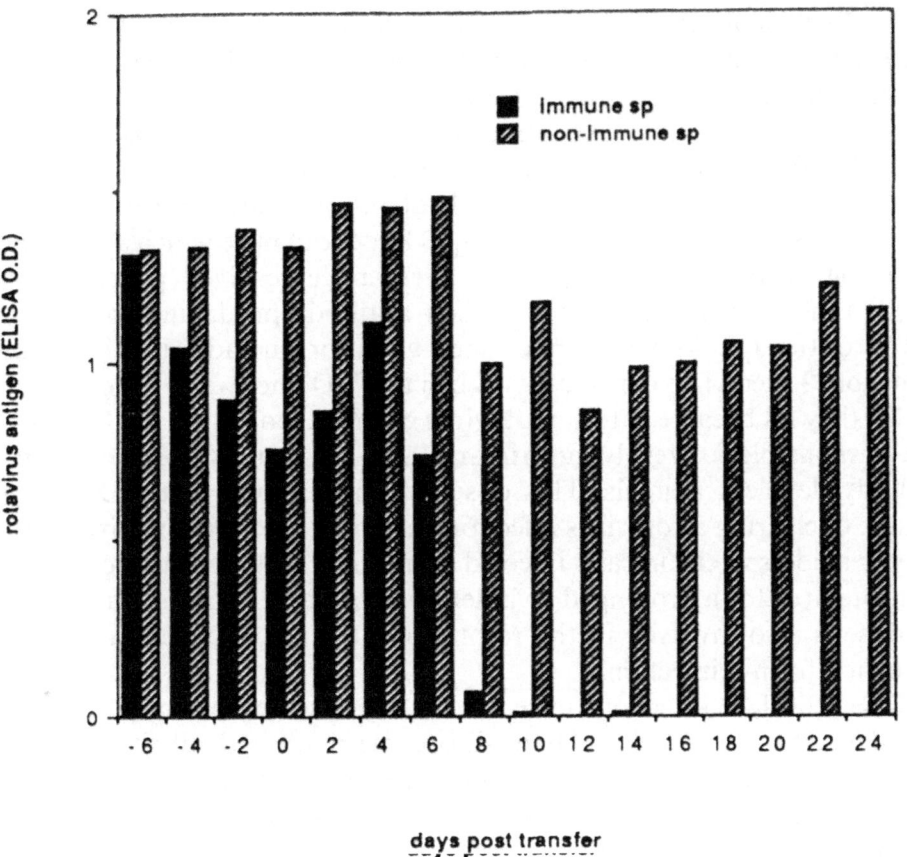

Fig. 2. Rotavirus shedding in stools of SCID mice before and after transfer of immune and nonimmune CD8 + spleen (sp) cells obtained from Balb/c mice immunized intraperitoneally with rotavirus. Rotavirus antigen was detected by ELISA. Each point represents a mean optical density of the data collected from five mice. For the immune cell transfer group, the standard error of the mean for values on days -6, -2, 0, 2 and 4 ranged from 9 to 15%; on days 6, 8, 10, 12, 14, 16, 18, 20, 22 and 24 it ranged from 19 to 32%. For the nonimmune cell transfer group, the standard errors of the mean for all values were less than 10%. Reproduced from reference [5] with permission

Because the β_2 microglobulin knockouts have been reported to retain some residual CD8 + T cell cytotoxic activity, we also treated these animals with anti-CD8 antibody [12]. β_2 microglobulin knockout mice, treated or not treated with the anti-CD8 monoclonal antibody, as do nude mice, resolve primary rotavirus infection in a slightly delayed fashion, although resolution is complete. Clearance of primary rotavirus infection in β_2 microglobulin knockout mice correlated with the production of intestinal rotavirus specific IgA. Hence, CD8 + T cells appear to play a role in clearance of primary infection but are not required for this function. We are currently investigating the mechanism by which CD8 + T cells carry out their antiviral effect. We also tested the ability of CD8 + T cell depleted mice, which had cleared primary infection, to resist rechallenge [12].

CD8 + T cell depleted mice rechallenged 6–8 weeks after primary infection were completely immune to reinfection. This immunity correlated with high levels of IgA antirotavirus antibody in the feces of immune animals. Hence, CD8 + T cells do not appear to play a significant role in the development of active protective immunity after homologous infection in the mouse.

In order to directly evaluate the role of antibody in rotavirus immunity, we examined the response to rotavirus in J_HD knockout mice which are incapable of producing antibody [4]. In general, these mice resolved primary rotavirus infection in an identical manner to their antibody producing controls. About 5–10% of the J_HD knockout mice shed virus for one additional day and one animal of 20 shed virus chronically. When the J_HD mice were depleted of CD8 + T cells, they all became chronic rotavirus carriers, indicating the CD8 + T cells were responsible for resolving primary infection in most but not all of these antibody deficient animals. This observation confirms that CD8 + cells are capable of clearing a rotavirus infection as shown by earlier passive CD8 + cell transfer studies in chronically infected SCID mice [5]. Because some J_HD mice shed rotavirus for a prolonged or indefinite period, it appears that an antibody response is also involved in the resolution of primary infection as well as in protection from reinfection.

When the J_HD knockout mice, which had resolved primary infection were rechallenged 6–8 weeks later, they all became reinfected, albeit at a somewhat lower level of shedding than did non-immune controls. We have interpreted this finding to indicate that complete resistance to reinfection is absolutely dependent on antibody (resumably IgA) and not on CD8 + T cells.

Studies of active vaccination in the mouse model

A substantial body of data has been published characterizing the factors that regulate immunity in the mouse passive protection model [7, 11, 20, 21, 24, 28]. These studies will not be further reviewed because their relevance to active immunity is not clearly established.

Homologous infection with any murine rotavirus fully protects against subsequent reinfection [3]. Since all the murine rotaviruses examined to date appear to have related VP7s, it is impossible to directly examine the role of G-serotype specific immunity in the mouse model of homologous infection. However, it is clear that to obtain complete protection from reinfection the original murine rotavirus and the murine virus used for challenge need not express the same P type if they share G types [3].

The nature of protective immunity following heterologous infection ("Jennerian vaccination") in mice has been studied [10]. Heterologous viruses vary considerably in their abilities to induce local immune responses in mice, with group A rotaviruses RRV (G3P5[3], simian origin) and SA11 (G3P[2], simian origin) appearing to be the most efficient. However, in all cases, heterologous infection is much less effective than homologous infection at stimulating a local intestinal humoral response [10]. The ability to stimulate a local response is not

A. Fecal IgA vs. Viral Antigen Shedding

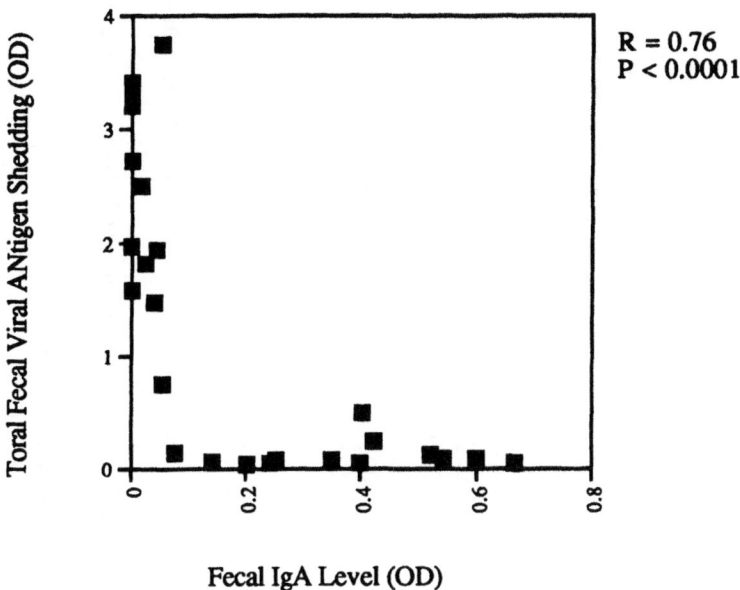

B. Serum IgG vs. Viral Antigen Shedding

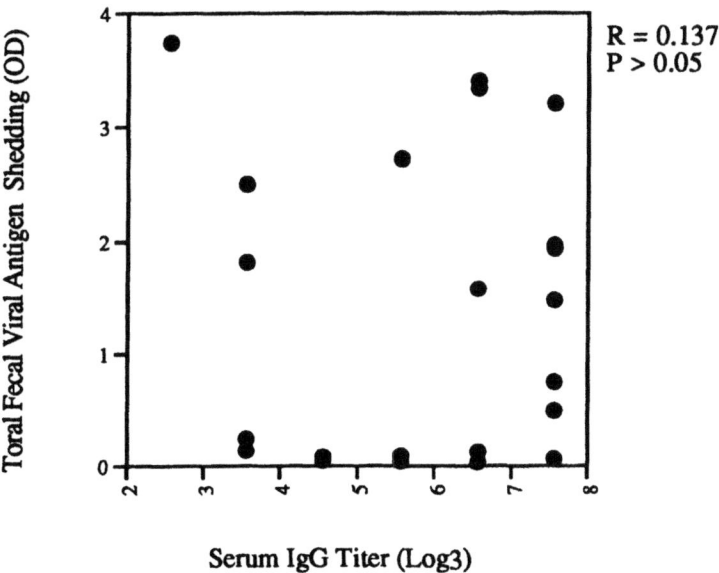

Fig. 3. Correlation between fecal IgA or serum IgG response and fecal viral antigen shedding after murine rotavirus challenge in mice immunized with a heterologous rotavirus. Six seven day-old mice were oral inoculated with a heterologous rotavirus, RRV (10^7 PFU – 10^5 PFU/mouse). Six weeks later mice were challenged with a virulent murine rotavirus, EC_w and viral antigen shedding in feces was measured by ELISA for 8 days. Total viral antigen shedding after challenge was correlated with levels of rotavirus-specific fecal IgA (A) or serum IgG (B) response prior to challenge

Table 1. Protection of suckling mice against challenge with simian (RRV) or murine (Eb) rotavirus following inoculation of dams with baculovirus-expressed VP4 protein

Immunizing antigen	Dam	HI titer against RRV[a]	HI titer of pup's stomach contents versus RRV	FRN[b] titer against:		No. of mice with diarrhea/no. of mice challenged with[c]:	
				RRV	Eb	RRV	Eb
AcNPV-expressed VP4	1	102 200	400	102 200		0/8	
	2	25 600	400	25 600		0/9	
	3	25 600	ND[d]	25 600			1/6
	4	102 200	ND	102 200		0/6	
Control AcNPV Protein	1	<100	<100	<100	<100	10/10	
	2	<100	<100	<100	<100	8/9	
	3	<100	ND	<100	<100		5/5
	4	<100	ND	<100	<100		7/9

[a] Hemagglutination inhibition (HI) titer are expressed as the reciprocal of the serum dilution required to inhibit hemagglutination of 4 to 8 hemagglutination units of RRV

[b] Focus reduction neutralization (FRN) data represent the reciprocal of the serum dilution resulting in a 60% reduction in the number of infected MA104 cells

[c] Pups were challenged 5 days after birth with 10^8 PFU of RRV or 10^4 infectious doeses of Eb

[d] ND Not determined

Adapted from [20]

directly related to virulence because attenuated murine viruses are more immunogenic than are virulent heterologous viruses [10]. Protection after heterologous infection is directly correlated with fecal IgA levels but not with serum IgG responses (Fig. 3). Preliminary studies indicate that local immunogenicity, rather than viral serotype, is the primary determinant of protection following heterologous infection (N. Feng, unpubl. data).

There have been limited studies of systemic immunization in mice with live or inactivated heterologous or homologous rotaviruses. The data available indicate that systemic immunization can induce protection from reinfection [22]. How this effect is mediated is unknown.

We have carried out a variety of immunization studies in mice employing several different expression vectors to produce rotavirus genes encoding VP4 and VP7. Prior studies in the passive model of rotavirus infection had clearly demonstrated that recombinant immunization of dams with either VP7 expressed by vaccinia or by an adenovirus or VP4 expressed by baculovirus could prevent illness in pups challenged with high doses of heterologous virus (Table 1) [7, 11, 20]. To date, our attempts to induce active immunity in the adult mouse model using vaccinia virus or adenovirus expressed VP4 or VP7 have been unsuccessful. Both parenteral and oral immunization protocols induced readily detectable levels of systemic antibody but very low or undetectable levels of local antibody even after oral immunization (N. Feng, unpubl. data). Presumably the inability of these recombinant vectors to efficiently stimulate the enteric immune system is responsible for their lack of efficacy. We are currently carrying out studies aimed at augmenting local responses to recombinant vectors. As well studies are now underway to evaluate the efficacy of DNA immunization in the mouse rotavirus model (J. Herrmann, pers. commun.).

Conclusion

The mouse model has provided us with several insights into immunity to rotaviruses. We have learned that antibody (presumably local IgA) appears to be the primary determinant of protection while both CD8 + T cells and antibody are involved in mediating resolution of primary infections. Experiments in mice indicate that heterologous rotaviruses are far less efficient stimulators of a local immune response and protective immunity than are homologous rotaviruses. Although heterologous infection can induce complete protection against challenge, protection is dependent on a large viral dose; a small decrease in immunization dose leads to considerable loss of efficacy. These observations appear relevant to current human vaccine trials and emphasize the need to carefully monitor local as well as systemic immune responses to vaccines.

Additional studies in the mouse have clearly demonstrated the development of heterotypic protective immunity after oral heterologous immunization. In fact, local immunogenicity, rather than serotype specificity of the immunizing heterologous virus, was the key determinant of protection in our studies (N. Feng, unpubl. data). The relevance of these observations to current "Jennerian" vaccine trials should be determined. The host factors that regulate the difference in immune response to homologous and heterologous infection also remain to be elucidated. Presumably, the degree of TH1 and TH2 activation in these two situations, as well as the location at which helper T cell activation occurs, must vary. The genetic basis of differences in the immune response to homologous and heterologous virus infection is currently under study. Preliminary data indicate a multigenic effect in which genes encoding VP4 or VP7 are not of primary importance (N. Feng, unpubl. data).

The mouse model of rotavirus infection and the availability of genetically altered mice with specific immune defects should make it easier to understand the reason(s) different vaccine strategies succeed or fail. Utilization of the mouse model to study rotavirus-like particles produced in baculovirus, DNA immunization and recombinant rotaviral proteins is currently underway. Preliminary failures using live recombinant expression vectors appear to be due to the difficulty of developing immunity to these vectors in the mouse gastrointestinal tract. The success of future strategies will be dependent, at least in part, on the ability of the immunization to induce antirotavirus antibody in the gastrointestinal tract.

Acknowledgments

This work was supported by grants RO1AI21632 and DK38707 from the NIH and by a grant from W.H.O. H. B. Greenberg is a medical investigator at the Palo Alto Veterans Administration Medical Center. M. A. Franco is funded by a Walter V. and Idun Berry fellowship.

References

1. Bernstein DI, Glass RI, Rodgers G, Davidson BL, Sack DA (1995) Evaluation of rhesus rotavirus monovalent and tetravalent reassortant vaccines in US children. US Rotavirus Vaccine Efficacy Group. JAMA 273: 1191–1196
2. Bruce MG, Campbell I, Xiong Y, Redmond M, Snodgrass DR (1994) Recognition of rotavirus antigens by mouse L3T4-positive T helper cells. J Gen Virol: 1859–1866
3. Burns JW, Krishnaney AA, Vo PT, Rouse RV, Anderson LJ, Greenberg HB (1995) Analyses of homologous rotavirus infection in the mouse model. Virology 207: 143–153
4. Chen J, Trounstine M, Alt FW, Young F, Kurahara C, Loring JF, Huszar D (1993) Immunoglobulin gene rearrangement in B cell deficient mice generated by targeted deletion of the J_H locus. International Immunol 5: 647–656
5. Dharakul T, Rott L, Greenberg H (1990) Recovery from chronic rotavirus infection in mice with severe combined immunodeficiency: virus clearance mediated by adoptive transfer of immune CD8+ T lymphocytes. J Virol 64: 4375–4382
6. Dunn SJ, Burns JW, Cross TL, Vo PT, Bremont M, Greenberg HB (1994) Comparisons of VP4 and VP7 of five murine rotavirus strains. Virology 203: 250–259
7. Dunn SJ, Fiore L, Werner RL, Cross TL, Broome RL, Ruggeri FM, Greenberg HB (1995) Immunogenicity, antigenicity, and protection efficacy of baculovirus expressed VP4 trypsin cleavage products, VP5(1)* and VP8* from rhesus rotavirus. Arch Virol 140: 1969–1978

8. Eiden J, Lederman HM, Vonderfecht S, Yolken R (1986) T-cell-deficient mice display normal recovery from experimental rotavirus infection. J Virol 57: 706–708

9. Eydelloth RS, Vonderfecht SL, Sheridan JF, Enders LD, Yolken RH (1984) Kinetics of viral replication and local and systemic immune response in experimental rotavirus infection. J Virol 50: 947–950

10. Feng N, Burns JW, Bracy L, Greenberg H (1994) Comparison of mucosal and systemic humoral immune responses and subsequent protection in mice orally inoculated with a homologous or a heterologous rotavirus. J Virol 68: 7776–7773

11. Fiore L, Dunn SJ, Ridolfi B, Ruggeri FM, Mackow ER, Greenberg HB (1995) Antigenicity, immunogenicity and passive protection induced by immunization of mice with baculovirus-expressed VP7 protein from rhesus rotavirus. J Gen Virol: 1981–1988

12. Franco MA, Greenberg HB (1995) Role of B cells and cytotoxic T lymphocytes in clearance of and immunity to rotavirus infection in mice. J Virol 69: 7800–7806

13. Franco MA, Lefevre P, Willems P, Tosser G, Lintermanns P, Cohen J (1994) Identification of cytotoxic T cell epitopes on the VP3 and VP6 rotavirus proteins. J Gen Virol 75: 589–596

14. Franco MA, Prieto I, Labbe M, Poncet D, Borras CF, Cohen J (1993) An immunodominant cytotoxic T cell epitope on the VP7 rotavirus protein overlaps the H2 signal peptide. J Gen Virol 74: 2579–2586

15. Glass RI, Gentsch J, Smith JC (1994) Rotavirus vaccines: success by reassortment? Science 265: 1389–1391

16. Greenberg HB, Vo PT, Jones R (1986) Cultivation and characterization of three strains of murine rotavirus. J Virol 57: 585–590

17. Ho M, Glass RI, Pinsky PF, Anderson L (1988) Rotavirus as a cause of diarrheal morbidity and mortality in the United States. J Infect Dis 158: 1112–1116

18. Institute of Medicine (1986) New vaccine development. Establishing priorities. Diseases of importance in developing countries, vol II. National Academy Press, Washington DC, pp 308–318

19. Kapikian AZ, Chanock RM (1996) Rotaviruses. In: Fields BN, Knipe DM, Howley PM (eds) Fields Virology, vol 2. Lippincott-Raven, Philadelphia, pp 1657–1708

20. Mackow ER, Vo PT, Broome R, Bass D, Greenberg HB (1990) Immunization with baculovirus-expressed VP4 protein passively protects against simian and murine rotavirus challenge. J Virol 64: 1698–1703

21. Matsui SM, Offit PA, Vo PT, Mackow ER, Benfield DA, Shaw RD, Padilla NL, Greenberg HB (1989) Passive protection against rotavirus-induced diarrhea by monoclonal antibodies to the heterotypic neutralization domain of VP7 and the VP8 fragment of VP4. J Clin Microbiol 27: 780–782

22. McNeal MM, Sheridan JF, Ward RL (1992) Active protection against rotavirus infection of mice following intraperitoneal immunization. Virology 191: 150–157

23. Offit PA, Boyle DB, Both GW, Hill NL, Svoboda YM, Cunningham SL, Jenkins RJ, McCrae MA (1991) Outer capsid glycoprotein VP7 is recognized by cross-reactive, rotavirus-specific, cytotoxic T lymphocytes. Virology 184: 563–568

24. Offit PA, Clark HF (1985) Protection against rotavirus-induced gastroenteritis in a murine model by passively acquired gastrointestinal but not circulating antibodies. J Virol 54: 58–64

25. Offit PA, Cunningham SL, Dudzik KI (1991) Memory and distribution of virus-specific cytotoxic T lymphocytes (CTLs) and CTL precursors after rotavirus infection. J Virol 65: 1318–1324

26. Offit PA, Dudzik KI (1989) Rotavirus-specific cytotoxic T lymphocytes appear at the intestinal mucosal surface after rotavirus infection. J Virol 63: 3507–3512
27. Offit PA, Dudzik KI (1990) Rotavirus-specific cytotoxic T lymphocytes passively protect against gastroenteritis in suckling mice. J Virol 64: 6325–6328
28. Offit PA, Shaw RD, Greenberg HB (1986) Passive protection against rotavirus-induced diarrhea by monoclonal antibodies to surface proteins VP3 and VP7. J Virol 58: 700–703
29. Ramig RF (1988) The effects of host age, virus dose, and virus strain on heterologous rotavirus infection of suckling mice. Microb Pathog 4: 189–202
30. Riepenhoff-Talty M, Dharakul T, Kowalski E, Michalak S, Ogra PL (1987) Persistent rotavirus infection in mice with severe combined immunodeficiency. J Virol 61: 3345–3348
31. Sereno MM, Gorziglia MI (1994) The outer capsid protein VP4 of murine rotavirus strain Eb represents a tentative new P type. Virology 199: 500–504
32. Shaw RD, Merchant AA, Groene WS, Cheng EH (1993) Persistence of intestinal antibody response to heterologous rotavirus infection in a murine model beyond 1 year. J Clin Microbiol 31: 188–191
33. Ushijima H, Morikawa S, Hasegawa A, Mukoyama A, Suzuki E, Yamamoto T, Nishio O (1995) Characterization of VP4 and VP7 of murine rotavirus (YR-1) Isolated in Japan. Fifth international symposium on double stranded RNA viruses. Abstract P 08
34. Ward RL, McNeal MM, Sheridan JF (1990) Development of an adult mouse model for studies on protection against rotavirus. J Virol 64: 5070–5075
35. Wolf JL, Cukor G, Blacklow NR, Dambrauskas R, Trier JS (1981) Susceptibility of mice to rotavirus infection: effects of age and administration of corticosteriods. Infect Immun 33: 565–574
36. Woode GN, Zheng S, Melendy DR, Ramig RF (1989) Studies on rotavirus homologous and heterologous active immunity in infant mice. Viral Immunol 2: 127–132

Authors' address: Dr. H. B. Greenberg, Stanford University, School of Medicine, Lab Surge, P 304, Stanford, CA 94304, U.S.A.

Arch Virol (1996) [Suppl] 12: 153–161

The gnotobiotic piglet as a model for studies of disease pathogenesis and immunity to human rotaviruses

L. J. Saif, L. A. Ward, L. Yuan, B. I. Rosen, and T. L. To

Food Animal Health Research Program,
Ohio Agricultural Research and Development Center,
The Ohio State University, Wooster, Ohio, U.S.A.

Summary. Gnotobiotic piglets serve as a useful animal model for studies of human rotavirus infections, including disease pathogenesis and immunity. An advantage of piglets over laboratory animal models is their prolonged susceptibility to human rotavirus-induced disease, permitting cross-protection studies and an analysis of active immunity. Major advances in rotavirus research resulting from gnotobiotic piglet studies include: 1) the adaptation of the first human rotavirus to cell culture after passage and amplification in piglets; 2) delineation of the independent roles of the two rotavirus outer capsid proteins (VP4 and VP7) in induction of neutralizing antibodies and cross-protection; and 3) recognition of a potential role for a nonstructural protein (NSP4) in addition to VP4 and VP7, in rotavirus virulence. Current studies of the pathogenesis of group A human rotavirus infections in gnotobiotic piglets in our laboratory have confirmed that villous atrophy is induced in piglets given virulent but not cell culture attenuated human rotavirus (G1, P1A, Wa strain) and have revealed that factors other than villous atrophy may contribute to the early diarrhea induced. A comprehensive examination of these factors, including a proposed role for NSP4 in viral-induced cytopathology, may reveal new mechanisms for induction of viral diarrhea. Finally, to facilitate and improve rotavirus vaccination strategies, our current emphasis is on the identification of correlates of protective active immunity in the piglet model of human rotavirus-induced diarrhea. Comparison of cell-mediated and antibody immune responses induced by infection with a virulent human rotavirus (to mimic host response to natural infection) with those induced by a live attenuated human rotavirus (to mimic attenuated oral vaccines) in the context of homotypic protection has permitted an analysis of correlates of protective immunity. Results of these studies have indicated that the magnitude of the immune response is greatest in lymphoid tissues adjacent to the local site of viral replication (small intestine). Secondly, there was a direct correlation between the degree of protection induced and the level of the intestinal immune response, with significantly higher local immune responses and complete protection induced only after primary exposure to virulent human rotavirus. These studies thus have established basic parameters related to immune protection in the piglet model of human rotavirus-induced disease, verifying the usefulness of this model to examine new strategies for the design and improvement of human rotavirus vaccines.

Introduction

Group A rotaviruses are a leading cause of dehydrating diarrheal infections in infants and young children worldwide [8]. Public health problems posed by rotaviruses have stimulated research on vaccination strategies. Vaccine development has focused on the use of live attenuated oral vaccines in a "Jennerian" approach using heterologous animal strains or human/animal reassortants as candidate vaccines [5, 8]. Unfortunately, these candidate vaccines have often failed in various aspects of safety, immunogenicity or efficacy, especially when tested in developing countries [8]. To facilitate and improve vaccine development, a more comprehensive understanding of rotavirus pathogenesis and mucosal immunity to rotaviruses is needed. These studies are most readily accomplished by the use of animal models to study disease pathogenesis and immunity *in vivo*.

Advantages of gnotobiotic piglets as models for studies of rotavirus pathogenesis and immunity

Although laboratory mice and rabbits serve as useful models for evaluating immune responses to rotaviruses, especially host-specific strains, these animal models are not conducive for studies of active immunity to clinical infections induced by human rotaviruses because human rotavirus infections in these species are usually subclinical [3, 13]. Moreover, older mice and rabbits are refractory to rotavirus disease and permit evaluation of active protection only against infection.

In comparison, gnotobiotic piglets remain susceptible to infection and disease induced by several human rotavirus strains for as long as 6 weeks of age (14, 22, 24, Saif, LJ, unpublished). Other advantages of the gnotobiotic piglet model include: 1) they closely resemble humans in gastrointestinal physiology (monogastrics) and mucosal immune development [9, 12]; 2) the placenta of pigs acts as a barrier to the transfer of maternal antibodies; hence colostrum-deprived gnotobiotic pigs are devoid of rotavirus maternal antibodies and are immunologically virgin but immunocompetent at birth, permitting analysis of true primary immune responses [9]; and 3) the derivation and maintenance of piglets in a gnotobiotic environment assures that exposure to extraneous rotaviruses or other enteric pathogens is eliminated as a confounding variable [1, 23].

The pathogenesis of a group A human rotavirus in neonatal gnotobiotic piglets

In spite of the host-specificity of rotaviruses, several researchers have found that gnotobiotic piglets or gnotobiotic calves were susceptible to infection and disease by heterologous rotaviruses, including human rotaviruses, under experimental conditions [10, 11, 14–16, 18, 21–24]. This observation agrees with antigenic and genetic data showing a close relationship between certain human and animal rotaviruses, suggesting that interspecies transmission of rotaviruses may occur under certain poorly defined circumstances in nature [19]. Gnotobiotic piglets were invaluable for initial studies of human rotaviruses. Passage of the Wa strain of human rotavirus (G1, P1A) from stool filtrates of infected infants into gnotobiotic piglets provided an amplified source of viable rotavirus, free of maternal or actively induced (in the early stages of infection) antibodies, and resulted in the first successful adaptation of a human rotavirus to serial propagation in cell culture [23].

In subsequent studies, the pathogenesis of the infant stool-passaged, virulent Wa strain of human rotavirus was analyzed in gnotobiotic piglets [21]. Piglets orally inoculated with 10^5 focus-forming units (FFU) (or 10^5 median infectious doses) of Wa rotavirus developed diarrhea within 13 post-inoculation (PI) hours, which correlated with the presence of rotavirus antigen within villous epithelial cells (Table 1). Mild to moderate villous atrophy was observed at PI hours 24–48 coincident with the peak of virus replication. Diarrhea and rotavirus shedding persisted between 4 to 7 PI days (PID). Recovery correlated with the presence of morphologically normal villi by PID 7. Thus the Wa human rotavirus induced lesions in gnotobiotic pigs, similar but less severe than those seen after infection with some (but not all) homologous porcine rotavirus strains [17]. Moreover, factors other than villous atrophy may contribute to the early diarrhea induced by PI hour 13, preceding the detection of villous atrophy.

Gnotobiotic piglets also proved useful in a recent study to identify the rotavirus genes associated with virulence and host range restriction [6]. The response of gnotobiotic piglets was analyzed after oral administration of a porcine X human reassortant rotavirus derived from a parental porcine rotavirus (G4, P9, SB1A strain) which caused diarrhea in piglets and a parental human rotavirus (G2, 1B, DS-1 strain) which was attenuated for piglets. The major conclusions were that replacing the VP3, VP4, VP7 or NSP4 genes of the attenuated human strain with the corresponding genes of the virulent porcine rotavirus yielded viral reassortants that failed to induce diarrhea. Similarly, reassortants possessing only one, two or three of these porcine rotavirus genes on the human rotavirus genetic background failed to induce diarrhea. These results suggest that replacement of any one of these four genes of a human rotavirus with that of an avirulent animal rotavirus could attenuate the human rotavirus, leading to a new rotavirus vaccine strategy.

Passive immunity to human rotaviruses in a gnotobiotic piglet model of disease

Gnotobiotic piglets have been used to evaluate the efficacy of passively administered bovine antibody to a human rotavirus for preventing human rotavirus-

Table 1. Summary of clinical signs, lesions, rotavirus detection and seroconversion after oral inoculation and challenge of gnotobiotic piglets[a] with Wa human rotavirus

Primary virus inoculum[a]	Primary inoculation[a]					Challenge (virulent Wa)[a]	
	Moderate to severe diarrhea	Fecal virus shedding	Viral antigen in gut	Villous atrophy	Sero-conversion	Moderate to severe diarrhea	Virus shedding
Virulent	Yes	Yes	Yes	Yes	Yes	No[b]	No
Attenuated	No[b]	Yes[c]	No[c]	No[c]	Yes	Partial[d]	Partial[d]
None	No[b]	No	No	No	No	Yes	Yes

[a] n = 8–18 piglets used for analysis of each response; piglets were orally inoculated with virulent, attenuated or no rotavirus (controls) and orally challenged with virulent Wa rotavirus at post-inoculation day 21

[b] 11–13% of piglets developed transient mild diarrhea

[c] Only 6% of piglets shed virus after inoculation with attenuated Wa rotavirus; none of the piglets examined for viral antigen or villous atrophy shed rotavirus prior to euthanasia

[d] 56% of piglets developed diarrhea (compared to 83% of controls) and 81% shed virus (compared to 100% of controls) after virulent Wa rotavirus challenge

induced diarrhea [15]. Cows were immunized with inactivated human rotavirus serotypes G1, P1A (Wa strain) and G2, P1B (S2 strain) and simian rotavirus (G3, [P2], SA11 strain) and the (immune) colostrum collected. Antibody concentrates from colostrum were fed three times daily to gnotobiotic piglets subsequently challenged with virulent Wa rotavirus. The immune colostrum feeding effectively reduced or eliminated both rotavirus shedding and diarrhea in a dose-dependent manner, confirming that a quantitative relationship exists between the protective antibody dose and the diarrheal disease response. Furthermore, piglets fed the immune colostrum and therefore protected against human rotavirus-induced disease, seroconverted to Wa rotavirus, indicative of the development of active immune responses in the presence of protective levels of passive colostral antibodies.

Active immunity to human rotaviruses in a gnotobiotic piglet model of disease

Previous studies of porcine rotavirus infections in gnotobiotic piglets confirmed that rotaviruses that share common VP4 (P) and VP7 (G) serotypes induced a high degree, or complete cross-protection against challenge with rotavirus strains bearing the common P or G types [7]. Little or no cross-protection was evident in the piglets inoculated and challenged with heterotypic (in both G and P type) serotypes [1].

We have expanded these studies to identify correlates of homotypic (common G and P types) protection in the gnotobiotic piglet model of human rotavirus-induced diarrhea [14, 22, 24]. In these studies, 3- to 5-day-old piglets were orally inoculated with the virulent (stood-passaged) or attenuated (cell culture-passaged) Wa strain of human rotavirus and challenged at PID 21 with the homologous virulent Wa rotavirus. These viruses were selected to mimic natural infection with virulent rotavirus or oral inoculation with a live attenuated candidate rotavirus vaccine. Piglets were examined for clinical signs of illness and for rotavirus shedding (by ELISA [4] and cell culture immunofluorescence assays [1]) after inoculation and challenge and intestinal lesions were evaluated in selected pigs [14, 21, 22, 24; Table 1]. Correlates of protective immunity were determined by ELISPOT (20, 24, B cell responses) and lymphoroliferative assays (2, 22, LPA, T cell responses) using intestinal (gut lamina propria; mesenteric lymph node) and systemic (blood; spleen) lymphoid tissues collected at various PID or post-challenge days (Table 2).

Piglets inoculated with virulent Wa rotavirus developed diarrhea and villous atrophy was evident within 24–72 PI hours [14, 21, 22, 24, Table 1]. All piglets shed virus in feces and seroconverted with neutralizing antibodies to Wa rotavirus. Upon challenge with homologus virulent Wa rotavirus, all piglets were protected from virus shedding and severe to moderate diarrhea. Piglets given attenuated Wa rotavirus developed transient mild or no diarrhea (like controls) and no villous atrophy was evident (Table 1). Fecal shedding was detected in only 6% of the pigs, but 96% of the pigs seroconverted to Wa

Table 2. Peak immune responses to Wa human rotavirus in intestinal lamina propria lymphoid tissues 21 days after primary inoculation or 4 days after challenge with Wa human rotavirus

Virus Inoculation	Primary inoculation (PID 21)[a]				Challenge (PCD 4)[a]			
	Mean ($\pm$SEM)[b] No. ASC[b]/5 $\times$ 10^5 MNC[b]			LPA[b]	Mean ($\pm$SEM)[b] No. ASC[b]/5 $\times$ 10^5 MNC[b]			LPA[b]
	IgG	IgA	IgG/IgA[c]	Mean ($\pm$SEM) CPM[b]	IgG	IgA	IgG/IgA[c]	Mean ($\pm$SEM) CPM[b]
Virulent	64 ($\pm$26)	53*[d] ($\pm$28)	1.2	3.2×10^4*[d] ($\pm 5.6 \times 10^3$)	108 ($\pm$43)	47 ($\pm$19)	2.3	2.7×10^4 ($\pm 4.8 \times 10^3$)
Attenuated	41 ($\pm$26)	6* ($\pm$4)	6.8	1.8×10^4* ($\pm 3.8 \times 10^3$)	270 ($\pm$52)	46 ($\pm$10)	5.9	2.6×10^4 ($\pm 2.6 \times 10^3$)

[a] Gnotobiotic piglets were orally inoculated with virulent or attenuated Wa rotavirus and orally challenged with virulent Wa rotavirus at post-inoculation day (PID) 21; PCD = post-challenge day

[b] SEM = standard error of the mean; ASC = antibody secreting cells; LPA = lymphoproliferative assay; MNC = mononuclear cells; CPM = counts per minute (minus background)

[c] IgG/IgA = ratio of Wa rotavirus-specific IgG ASC to IgA ASC based on mean numbers of ASC per 5 $\times$ 10^5 MNC

[d] "*" denotes significantly different (p < 0.05) numbers of ASC or LPA responses (following rank transformation of mean CPM) between the virulent and attenuated rotavirus-inoculated groups

rotavirus. Piglets were only partially protected from diarrhea (56% with diarrhea) and virus shedding (81% shed virus) after challenge exposure.

Assessment of the immune responses in these pigs revealed that the highest numbers of antibody secreting cells (ASC) (measured by ELISPOT) [14, 24] and LPA responses [22] [measured by virus-stimulated counts per minute (CPM) minus background CPM] were in intestinal tissues (adjacent to the site of rotaviral replication) of both groups of pigs (Table 2). The number of ASC in intestinal tissues was at least 5-fold higher than the number of ASC in systemic tissues (data not shown) before challenge. The number of IgA ASC and the LPA responses (CPM) were significantly higher ($p < 0.05$) at challenge (PID 21) in the intestinal lymphoid tissues of the virulent-Wa rotavirus-inoculated pigs compared to the attenuated Wa rotavirus-inoculated pigs (Table 2). Moreover the mean IgG/IgA ratios were ~ 1 in the virulent Wa rotavirus-inoculated pigs, but were ~ 7 in the attenuated Wa rotavirus- inoculated pigs, reflecting the predominance of IgG ASC in the latter group of pigs. After challenge of the virulent Wa rotavirus-inoculated pigs, only transient low ($\leqslant 2$ fold) or no increases occurred in numbers of IgA and IgG ASC and LPA responses (Table 2) reflecting the limited viral replication and antigenic stimulation which coincided with complete protection. The lower numbers of ASC (particularly IgA ASC) and LPA immune responses seen in the attenuated Wa rotavirus-inoculated pigs at challenge exposure (Table 2, PID 21) correlated with induction of only partial protection against diarrhea and virus shedding after challenge (Table 1). Furthermore, these pigs developed greatly increased ASC numbers (6–7-fold) and LPA responses (2–4-fold) after challenge, consistent with virus infection. Thus it appears that the magnitude of the immune response is greatest in lymphoid tissues adjacent to the site of rotavirus replication and tissue destruction (small intestine) and that the level (and IgA antibody isotype) of the local immune response may correlate with the degree of protection induced.

Acknowledgements

Salaries and research support provided by state and federal funds appropriated to the Ohio Agricultural Research and Development Center, The Ohio State University. Supported in part by Grants R01AI33561 and R01AI37111 from the National Institute of Allergy and Infectious Diseases, National Institutes of Health and Grant GPV/V27/181/24 from the World Health Organization.

References

1. Bohl EH, Theil KW, Saif LJ (1984) Isolation and serotypes of porcine rotaviruses and antigenic comparison with other rotaviruses. J Clin Microbiol 19: 105–111
2. Brim TA, Van Cott JL, Lunney JK, Saif LJ (1994) Lymphocyte proliferative responses of pigs inoculated with transmissible gastroenteritis virus or porcine respiratory coronavirus. Am J Vet Res 55: 494–501
3. Conner ME, Gilger MA, Estes MK, Graham DY (1991) Serologic and mucosal immune response to rotavirus infection in the rabbit model. J Virol 65: 2562–2571

4. Hoblet KH, Saif LJ, Kohler EM, Theil KW, Bech-Nielsen S, Stitzlein GA (1986) Efficacy of an orally administered modified-live porcine-origin rotavirus vaccine against post-weaning diarrhea in pigs. Am J Vet Res 47: 1697–1703

5. Hoshino Y, Kapikian AZ (1994) Rotavirus vaccine development for the prevention of severe diarrhea in infants and young children. Trends Microbiol 2: 242–249

6. Hoshino T, Saif LJ, Kang SY, Sereno M, Chen WK, Kapikian AZ (1995) Identification of group A rotavirus genes associated with virulence of a porcine rotavirus and host range restriction of a human rotavirus in the gnotobiotic piglet model. Virology 209: 274–280

7. Hoshino Y, Saif LJ, Sereno MM, Chanock RM, Kapikian AZ (1988) Infection immunity of piglets to either VP3 or VP7 outer capsid protein confers resistance to challenge with a virulent rotavirus bearing the corresponding antigen. J Virol 62: 744–748

8. Kapikian AZ, Chanock RM (1990) Rotaviruses. BN Fields, Knipe DM, Chanock RM, Hirsch MS, Melnick JL, Monath TP, Roizman B, Straus SE (eds) Virology, Raven Press, New York, pp 1353–1403

9. Kim YB (1975) Developmental immunity in the piglet. Birth Defects 11: 549

10. Mebus CA, Wyatt RG, Sharpee RL, Sereno MM, Kalica AR, Kapikian AZ, Twiehaus MJ (1976) Diarrhea in gnotobiotic calves caused by the reovirus-like agent of human infantile gastroenteritis. Infect Immun 14: 471–474

11. Middleton PJ, Petric M, Szymanski MT (1975) Propagation of infantile gastroenteritis virus (orbigroup) in conventional and germfree piglets. Infect Immun 12: 1 276–1 280

12. Phillips RW, Tumbleson ME (1986) Models. In: Tumbleson ME (ed) Swine in Biomedical Research. Plenum Press, New York, pp 437–440

13. Ramig F (1988) The effects of host age, virus dose, and virus strain on heterologous rotavirus infection of suckling mice. Microb Pathog 4: 189–202

14. Saif LJ, Ward L, Yuan L, To TL (1996) Studies of the pathogenesis and immunity to a human rotavirus in a gnotobiotic pig model of enteric disease. Proc First Intl Rushmore Conf on Mechanisms in the Pathogenesis of Enteric Diseases, Rapid City, SD, September 27–30, 1995. Plenum Press, New York

15. Schaller JP, Saif LJ, Cordle CT, Candler E, Winship TR, Smith KL (1992) Prevention of human rotavirus induced diarrhea in gnotobiotic piglets using bovine antibody. J Inf Dis 165: 623–630

16. Steel R, Torres-Medina A (1984) Effects of environmental and dietary factors on human rotavirus infection in gnotobiotic piglets. Infect Immun 43: 906–911

17. Theil KW, Bohl E, Cross R, Kohler E, Agnes A (1978) Pathogenesis of porcine rotaviral infection in experimentally inoculated gnotobiotic pigs. Am J Vet Res 39: 213–220

18. Torres-Medina A, Wyatt RG, Mebus CA, Underdahl NR, Kapikian AZ (1976) Diarrhea in gnotobiotic piglets caused by the reovirus-like agent of human infantile gastro-enteritis. J Infect Dis 133: 22–27

19. Urasawa S, Hasegawa A, Urasawa T, Taniguchi K, Wakasugi F, Suzuki H, Inouye S, Pongprot B, Supawadee J, Suprasert S, Rangsiyanond, Tonusin S, Yamazi Y (1992) Antigenic and genetic analysis of human rotaviruses in Chiang Mai Thailand: evidence for a close relationship between human and animal rotaviruses. J Inf Dis 166: 227–234

20. Van Cott J, Brim T, Lunney J, Saif LJ (1994) Contribution of antibody secreting cells induced in mucosal lymphoid tissues of pigs inoculated with respiratory or enteric strains of coronavirus to immunity against enteric coronavirus challenge. J Immunol 152: 3980–3990

21. Ward LA, Rosen BI, Yuan L, Saif LJ (1996) Pathogenesis of an attenuated and a virulent strain of group A human rotavirus in neonatal gnotobiotic pigs. J Gen Virol 77: 1431–1441

22. Ward LA, Yuan L, Rosen BI, Saif LJ (1996) Local and systemic T cell immunity to human group A rotavirus in neonatal gnotobiotic pigs. Clin Diag Lab Immunol 3: 342–350
23. Wyatt RG, James WD, Bohl EH, Theil KW, Saif LJ, Kalica AR, Greenberg HB, Kapikian AZ, Chanock RM (1980) Human rotavirus type 2: cultivation *in vitro*. Science 207: 189–191
24. Yuan L, Ward LA, Rosen BI, To TL, Saif LJ (1996) Evaluation of systemic and intestinal antibody-secreting cell responses to human rotavirus in a gnotobiotic piglet model of disease. J Virol 70: 3075–3083

Authors' address: Dr. L. J. Saif, Ohio Agricultural Research and Development Center, The Ohio State University, Wooster, Ohio 44691, U.S.A.

Arch Virol (1996) [Suppl] 12: 163–175

Jennerian and modified Jennerian approach to vaccination against rotavirus diarrhea using a quadrivalent rhesus rotavirus (RRV) and human-RRV reassortant vaccine

A. Z. Kapikian[1], Y. Hoshino[1], R. M. Chanock[1], and I. Perez-Schael[2]

[1] Laboratory of Infectious Diseases, National Institute of Allergy and Infectious Diseases, National Institutes of Health, Bethesda, MD
[2] Central University of Venezuela, Caracas, Venezuela

Summary. Rotaviruses are the single most important cause of severe diarrhea of infants and young children world-wide. Deaths from rotavirus diarrhea occur infrequently in developed countries; however, in developing countries, rotaviruses are estimated to cause over 870 000 deaths in the under five-year age group. There is, therefore, a vital need for a vaccine to prevent severe rotavirus diarrhea in infants and young children. The most extensively evaluated strategy for rotavirus vaccination has been the "Jennerian" approach in which an antigenically related rotavirus strain from an animal host (bovine or simian [rhesus monkey]) is used as the immunogen to induce protection against the four epidemiologically important group A human rotavirus serotypes. These orally administered vaccines were safe and immunogenic but had only limited success because serotype-specific immunity was not induced consistently in the under six-month age group. Therefore, a modified "Jennerian" approach was adopted with the goal of attaining broader antigenic coverage. In this approach, four serotypes are combined to form a quadrivalent vaccine comprised of: (i) rhesus rotavirus (RRV) which provides coverage for VP7 serotype 3, and (ii) three human-RRV reassortants each with ten RRV genes and a single human rotavirus gene that encodes VP7 serotype 1, 2, or 4 specificity. This modified "Jennerian" approach appears to be quite promising in preventing severe diarrhea in field trials. However, if this approach fails to yield an optimal level of protection consistently, additional modified "Jennerian" strategies are under development that consider not only human rotavirus VP7 but also human rotavirus VP4, the other outer capsid protein. In addition, a non- "Jennerian" approach includes the development of cold-adapted human rotavirus strains or cold-adapted human rotavirus reassortants as vaccine candidates.

Introduction

The need for an effective rotavirus vaccine has received firm international endorsement [28]. The importance of rotavirus is highlighted by studies showing that they are the single most important etiological agents of severe diarrheal illnesses in infants and young children in the developed and developing countries, being responsible for approximately 35–50% of such illnesses [13]. However, the consequences of such illness are radically different in developing countries, where rotaviruses cause more than 870 000 deaths in the less than five year age group, annually, whereas in a developed country, such as the U.S., ∼150 children die from this affliction [5, 10]. Therefore, the goal of a rotavirus vaccine is the prevention of severe diarrhea during the first two years of life, when such illnesses are most debilitating and may lead to death.

Before considering vaccination strategies, it is important to summarize a few virological features of rotaviruses that are relevant for vaccine development. Rotaviruses are 70 nm in diameter, icosahedral, non-enveloped, and possess a distinctive double-shelled outer capsid (Fig. 1) [13]. Within the inner shell is the core (constituting a third layer) which encloses the genome, comprised of 11 segments of double-stranded RNA [19]. During coinfection, the segmented genome readily undergoes genetic reassortment.

Rotaviruses have three important antigenic specificities: group, subgroup and serotype, which are mediated by various proteins (Fig. 1) [6, 13, 19]. Group specificity is determined predominantly by VP6 (encoded by gene 6); seven distinct groups are recognized and are designated by the letters A-G. Groups A, B and C have been recovered from both humans and various animal species. However, most human rotaviruses of clinical relevance belong to group A and for this reason, vaccine development has been limited exclusively to this group. Subgroup specificity is mediated by VP6, with most human strains belonging to subgroup I or II. Serotype specificity is defined by two proteins: VP7, which is one of the two major neutralization antigens located on the outer capsid (encoded by gene 7, 8 or 9 depending on the strain); and by VP4 (encoded by gene 4), the other outer capsid protein, which also induces neutralizing antibodies and which protrudes from the outer surface in the form of 60 spikes of approximately 10–12 nm length [23]. Although the immune mechanisms of protection are not clearly established, antibodies to both VP4 and VP7 are independently associated with protection against rotavirus illness in studies in animal models [6]. Until recently, serotypes were defined exclusively according to VP7 or "G" specificity. However, a serotyping scheme which considers VP4 or "P" specificity is also under development, with a proposed numbering system based on neutralization [6]. Fourteen group A rotavirus G serotypes are recognized: ten in humans and thirteen in animals with nine of the fourteen having been recovered from both humans and animals. Only four of the ten human rotavirus "G" serotypes (numbered 1, 2, 3, or 4) are of epidemiological importance, making vaccine development a manageable goal.

However, there are certain other, more formidable, obstacles to successful immunization against severe rotavirus diarrhea: (i) infants should be immunized shortly after birth because many infants in developing countries are seen by a healthcare provider only during the newborn period; (ii) infants are born with high levels of maternally derived rotavirus neutralizing antibodies that may

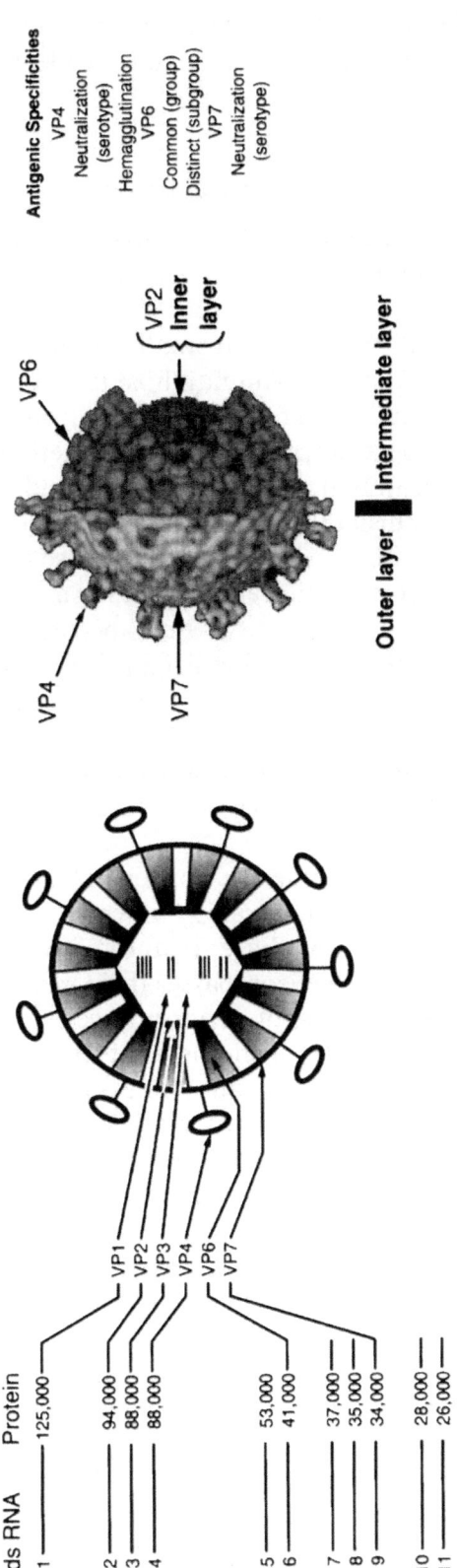

Fig. 1. Rotavirus gene products. *Left*: Schematic representation of the rotavirus particle. *Right*: Surface representations of the three-dimensional structures of the outer layer of the complete particle (left) and a particle (right) in which the outer layer and a small triangular portion of the intermediate layer have been removed exposing the inner layer. Modified from Kapikian and Chanock, Ref [13]. The three-dimensional figure on the right is by courtesy of B.V.V. Prasad

interfere with the immunological response to the vaccine; and (iii) immunological immaturity during the neonatal period may also adversely affect the response to the vaccine [11, 13].

Jennerian approach to vaccination

Approaches to rotavirus vaccine development range from conventional growth in cell culture of human or animal rotavirus strains to the use of molecular biological techniques. The most extensively studied strategy is the concept pioneered by Edward Jenner in 1798 for human smallpox vaccination in which a related, live, attenuated agent from a non-human host is used as the immunogen [11, 15, 16]. Early serologic studies were instrumental in suggesting that the Jennerian approach might provide a simple and practical method of immunoprophylaxis: (i) human and certain animal rotaviruses were found to share a group antigen that made them indistinguishable in various serological assays (such as complement fixation) using hyperimmune sera raised to these strains in laboratory animals; and (ii) infants and young children undergoing a rotavirus infection developed a serologic response not only to the human rotavirus but also to animal rotaviruses of bovine, simian or murine origin [14]. Furthermore, the feasibility of the Jennerian approach, in which an animal rotavirus is used as a substitute antigen for the human virus, was demonstrated when gnotobiotic calves that were inoculated *in utero* with a bovine rotavirus (NCDV) failed to develop diarrhea following homotypic (NCDV) or heterologous challenge with a human rotavirus at birth [30].

Because of these observations, the Jennerian approach became the keystone of our vaccine studies, replacing a previous strategy in which a human rotavirus strain Wa (G1) had been evaluated in phase 1 studies in adult volunteers [11, 15–17, 29]. The Jennerian approach encompasses three important principles: (i) the antigenically related animal virus should not induce significant illness in humans; (ii) infection with the animal strain should protect against the virulent human virus; and (iii) the vaccine should not cause illness in contacts. Although it had been demonstrated that the bovine virus NCDV could induce heterologous protection in the gnotobiotic calf model, we did not pursue studies in humans with this virus because of a lack of clarity regarding its cell culture passage history; however, we undertook limited phase 1 studies in adult volunteers with bovine rotavirus strain UK (which shares VP7 but not VP4 specificity with NCDV [6]) but which had a proven cell culture pedigree [11, 15, 28].

We therefore pursued the Jennerian approach with a different animal rotavirus, rhesus rotavirus MMU18006, which shares neutralization specificity with a reference strain belonging to human rotavirus VP7 serotype 3 [6, 12, 15, 28]. The RRV strain grew efficiently in primary simian cell culture and was adapted to replicate in FRhL2 cells, a semicontinuous simian diploid cell strain developed by the forerunner of the Food and Drug Administration as a potential cell substrate for vaccine production [27].

The RRV vaccine was evaluated for safety and antigenicity in stepwise fashion in adults, older children, younger children, older infants, and finally in the target population of infants less than six months of age [15, 16]. At a 10^5 pfu dose it induced an unacceptably high rate of reactions in infants who were predominantly six months of age and older whereas in two to five months old infants (the target population) the vaccine proved to be safe and antigenic at a 10^4 pfu dose, although, in some studies, it induced a mild, transient febrile response on day three or four after vaccination in about one-third of vaccinees in this age group [12, 15, 16]. The decrease in the rate of vaccine reactions appeared to be a function of passively acquired maternal antibodies in the less than six month age group rather than the lower titer of the vaccine.

The RRV vaccine or placebo was administered orally in a single 10^4 or 10^5 pfu dose to more than 1 500 infants and young children 1 to 20 months of age in double-masked placebo-controlled field trials in the USA, Sweden, Finland, and Venezuela. The protective efficacy of this vaccine was variable, ranging from nil to 64% against any rotavirus diarrhea and nil to 85% against clinically significant rotavirus diarrhea [12]. Because the RRV vaccine achieved its greatest degree of efficacy in the Venezuelan study in which the homotypic serotype 3 strain was predominant, it was proposed that the inconsistency in protective efficacy resulted from the failure of the vaccine to protect against rotavirus diarrhea caused by strains that differed in serotype from that of the vaccine strain [22]. In addition, it was considered that in the vaccine failures, the target population had not experienced a prior rotavirus infection and thus had responded only homotypically rather than both homotypically and heterotypically to the RRV vaccine [16, 22].

Modified Jennerian approach to vaccination

We therefore altered our strategy and developed a modified Jennerian approach in order to achieve broader antigenic coverage [15, 16]. Single gene substitution reassortant viruses had been generated that contained: (i) the human rotavirus gene that encodes the outer capsid protein VP7 which confers antigenic specificity of serotype 1, 2 or 4, and (ii) the remaining 10 genes from RRV (Fig. 2) [20, 21]. The RRV itself provided coverage for VP7 serotype 3. It was anticipated that these four strains would be combined into a quadrivalent vaccine which covered the antigenic specificity of each of the four epidemiologically important serotypes.

Phase I studies with individual reassortant vaccine strains representing VP7 serotype 1 (D × RRV), 2 (DS-1 × RRV), or 4 (ST-3 × RRV) indicated that each resembled the RRV vaccine with regard to attenuation and immunogenicity, thus paving the way for efficacy trials with individual reassortants and ultimately with the quadrivalent formulation [16]. Protective efficacy of individual reassortants with VP7 serotype 1 or 2 specificity was promising in Tampere, Finland or in Rochester, NY, U.S.A. [18, 25]. Unexpectedly, heterotypic protection was observed in the USA study in which RRV vaccine was evaluated along with

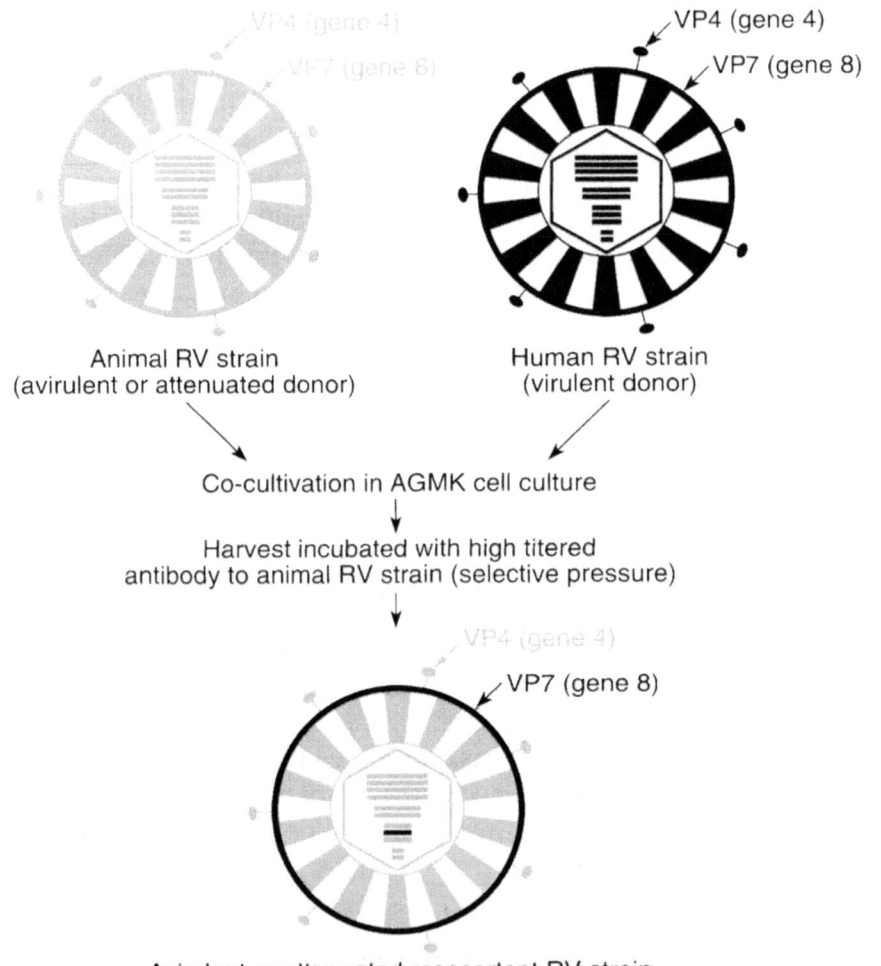

Fig. 2. Production of reassortant rotavirus (RV) vaccine (adapted from Kapikian *et al.* Ref. [15])

a serotype 1 reassortant vaccine, and in the Finnish study in which two reassortant vaccines (types 1 and 2) were under study. Priming by prior exposure was a distinct possibility in the USA study but unlikely in Finland.

Next, the three reassortants were combined together with RRV into a single quadrivalent vaccine, which was evaluated in phase I studies for reactogenicity and antigenicity. The quadrivalent vaccine was similar to the RRV parental strain or individual reassortants with regard to attenuation [3, 4]. However, as measured by serum neutralization assay, the rotavirus antibody response ("take rate") induced by the quadrivalent vaccine for each of the four VP7 serotypes was disappointing. Therefore, after various adjustments, a two-dose regimen of 10^5 PFU of each serotype in the quadrivalent vaccine (*i.e.*, a total of 4×10^5 PFU) was found to be acceptable because it achieved a "take rate" approaching or exceeding 50% for each of the four serotypes [4, 16].

Efficacy trials with quadrivalent rotavirus vaccine

The results of two recently completed studies in the USA indicate that the quadrivalent formulation provided a high level of protective efficacy against severe rotavirus diarrheal disease. A multi-center (23 sites), three cell, prospective, double-masked, placebo-controlled efficacy trial covering a period of two rotavirus seasons was performed [1]. In this clinical trial, 898 infants and young children received three doses of 4×10^4 PFU of quadrivalent vaccine (1×10^4 PFU of each component) or three doses of 4×10^4 PFU of D $\times$ RRV reassortant vaccine (VP7 serotype 1) or a placebo at approximately two, four and six months of age. Reactions were mild and limited after vaccination. Following the first dose, infants who received the quadrivalent vaccine developed a significantly greater number of febrile episodes ($> 38\,^{\circ}\mathrm{C}$ [rectal]) than did the placebo group on day 4 ($\sim 7\%$ vs. $\sim 3\%$) and day 5 ($\sim 5\%$ vs. $\sim 1\%$). Significant differences between the quadrivalent vaccine and control groups in the occurrence of diarrhea or vomiting were not observed during any of the five days following the first dose. With regard to antigenicity, 93% of the quadrivalent vaccine recipients and 16% of the placebo recipients tested, developed a four-fold or greater serum antibody response to rotavirus by ELISA or neutralization. The neutralizing antibody response to human rotavirus serotypes 1–4 ranged from 29% vs. serotype 1 to 43% to serotype 4. Neutralizing antibody rises reached 88% against RRV reflecting a vigorous response to its VP4 component [1].

The efficacy against rotavirus diarrhea of any severity over the two "season" period was 40% for the monovalent vaccine and 57% for the quadrivalent vaccine (Table 1 [for quadrivalent vaccine]). However, the efficacy against very severe gastroenteritis (a severity score of 15–20 with 20 being the maximal score) reached 73% and 82% for the monovalent and quadrivalent vaccines, respectively. The efficacy of the monovalent vaccine against diarrheal illness of greater than three days duration was 36%, whereas that for the quadrivalent vaccine was 92%. In addition, there was a reduction in medical visits for rotavirus diarrhea of 67% and 78% for the monovalent and quadrivalent vaccine recipients, respectively. In view of our strategy aimed at inducing broader antigenic coverage with the modified Jennerian approach, it was of particular interest that although both vaccines protected against serotype 1 rotavirus diarrhea (the predominant serotype), only the quadrivalent vaccine was associated with protection against non-serotype 1 rotavirus diarrhea during the second season. However, it was not clear whether this was due to serotype-specific protection or to the duration of protection [1].

These encouraging observations were confirmed and expanded during a second prospective, double-masked placebo-controlled efficacy study at 24 centers in the USA [24]. Three doses of a preparation of the quadrivalent vaccine containing ten-fold more virus (1×10^5 pfu of each component), or ten-fold more

Table 1. Protective efficacy of quadrivalent rotavirus (RV) vaccine (4×10^4 PFU) against the occurrence of RV diarrhea of varying severity over two RV seasons in U.S.A.*

Parameter	Number with Indicated Parameter who Received		Protective Efficacy
	Quadrivalent Vaccine** (N = 305)	Placebo (N = 296)	
RV Diarrhea**	29 (10%)	65 (22%)	57%[a]
Severity Score of			
1–8	14 (5%)	24 (9%)	49%[b]
9–14	11 (4%)	25 (9%)	59%[c]
15–20	2 (1%)	11 (4%)	82%[d]
RV Diarrhea > 3 Days Duration	2 (1%)	25 (9%)	92%[e]
Medical Visits	6 (2%)	27 (9%)	78%[f]

*Adapted from Bernstein *et al*, Reference [1]

**Severity score not available from two vaccinees and five controls. Individuals with score >8 excluded from 1–8 efficacy analysis and those with score >14 from 9–14 efficacy analysis

[a] $P < 0.0001$ (Fisher Exact Test), 95% Cl 35, 71%

[b] $P < 0.05$, Cl 4, 73%

[c] $P < 0.001$, Cl 20, 79%

[d] $P < 0.02$, Cl 31, 96%

[e] $P < 0.001$, Cl 71, 98%

[f] $P < 0.001$, Cl 50, 91%

D × RRV (VP7:1) reassortant vaccine (4×10^5 pfu) or a placebo were administered orally to 1 278 infants at approximately two, four and six months of age. Surveillance of 1 187 children was maintained for one rotavirus "season". Once again, reactions were mild and limited after vaccination. Following the first dose, infants who received the quadrivalent vaccine developed a significantly greater number of febrile episodes (> 38 °C axillary) than did placebo recipients on day four following vaccination (2.2% vs. 0.2%). The lower number of febrile reactions in this study probably reflected the measurement of axillary temperature. With regard to antigenicity, 92% of the quadrivalent vaccine recipients and 4% of the placebo recipients tested developed a four-fold or greater serum antibody response to rotavirus by ELISA or neutralization. The neutralizing antibody responses ranged from 14% to serotype 1 to 31% to serotype 4. Neutralizing antibody responses reached 90% to RRV, once again reflecting the efficient response to the VP4 component of this strain.

Each of the vaccines induced significant resistance to the occurrence of rotavirus diarrhea of any severity; protective efficacy was 54% and 49% for the monovalent and quadrivalent vaccines, respectively (Table 2 [for quadrivalent vaccine]). However, protective efficacy increased further when measured accord-

Table 2. Protective efficacy of quadrivalent rotavirus (RV) vaccine (4×10^5 PFU) against the occurrence of RV diarrhea of varying severity over one RV season in U.S.A.*

Parameter	Number with Indicated Parameter who Received		Protective Efficacy
	Quadrivalent Vaccine (N = 398)	Placebo (N = 385)	
RV Diarrhea	51 (13%)	97 (25%)	49%[a]
Severity Score of RV Diarrhea			
>8	24 (6%)	72 (19%)	68%[b]
>14	7 (2%)	34 (9%)	80%[c]
Medical Intervention	16 (4%)	58 (15%)	73%[d]
Dehydration	0	13 (3%)	100%[e]

*Adapted from Rennels et al, Reference [24]
[a]P < 0.0001 (Fisher Exact Test) 95% Cl 31, 63%
[b]P < 0.0001, Cl 50, 79%
[c]P < 0.0001, Cl 56, 91%
[d]P < 0.0001, Cl 54, 84%
[e]P < 0.001, Cl 73, 100%

ing to various clinical parameters: (i) the quadrivalent vaccine achieved a protective efficacy of 80% and the monovalent vaccine, 69% against severe rotavirus diarrhea (i.e., severity score > 14 points); (ii) the monovalent and quadrivalent vaccines reduced the need for physician intervention by 67% and 73%, respectively; and (iii) most striking was the reduction in the number of dehydrating diarrheal episodes as thirteen were observed in the placebo group, only two in the monovalent vaccine group, and none in the recipients of the quadrivalent vaccine. Both vaccines provided comparable protection against serotype 1 rotavirus diarrhea, the most frequently detected serotype (77%), with 44% efficacy for the quadrivalent vaccine and 55% for monovalent vaccine. However, with regard to protection against serotype 3 rotavirus diarrhea, which was the second most frequently detected serotype (19%), the quadrivalent vaccine achieved a protective efficacy of 77% and the monovalent vaccine 45%, a trend that was not statistically significant (P = 0.14) when the comparison was made between vaccine groups. However, this level of serotype-specific protection was statistically significant for the quadrivalent, but not the monovalent vaccine when the comparison was made with the placebo group (P < 0.01) [24].

Other modified Jennerian approaches to vaccination

It should be noted that reassortant rotaviruses containing the VP7 gene of serotype 1, 2, 3, or 4 and 10 genes from the bovine UK strain have been generated

[20, 21] and evaluated in phase 1 studies at Johns Hopkins University or Vanderbilt University (Makhene *et al.* [Abstract SV-39 Fifth Rotavirus Vaccine Workshop: Current Issues and Future Developments, Atlanta, GA.] 1995). It is planned to combine these strains into a quadrivalent vaccine for phase 1 and phase 2 studies. In this regard, other reassortant strains with: (i) the VP4 of the Wa virus (P1A:G1) and 10 genes from the bovine UK strain, and (ii) VP4 of DS-1 virus (P1B:G2) and 10 genes from UK are available for formulation of a pentavalent or hexavalent vaccine (in combination with the quadrivalent bovine rotavirus-based reassortant vaccine) if this should prove necessary [6]. Similar but not identical approaches for the quadrivalent RRV-based vaccine described above are also available and could be implemented if a pentavalent or hexavalent vaccine with VP4 from human rotavirus Wa (P1A) or DS-1 (P1B) is required for optimal protection with the RRV-based formulation (Y. Hoshino, unpubl. studies).

In a further modification of the Jennerian approach, reassortant rotaviruses have been generated that possess human rotavirus P1A and G serotype 1, 2, 3, or 4 specificities and the remaining genes from bovine rotavirus UK [6, 8]. These strains, which have both VP4 and VP7 from human rotavirus, may display the attenuation phenotype because it was shown recently in the gnotobiotic piglet model that gene 3 (VP3), or 4 (VP4), or 9 (VP7), or 10 (NS28, NSP4) each independently play a role in viral virulence [8]; elimination of any one of these genes led to loss of virulence.

Non-Jennerian approach to rotavirus vaccination

Two additional rotavirus vaccination approaches, both of them non-Jennerian, deserve special attention because they have been evaluated in phase 1 or phase 2 studies in humans. The first involved the use of a VP7 serotype 1 rotavirus strain, M37, that was recovered from the stool of an asymptomatic neonate in Venezuela [3]. The impetus for using this strain came from the observation in another study that neonates who developed subclinical rotavirus infection during the first 14 days of life were protected against severe rotavirus diarrhea (but not infection) for up to three years [2]. However, during a preliminary study in Finland, this vaccine failed to induce protection against rotavirus diarrhea [26].

The second non-Jennerian approach involves the development of cold-adapted human rotavirus strains with VP7 serotype 1, 2, 3, or 4 specificity or cold-adapted human rotavirus × human rotavirus reassortant strains [7, Y. Hoshino, unpubl. studies]. An important attribute of these strains is that they possess the VP4 and VP7 of human rotavirus. Phase 1 safety studies of a 30 °C cold-adapted vaccine strain have been initiated in adults at the University of Rochester with plans to proceed stepwise in children and infants if the vaccine proves to be safe and antigenic. (Treanor *et al.* [Abstract SV-40 Fifth Rotavirus Vaccine Workshop. Current Issues and Future Developments, Atlanta, GA 1995.]).

Conclusion

As a result of the successful field trials with the quadrivalent vaccine in the USA, the prospects for the availability of an effective rotavirus vaccine for general pediatric use in the not-too-distant future appear to be quite promising. The need for second generation vaccines will depend on the acceptability and overall efficacy of the current formulation.

It is clear from this short summary that the development of a rotavirus vaccine has required an immense collaborative effort involving numerous investigators at NIH and at many other institutions. This becomes especially clear from perusal of the numerous collaborators participating in the non-review publications cited in the references. It would have been impossible to reach the current stage of vaccine development without the superb collaborative efforts of so many colleagues and their respective institutions.

References

1. Bernstein DI, Glass RI, Rogers G, Davidson BL, Sack DA, for the U.S. Rotavirus Vaccine Efficacy Group (1995) Evaluation of rhesus rotavirus monovalent and tetravalent reassortant vaccines in US children. JAMA 273: 1191–1196
2. Bishop RF, Barnes GL, Cipriani E, Lund JS (1983) Clinical immunity after neonatal rotavirus infection. A prospective longitudinal study in young children. N Engl J Med 309: 72–76
3. Flores J, Perez-Schael I, Blanco M, White L, Garcia D, Vilar M, Cunto W, Gonzalez R, Urbina C, Boher J, Mendez M, Kapikian AZ (1990) Comparison of reactogenicity and antigenicity of M37 rotavirus vaccine and rhesus-rotavirus-based quadrivalent vaccine. Lancet 336: 330–334
4. Flores J, Perez-Schael I, Blanco M, Rojas AM, Alfonzo E, Crespo I, Cunto W, Pittman A, Kapikian AZ (1993) Reactogenicity and immunogenicity of a high-titered rhesus rotavirus-based quadrivalent rotavirus vaccine. J Clin Microbiol 31: 2439–2445
5. Ho M-S, Glass RI, Pinsky PF, Young-Okoh N, Sappenfield WM, Buehler JW, Gunter N, Anderson LJ (1988) Diarrheal deaths in American children. Are they preventable? JAMA 260: 3281–3285
6. Hoshino Y, Kapikian AZ (1994) Rotavirus vaccine development for the prevention of severe diarrhea in infants and young children. Trends Microbiol 2: 242–249
7. Hoshino Y, Kapikian AZ, Chanock RM (1994) Selection of cold-adapted mutants of human rotaviruses that exhibit various degrees of growth restriction in vitro. J Virol 68: 7598–7602
8. Hoshino Y, Saif L, Kang S-Y, Sereno MM, Chen W-K, Kapikian AZ (1995) Identification of group A rotavirus genes associated with virulence of a porcine rotavirus and host range restriction of a human rotavirus in the gnotobiotic piglet model. Virology 209: 274–280
9. Institute of Medicine (1986) Prospects for immunizing against rotavirus. In: New Vaccine Development. Establishing Priorities. Diseases of Importance in the United States, vol 1. National Academy Press, Washington, DC, pp 410–423
10. Institute of Medicine (1986) The prospects for immunizing against rotavirus. In: New Vaccine Development. Establishing Priorities. Diseases of Importance in Developing Countries, vol 2. National Academy Press, Washington, DC, pp 308–316

11. Kapikian AZ (1994) Jennerian and modified Jennerian approach to vaccination against rotavirus diarrhea in infants and young children: an introduction. In: Kapikian AZ (ed) Viral Infections of the Gastrointestinal Tract. Marcel Dekker, New York, pp 409–417

12. Kapikian AZ (1994) Rhesus rotavirus-based human rotavirus vaccines and observations on selected non-Jennerian approaches to rotavirus vaccination. In: Kapikian AZ (ed) Viral Infections of the Gastrointestinal Tract. Marcel Dekker, New York, pp 443–470

13. Kapikian AZ, Chanock RM (1996) Rotaviruses. In: Fields BN, Knipe DM, Howley PM, Chanock RM, Melnick JL, Monath TP, Roizman B, Straus SE (eds) Virology. Lippincott-Raven Publishers, Philadelphia, pp 1657–1708

14. Kapikian AZ, Cline WL, Kim HW, Kalica AR, Wyatt RG, VanKirk DH, Chanock RM, James HD Jr, Vaughn AL (1976) Antigenic relationships among five reovirus-like (RVL) agents by complement-fixation (CF) and development of new substitute CF antigens for the human RVL agent of infantile gastroenteritis. Proc Soc Exp Biol Med 152: 535–539

15. Kapikian AZ, Hoshino Y, Flores J, Midthun K, Glass RI, Nakagomi O, Nakagomi T, Chanock RM, Potash L, Levine MM, Dolin R, Wright PF, Belshe RE, Anderson EL, Vesikari T, Gothefors L, Wadell G, Perez-Schael I (1986) In: Holmgren J, Lindberg A, Mollby R (eds) Development of Vaccines and Drugs Against Diarrhea. 11th Nobel Conference, Stockholm 1985. Student litteratur, Lund, Sweden, pp 192–214

16. Kapikian AZ, Vesikari T, Ruuska T, Madore HP, Christy C, Dolin R, Flores J, Green KY, Davidson BL, Gorziglia M, Hoshino Y, Chanock RM, Midthun K, Perez-Schael I (1992) An update on the "Jennerian" and modified "Jennerian" approach to vaccination of infants and young children against rotavirus diarrhea. Adv Exp Med Biol 327: 59–69

17. Kapikian AZ, Wyatt RG, Levine MM, Black RE, Greenberg HB, Flores J, Kalica AR, Hoshino Y, Chanock RM (1983) Studies in volunteers with human rotaviruses. Dev Biol Stand 53: 209–218

18. Madore HP, Christy C, Pichichero M, Long C, Pincus P, Vosefsky P, Kapikian AZ, Dolin R, Elmwood, Panorama, and Westfall Pediatric Groups (1992) Field trial of rhesus rotavirus or human-rhesus reassortant vaccine of VP7 serotype 3 or 1 specificity in infants. J Infect Dis 166: 235–243

19. Mattion NM, Cohen J, Estes MK (1994) The rotavirus proteins. In: Kapikian AZ (ed) Viral Infections of the Gastrointestinal Tract. Marcel Dekker, New York, pp 169–249

20. Midthun K, Greenberg HB, Hoshino Y, Kapikian AZ, Wyatt RG, Chanock RM (1985) Reassortant rotaviruses as potential live rotavirus vaccine candidates. J Virol 53: 949–954

21. Midthun K, Hoshino Y, Kapikian AZ, Chanock RM (1986) Single gene substitution rotavirus reassortants containing the major neutralization protein (VP7) of human rotavirus serotype 4. J Clin Microbiol 24: 822–826

22. Perez-Schael I, Garcia D, Gonzalez M, Gonzalez R, Daoud N, Perez M, Cunto W, Kapikian AZ, Flores J (1990) Prospective study of diarrheal diseases in Venezuelan children to evaluate the efficacy of rhesus rotavirus vaccine. J Med Virol 30: 219–229

23. Prasad BV, Burns JW, Marietta E, Estes MK, Chiu W (1990) Localization of VP4 neutralization sites in rotavirus by three-dimensional cryo-electron microscopy. Nature 343: 476–479

24. Rennels MB, Glass RI, Dennehy PH, Bernstein DI, Pichichero ME, Zito ET, Mack ME, Davidson BL, Kapikian AZ, for the United States Rotavirus Efficacy Group (1996) Safety and efficacy of high-dose rhesus-human reassortant rotavirus vaccines-report of the national multicenter trial. Pediatrics 97: 7–13

25. Vesikari T, Ruuska T, Green KY, Flores J, Kapikian AZ (1992) Protective efficacy against serotype 1 rotavirus diarrhea by live oral rhesus-human reassortant rotavirus

vaccines with human rotavirus VP7 serotype 1 or 2 specificity. Pediatr Inf Dis J 11: 535–542

26. Vesikari T, Ruuska T, Kouvu HP, Green KY, Flores J, Kapikian AZ (1991) Evaluation of the M37 human rotavirus vaccine in 2- to 6-month old infants. Pediatr Inf Dis J 10: 912–917

27. Wallace RE, Vasington PJ, Petricciani JC, Hopps HE, Lorenz DE, Kadanka Z (1973) Development of a diploid cell line from fetal rhesus monkey lung for virus vaccine production. In Vitro 8: 323–332

28. World Health Organization Scientific Working Group (1980) Rotavirus and other viral diarrheas. Bull WHO 58: 183–198

29. Wyatt RG, Kapikian AZ, Hoshino Y, Flores J, Midthun K, Greenberg HB, Glass RI, Askaa J, Levine MM, Black RE, Clements ML, Potash L, London WT (1985) In Control and Eradication of Infectious Diseases. An International Symposium PAHO copublication series no. 1. Washington, DC, Pan American Health Organization, pp 17–28

30. Wyatt RG, Mebus CA, Yolken RH, Kalica AR, James HD Jr, Kapikian AZ, Chanock RM (1979) Rotaviral immunity in gnotobiotic calves: Heterologous resistance to human virus induced by bovine virus. Science 203: 548–550

Authors' address: Dr. A. Z. Kapikian, Laboratory of Infectious Diseases, National Institute of Allergy and Infectious Diseases, National Institutes of Health, Bethesda, MD 20892, U.S.A.

Arch Virol (1996) [Suppl] 12: 177–186

Trials of oral bovine and rhesus rotavirus vaccines in Finland: a historical account and present status

T. Vesikari

University of Tampere, Medical School, Tampere, Finland

Summary. Live oral rotavirus vaccine strain RIT 4237, derived from group A bovine rotavirus NCDV, was given to human volunteers in Tampere, Finland in 1982. Efficacy studies of this vaccine in 6–12 month-old children gave results characteristic of the performance of oral rotavirus vaccines in general: 58% protective efficacy against any rotavirus gastroenteritis and 82% against "clinically significant" gastroenteritis. Four trials of RIT 4237 bovine rotavirus vaccine, one trial of group A RRV-1 rhesus rotavirus vaccine, and one trial of rhesus-human reassortant rotavirus vaccines D × RRV and DS1 × RRV were carried out between 1983–1989. A meta-analysis of the protective efficacy of these vaccines indicated a 67% (95% C.I. 55–77%) efficacy against moderately severe rotavirus disease and an 81% (95% C.I. 60–91%) efficacy against severe rotavirus disease. There was no apparent difference between bovine and rhesus-based rotavirus vaccines in the protective efficacy against severe rotavirus gastroenteritis. Problems associated with the use of any oral rotavirus vaccine include acid lability of the vaccine virus, which requires buffering, and a slight but significant interference of oral poliovirus vaccine with the uptake of rotavirus vaccine. In the near future, oral heterologous rotavirus vaccines may be available for prevention of severe rotavirus gastroenteritis.

Introduction

Live attenuated group A bovine rotavirus vaccine strain RIT 4237, derived from Nebraska Calf Diarrhea Virus (NCDV), became available for human studies in 1982 [6]. The new oral vaccine was first tested in Finland for safety and immunogenicity, and found to be both innocuous and immunogenic [22]. Thereafter, three separate placebo-controlled efficacy trials were conducted from 1983 to 1987 [17, 21, 24, 27–28]. Rhesus group A rotavirus vaccine, strain RRV-1 was tested for immunogenicity and safety in 1984 [25] and for efficacy in one trial from 1985 to 1987 [26]. Rhesus-human reassortant rotaviruses D × RRV and DS1 × RRV, expressing the VP7 surface protein of rotavirus G-serotypes 1 and 2, were tested for safety and immunogenicity in 1987 [31] and for efficacy in another trial from 1987 to 1989 [29].

The efficacy trials were conducted in the same clinical setting and had a comparable design in that the infants, aged between 0 and 12 months, were vaccinated in the autumn before the usual rotavirus season and followed through two successive winter and spring epidemic seasons of rotavirus. This paper summarizes the principal findings of these efficacy trials of the heterologous oral candicate rotavirus vaccines in Finland, and describes experience with relevant studies on immunogenicity of the vaccines.

Materials and methods

Details of the vaccine trials have been published previously and are only summarized.

The bovine rotavirus strain RIT 4237 (P6 [1], G6), comprising the vaccine, was attenuated in primary bovine kidney cells and prepared for use as vaccine at the 154th passage level in primary monkey kidney cells [6]. The vaccine lots used in clinical trials titered between $10^{8.1}$ and $10^{8.3}$ tissue culture infective doses (TCID$_{50}$). The lyophilized vaccine preparation (made by Smith Kline-RIT, Rixensart, Belgium) was supplied in individual vials and reconstituted with an appropriate diluent immediately before vaccination. An indistinguishable placebo preparation made from uninfected cells was supplied by the manufacturer for each efficacy trial [21, 24, 28].

The RRV-1 (strain MMU-18006, P5 [3], G3) rhesus rotavirus vaccine was prepared from a 16th passage rhesus rotavirus, produced in fetal rhesus lung diploid cells by Flow Laboratories (McLean, Virginia). This vaccine was supplied as frozen bulk, which was diluted and divided into doses under sterile conditions at the site of studies. Tissue culture medium used for dilution was applied as placebo vaccine. The vaccine bulk had a titre of 10^6 PFU, and 1:100 dilutions of the bulk were made so that each 1 ml dose contained 10^4 PFU. Individual dose vials of diluted vaccine were kept at $-70\,°C$ until use, and thawed and kept on crushed ice until the vaccine was administered [26].

The D × RRV and DS1 × RRV vaccines were also produced at the Flow Laboratories and supplied as frozen bulk. The D × RRV vaccine of 10^6 PFU was diluted 1:100 for use in the trial, following a similar procedure as described above for RRV-1. The DS1 × RRV vaccine bulk was used undiluted and the dose of vaccine given in the efficacy trial was 10^5 PFU [29].

In all studies the vaccines were administered orally using a 1 ml tuberculin syringe. For efficacy studies of the RIT 4237 vaccine, vaccine was given after a meal of breast milk or, if the infant was not breast-fed, formula milk [21, 24, 28]. The RRV-1, D × RRV and DS1 × RRV vaccines were administered following a 30 ml dose of bicarbonate-buffered soy milk [26, 29, 31].

All efficacy trials were randomized, double blind, and placebo-controlled. The basic design was similar in that the participating infants were vaccinated before the expected rotavirus epidemic season and followed for acute diarrhea for two successive epidemic seasons. In all studies the follow-up was done at the Tampere University Hospital, and the catchment area was the same in all trials.

In all studies the parents of participating children were asked to contact the investigator whenever the child had diarrhea or gastrointestinal upset; thus there was no preselection of cases according to severity. A stool specimen was collected in all cases of diarrhea (defined as 3 or more loose or watery stool in a 24 hour period), regardless of clinical severity. Diagnosis was based on demonstration of rotavirus antigen in stools.

In the first two efficacy trials of the RIT 4237 vaccine, the more severe episodes were described as "clinically significant" if they required active treatment, i.e., rehydration on an outpatient basis or admission to hospital. In subsequent trials, clinical information about the diarrheal episodes was collected in sufficient detail to allow analysis on a 0–20 point clinical score [18].

Results

Completed efficacy studies of heterologous rotavirus vaccines

All of the tested vaccines were efficacious in reducing rotavirus gastroenteritis (Table 1). Only one trial may be regarded as a failure; in this study newborn infants received a single dose of RIT 4237 vaccine in June and, presumably, did not contract wild rotavirus until the next epidemic season half a year later [17]; this study is not included in Table 1. In all other studies, the vaccine was given shortly before the epidemic season; in these studies the efficacy ranged from 8% (in neonates) to 67%, with a mean of 45%. There was no apparent difference between the bovine and rhesus-based rotavirus vaccines in this respect (Table 1).

When vaccine efficacy was calculated in relation to the severity of rotavirus episodes (Table 1), the RIT 4237 vaccine induced 56–89% protection against moderately severe ("clinically significant") and 89–100% protection against severe rotavirus diarrhea, respectively. Of the rhesus-derived vaccines, the DS1 × RRV vaccine with 10^5 PFU induced at least as good protection against moderately severe and severe rotavirus diarrhea as did the RIT 4237 vaccine. The efficacy of D × RRV and the RRV-1 vaccine, each titering 10^4 PFU, was somewhat lower. In a meta-analysis of all studies, the mean protection rate for cases of rotavirus diarrhea with a severity score $\geqslant 7$ was 67% (95% C.I. 55–77%) and for cases with a severity score $\geqslant 11$, the protection rate was 81% (95% C.I. 60–91%) [20].

Each of the four vaccines reduced significantly all severe diarrhea regardless of etiology. In fact, vaccine efficacy against all diarrhea with a severity score of 11 or greater in the five evaluable studies was 73% (95% C.I. 54–84%), close to the efficacy in cases with proven rotavirus etiology and corresponding clinical severity. Thus, all vaccines showed apparent efficacy against "non-rotavirus" diarrhea. An explanation for this effect is not known at present. However, the sensitivity of enzyme immunoassay may be insufficient to detect all cases of rotavirus diarrhea, and, in fact, reverse transcriptase polymerase chain reaction may reveal 15–20% more cases [33]. Therefore, protective efficacy against "non-rotavirus" diarrhea may, at least in part, represent protection against undiagnosed cases of rotavirus diarrhea.

Current status of heterologous oral rotavirus vaccines

None of the vaccines tested in Finland between 1982 and 1989 are currently regarded as active candidate vaccines for human use. However, much of the experience obtained in the Finnish studies may be relevant with regard to the immediate successors of these vaccines.

The RIT 4237 strain was withdrawn because of its insufficient efficacy in developing countries [7, 10, 14]. Likewise, bovine rotavirus strain WC3 (P7 [5]

Table 1. Summary of protective efficacy against rotavirus gastroenteritis of varying clinical severity found in four trials of bovine rotavirus vaccine and three trials of rhesus rotavirus vaccine and its reassortant derivatives

Study year	Vaccine	Titer (log 10)	Age of vaccinees (months)	Protective efficacy (95% confidence interval)					Reference
				Any rotavirus diarrhea %	Clinically significant rotavirus diarrhea %	(C.I.)	Severe rotavirus diarrhea %	(C.I.)	
1983–84	RIT 4237	8.1	8–11	55	89	(56–97)	N.A.		3
1989–85	RIT 4237	8.1*	6–12	62	80	(54–92)	N.A.		4
1984–87	RIT 4237	8.3	0	8	73	(36–89)	100		6
1985–87	RIT 4237	8.3*	0, 7	40	56	(7–79)	89	(11–99)	7
1985–87	RRV-1	4	2–5	38	60	(−23–87)	75	(−120–97)	9
1987–89	D × RRV	4	2–5	39	27	(−53–65)	44	(−61–81)	11
1987–89	DS1 × RRV	5	2–5	67	66	(10–87)	89	(13–99)	11
Mean of all studies				45	67	(55–77)	81	(60–91)	

*Two doses of vaccine

G6), which was tested in the USA, China, and Africa, was withdrawn because of inconsistant efficacy [2, 8]. However, the successor of the latter is a bovine-human reassortant vaccine WI79-9, which expresses human serotype 1 VP7 on the surface antigen of WC3 vaccine virus [5]. This vaccine was tested for efficacy in Rochester, New York, with results very comparable to those obtained with bovine rotavirus vaccine in Finland [19].

The successor of single serotype rhesus and rhesus-human reassortant rotavirus vaccines is a tetravalent combination vaccine (RRV-TV), which includes reassortants for G serotypes 1, 2 and 4 plus RRV for G serotype 3. This vaccine, now manufactured by Wyeth-Ayerst Research (Philadelphia, Pennsylvania), is undergoing extensive clinical evaluation in the USA and Finland and in several developing countries. In the USA, RRV-TV at a dose of 4×10^4 PFU provided 57% protection against all rotavirus gastroenteritis, and a greater efficacy for more severe cases [1].

RRV-TV vaccine is now being studied at a dose level of 4×10^5 PFU. In Finland, a double-blind placebo controlled efficacy trial involving 2388 infants started in September 1993 and ended in June 1995. Results of EIA tests of coded sera indicate that 47% of the entire study group produced antibody after vaccination, which, considering that one half of the children received a placebo means that the actual "take" rate of the vaccine was >90%. This is better than in previous studies and appears promising with regard to vaccine efficacy. The complete results will not be available until late 1996, but preliminary results indicate that RRV-TV has similar or greater efficacy than DS1 $\times$ RRV vaccine at the same dose level (10^5 PFU).

Factors influencing the take of oral heterologous rotavirus vaccines

While the mechanism of protection induced by oral heterologous vaccines remains largely unknown, it is clear that the vaccine will have to "take" and to initiate an intestinal infection. The "take" is reflected as a serological response, and therefore serum antibody response may be used as a proxy indicator of a successful vaccination, which, in turn, is a prerequisate of protective immunity. The experience accumulated in the immunogenicity studies of the prior candidate vaccines is likely to be applicable to the present oral rotavirus vaccines as well (Table 2).

Buffering

Early studies in Finland showed that a milk meal, which neutralizes gastric acid, given prior to vaccination increased the serological response to RIT 4237 vaccine from 50% to 88% [23]. Since then, buffering against gastric acidity has been regarded as an essential part of oral rotavirus vaccination.

More recently, a buffer has been combined with the RRV-TV vaccine, increasing the dose volume to 2.5 ml [12]. This may eliminate the need for a separate milk meal before vaccination.

Table 2. Some factors affecting the "take" and immunogenicity of oral heterologous
rotavirus vaccines

Study issue	Vaccine	Feeding with vaccine	Rotavirus antibody response %	Reference
Buffering	RIT 4237	No milk	50	22
		Milk	88	
Breast milk	RIT 4237	Breast milk	81	24
		Formula milk	86	
	RRV-TV	Breast milk	74	26
		Buffered soy milk	70	
Oral immuno-adjuvant	D × RRV	Lactobacillus GG	93	27
		No Lactobacillus GG	74	

Breast milk

In the Finnish studies, a breast milk meal before vaccination did not significantly
suppress the take of RIT 4237 rotavirus vaccine (Table 2) [30]. However,
Pichichero concluded from a meta-analysis that breast-fed infants on average
showed lower (48%) responses to rhesus rotavirus vaccine than non-breast-fed
infants (70%) [16]. In Turkish infants, however, the serological response rate to
RRV-TV vaccine was 74% when fed with mother's milk and 70% when fed with
buffered soy milk prior to vaccination [3]. Therefore, it seems that if breast milk
has any suppressing effect at all on the take of oral heterologous rotavirus
vaccines, this effect is likely to be small.

Lactobacilli

We recently observed that oral administration of *Lactobacillus casei* strain GG
concomitant with D × RRV rotavirus vaccine may enhance immune responses
to the vaccine. In a small trial, 93% of infants given *Lactobacillus* GG vs. 74% of
those not given *Lactobacillus* GG showed an IgA seroconversion after a single
dose of D × RRV vaccine [13]. This preliminary finding is encouraging and
further studies on immunoadjuvants to improve the immunogenicity and,
possibly, efficacy of heterologous oral rotavirus vaccines are warranted.

Oral poliovirus vaccine

Possible interference of oral poliovirus vaccine (OPV) and oral heterologous
rotavirus vaccines was not investigated in Finland, because OPV is not routinely
used in this country. Studies elsewhere have established that bovine rotavirus
vaccine RIT 4237 is sensitive to interference by OPV, and multiple doses of

rotavirus vaccine may be required to overcome this effect and reach full serological response [9, 32]. The experience with the rhesus-based vaccines is mixed [11, 12], but a trial in Thailand unequivocally indicated that the take of RRV-TV at a dose level of 4×10^4 PFU was partially suppressed by OPV [15]. A higher titering vaccine (4×10^5 PFU) and multiple doses may be required to overcome this problem, which appears inherent to all oral rotavirus vaccines.

Discussion

Live oral heterologous vaccines, whether of bovine or rhesus origin, offer significant clinical protection against severe gastroenteritis in the first two or three years of life. In older children rotavirus gastroenteritis is much less frequent and does not usually constitute a significant medical problem. Therefore, protection over two or three rotavirus seasons may be regarded as sufficient from the clinical point of view.

All oral heterologous vaccines tested in Finland were particularly effective for prevention of severe rotavirus gastroenteritis. This is important, as the goal of rotavirus vaccination should be protection against severe disease. Moreover, it is unrealistic to assume that a live attenuated rotavirus vaccine would induce protection against subclinical rotavirus infection, because natural rotavirus infection does not. Rather, rotavirus reinfections are frequent, usually subclinical, and probably important in maintaining protective immunity [35]. It may be assumed that naturally occurring subclinical reinfections are likely to contribute to the efficacy of the rotavirus vaccines given in the autumn before the rotavirus epidemic season, as was the case in all efficacy studies in Finland.

The experience in Finland does not provide any clear evidence that G-serotype-specific immunity is operative in the protection induced by bovine or rhesus rotavirus vaccines. All vaccines showed a roughly similar protection against rotavirus disease, associated mostly with group A G-serotype 1 rotavirus, regardless of the G-serotype specificity of the vaccine virus. The D × RRV vaccine which expresses the human G-serotype 1 surface antigen, was, in fact, less protective than the G-serotype unrelated bovine or rhesus rotavirus vaccines. G-serotype-specific neutralizing antibodies certainly have a role in protective immunity [2, 4], but protection induced by heterologous rotavirus vaccines may largely operate through other mechanisms. Conceivably, cell-mediated immunity might play an role in this protection. Studies of Yosukawa et al. have demonstrated that human adult, but not cord blood, T-lymphocytes have epitopes common to both human and bovine rotaviruses, suggesting the occurrence of cross-protective cell-mediated immunity between human and bovine rotaviruses [34]. Therefore, heterologous rotavirus vaccines might evoke a T-cell memory, and, on reinfection, cytotoxic lymphocytes may recognize peptides of conserved internal rotavirus proteins that are similar in human and animal rotaviruses.

In summary, the properties of oral heterologous rotavirus vaccines in humans have been learned empirically through clinical trials in infants, many of which were conducted in Finland. The accumulated experience indicates that several candidate oral rotavirus vaccines could successfully prevent a major proportion of severe rotavirus gastroenteritis, which typically occurs in the cold season in countries with temperate climate. The first candidate vaccine to gain licensure is likely to be RRV-TV, which may become available by 1997.

References

1. Bernstein DI, Glass RI, Rodgers G, Davidson BL, Sack DA, for the US Rotavirus Vaccine Efficacy Group (1995) Evaluation of rhesus rotavirus monovalent and tetravalent reassortant vaccines in US children. JAMA 273: 1191–1196
2. Bernstein DI, Smith VE, Sander DS, Pax KA, Schiff GM, Ward RL (1990) Evaluation of WC3 rotavirus vaccine and correlates of protection in healthy infants. J Infect Dis 162: 1055–1062
3. Ceyhan M, Kanra G, Seçmeer G, Midthun K, Davidson BL, Zito ET, Vesikari T (1993) Take of rhesus-human reassortant tetravalent rotavirus vaccine in breast-fed infants. Acta Paediatr 82: 223–227
4. Chiba S, Yokoyama T, Nakata S, Morita Y, Urasawa T, Taniguchi K, Urasawa S, Nakao T (1986) Protective effect of naturally acquired homotypic and heterotypic rotavirus antibodies. Lancet 2: 417–421
5. Clark HF, Borian FE, Plotkin SA (1990) Immune protection of infants against rotavirus gastroenteritis by a serotype 1 reassortant of bovine rotavirus WC 3. J Infect Dis 161: 1099–1104
6. Delem A, Lobmann M, Zygraich N (1984) A bovine rotavirus developed as a candicate vaccine for use in humans. J Biol Standard 12: 443–445
7. DeMol P, Zissis G, Butzler J-P, Mutwewingabo A, André FE (1986) Failure of live attenuated oral rotavirus vaccine. Lancet 2: 108
8. Georges-Courbot MC, Monges J, Siopathis MR, Roungon JB, Gresenguet G, Bellec L, Bouquety JC, Lanckriet C, Cadoz M, Hessel L, Gouvea V, Clark F, Georges AJ (1991) Evaluation of the efficacy of a low-passage bovine rotavirus (strain WC3) vaccine in children in Central Africa. Res Virol 142: 405–411
9. Giammanco G, De Grandi V, Lupo L, Mistretta A, Pignato S, Teuween D, Bogaerst H, Andre FE (1988) Interference of oral poliovirus vaccine on RIT 4237 oral rotavirus vaccine. Eur J Epidemiol 4: 121–123
10. Hanlon P, Hanlon L, Marsh V, Byass P, Shenton F, Hassan-King M, Jobe O, Sillah H, Hayes R, M'Boge BH, Whittle HC, Greenwood BM (1987) A trial of an attenuated bovine rotavirus vaccine (RIT 4237) in Gambian children. Lancet 2: 1342–1345
11. Ho M-S, Floyd RL, Glass RI, Pallansch MA, Jones B, Hamby B, Woods P, Penaranda ME, Kapikian AZ, Bohan G, Wilcox WD, Blumberg R (1989) Simultaneous administration of rhesus rotavirus vaccine and oral poliovirus vaccine: immunogenicity and reactogenicity. Pediatr Infect Dis J 8: 692–996
12. Ing DJ, Glass RI, Woods PA, Simonetti M, Pallansch MA, Wilcox WD, Davidson BL, Sievert AJ (1991) Immunogenicity of tetravalent rhesus rotavirus vaccine administered with buffer and oral poliovirus vaccine. Am J Dis Child 145: 892–897
13. Isolauri E, Joensuu J, Suomalainen H, Luomala M, Vesikari T (1995) Improved immunogenicity of oral D × RRV reassortant rotavirus vaccine by Lactobacillus casei GG. Vaccine 13: 310–312
14. Lanata CF, Black RE, del Aguila R, Gil A, Verastegui H, Gerna G, Flores J, Kapikian AZ, André FE (1989) Protection of Peruvian children gainst rotavirus diarrhea of specific serotypes by the RIT 4237 attenuated bovine rotavirus vaccine. J Infect Dis 159: 452–459

15. Migasena S, Simasathien S, Samakoses R, Pitisuttitham P, Sangaroon P, van Steenis G, Beuvery EC, Bugg H, Bishop R, Davidson BL, Vesikari T (1995) Simultaneous administration of oral rhesus-human reassortant tetravalent (RRV-TV) rotavirus vaccine and oral poliovirus vaccine (OPV) in Thai infants. Vaccine 13: 168–174

16. Pichichero ME (1990) Effect of breast-feeding on oral rotavirus vaccine seroconversion;a metaanalysis. J Infect Dis 162: 753–755

17. Ruuska T, Vesikari T, Delem A, André FE, Beards GM, Flewett TH (1990) Evaluation of RIT 4237 bovine rotavirus vaccine in newborn infants: Correlation of vaccine efficacy to season of birth in relation to rotavirus epidemic period. Scand J Infect Dis 22: 269–278

18. Ruuska T, Vesikari T (1990) Rotavirus disease in Finnish children: use of numerical scores for clinical severity of diarrhoeal episodes. Scand J Infect Dis 20: 259–267

19. Treanor JJ, Clark F, Pichichero M, Christy C, Gouvea V, Shrager D, Palazzo S, Offit P (1995) Evaluation of the protective efficacy of a serotype 1 bovine-human rotavirus reassortant vaccine in infants. Pediatr Infect Dis J 14: 301–307

20. Vesikari T (1993) Clinical trials of live oral rotavirus vaccines: the Finnish experience. Vaccine 11: 255–261

21. Vesikari T, Isolauri E, Delem A, D'Hondt E, André FE, Beards GM, Flewett TH (1985) Clinical efficacy of the RIT 4237 live attenuated bovine rotavirus vaccine in infants vaccinated before a rotavirus epidemic. J Pediatr 107: 189–194

22. Vesikari T, Isolauri E, Delem A, D'Hondt E, André FE, Zissis G (1983) Immunogenicity and safety of live oral attenuated bovine rotavirus vaccine strain RIT 4237 in adults and young children. Lancet 2: 807–811

23. Vesikari T, Isolauri E, D'Hondt E, Delem A, André FE (1984) Increased "take" rate of oral rotavirus vaccine in infants after milk feeding. Lancet ii: 700

24. Vesikari T, Isolauri E, D'Hondt E, Delem A, André FE, Zissis G (1984) Protection of infants against rotavirus diarrhoea by RIT 4237 attenuated bovine rotavirus strain vaccine. Lancet 1: 977–981

25. Vesikari T, Kapikian AZ, Delem A, Zissis G (1986) A comparative trial of rhesus monkey (RRV-1) and bovine (RIT 4237) oral rotavirus vaccines in young children. J Infect Dis 153: 832–839

26. Vesikari T, Rautanen T, Varis T, Beards GM, Kapikian AZ (1990) Clinical trial of rhesus rotavirus candidate vaccine (strain MMU 18006) in children vaccination between 2 and 5 months of age. Am J Dis Child 14: 285–289

27. Vesikari T, Ruuska T, Delem A, André FE, Beards GM, Flewett TH (1991) Efficacy of two doses of RIT 4237 bovine rotavirus vaccine for prevention of rotavirus diarrhoea. Acta Paediatr Scand 80: 173–180

28. Vesikari T, Ruuska T, Delem A, André FE (1987) Neonatal rotavirus vaccination with RIT 4237 bovine rotavirus vaccine: a preliminary report. Pediatr Infect Dis J 6: 164–169

29. Vesikari T, Ruuska T, Delem A, André FE (1986) Oral rotavirus vaccination in breast- and bottle-fed infants aged 6 to 12 months. Acta Paediatr Scand 75: 573–578

30. Vesikari T, Ruuska T, Green K, Flores J, Kapikian AZ (1992) Protective efficacy against serotype 1 rotavirus diarrhea by live oral rhesus-human reassortant rotavirus vaccines with human rotavirus VP7 serotype 1 or 2 specificity. Pediatr Infect Dis J 11: 535–542

31. Vesikari T, Varis T, Green J, Kapikian AZ (1991) Immunogenicity and safety of rhesus-human reassortant vaccines with serotype 1 or 2 VP7 specificity. Vaccine 9: 334–339

32. Vodopija I, Baklaic Z, Vlatkovic R, Bogaerts H, Delem A, André FE (1986) Combined vaccination with live oral poliovaccine and the bovine rotavirus RIT 4237 strain. Vaccine 4: 233–236

33. Wilde J, Eiden J, Willoughby J (1991) Improved detection of rotavirus shedding by polymerase chain reaction. Lancet 337: 323–326
34. Yosukawa M, Nakagomi O, Dobayaski Y (1990) Rotavirus induces proliferative response and augments non-specific cytotoxic activity of lymphocytes in humans. Clin Exp Immunol 80: 49–55
35. Zheng BJ, Lo SKF, Tam JSL, Lo M, Yeung CY, Ng MH (1989) Prospective study of community acquired rotavirus infection. J Clin Microbiol 27: 1939–1945

Author's address: Dr. T. Vesikari, University of Tampere, Medical School, P.O. Box 607, FIN-33101 Tampere, Finland.

Arch Virol (1996) [Suppl] 12: 187–198

WC3 reassortant vaccines in children

Brief review

H. F. Clark[1], P. A. Offit[1], R. W. Ellis[2], D. Krah[2], A. R. Shaw[2], J. J. Eiden[2], M. Pichichero[3], and J. J. Treanor[3]

[1] Department of Pediatrics, University of Pennsylvania, Philadelphia, U.S.A.
[2] Merck Research Laboratories, West Point, PA, U.S.A.
[3] School of Medicine, University of Rochester, New York, U.S.A.

Summary. Bovine rotavirus strain WC3 (P7[5], G6) administered at the 12th passage level was well tolerated clinically in infants and efficiently induced serum virus neutralizing antibody (VNA) with bovine rotavirus G6 specificity. The protective efficacy of WC3 vaccine against all rotavirus disease was inconsistent, varying in four separate trials from 76% to 0%; some selective protection against severe disease was seen in all trials. WC3 reassortants containing the gene for an individual human rotavirus VP7 (G) or VP4 (P) surface antigen were also well tolerated, but preferentially induced VNA to the WC3 parent. Efficacy trials of human G1 VP7 reassortant WI79-9 (P7[5], G1) consistently led to $>60\%$ protection against all rotavirus disease. A quadrivalent WC3 reassortant vaccine was developed to contain four separate monovalent reassortants expressing human rotaviruses surface proteins G1, G2, G3, and P1A [8] respectively. In a multicenter trial including 439 infants, this vaccine induced 67.1% protection against all rotavirus disease (defined as positive for rotavirus antigen by ELISA only [$p = <0.001$]) and 72.6% protection when the standard for rotavirus diagnosis was a positive test of stool for both rotavirus antigen by ELISA and rotavirus RNA by electropherotype analysis ($p = <0.001$). In this trial, episodes of the most severe rotavirus disease (clinical severity score >16.0, eight cases) occurred only in placebo recipients.

Introduction

Rotaviruses are double-stranded RNA segmented genome viruses comprising the genus *Rotavirus* of the family *Reoviridae*. The rotaviruses are grouped according to the sharing of common core antigens; the group A rotaviruses are the most common rotavirus pathogens of man and animals.

Group A rotaviruses were first identified as a cause of gastroenteritis in cattle; isolation of a bovine rotavirus was reported in 1969 [21]. A live oral rotavirus vaccine for calves was described in 1972 [22], but has never been proven protective in a placebo-controlled efficacy trial [1]. A rotavirus of group A was first associated with human disease in 1973 [4]. Human rotaviruses are now recognized as the leading cause of severe dehydrating diarrhea of infants, with annual mortality estimates approaching one million [19].

It was recognized early that group A rotaviruses of human or of mammalian origin share common virion core antigens but also express type-specific surface antigens detectable by virus neutralization tests [14]. However, vaccine-challenge experiments in animals, designed to assess the importance of type-specific immunity in protection against disease, gave conflicting results (reviewed in [10]).

Therefore, first generation rotavirus vaccines consisted of animal origin rotaviruses, used in the hope that A rotavirus group-specific immunity might be cross-protective [5, 13]. More recently developed candidate vaccines have consisted of animal rotaviruses with genomes "reassorted" to include genes expressing type-specific surface proteins of human rotaviruses [23]. We describe here clinical studies of a bovine rotavirus strain WC3 evaluated as a candidate vaccine for infants. Initial studies of WC3 vaccine have been followed by a series of clinical trials of WC3 reassortant viruses expressing type-specific human rotavirus surface protein VP7 (G serotype specificity) and/or human rotavirus surface protein VP4 (P genotype specificity). These reassortants have been employed as monovalent vaccines or in a polyvalent combination.

The terminology for designating the specific types of rotavirus surface proteins has been established by the Rotavirus Nomenclature Working Group. For example, with rotavirus strain WI79 of surface protein structure P1A[8], G1, PIA = the VP4 serotype, [8] = the VP4 genotype and G1 = the VP7 serotype [15].

WC3 rotavirus vaccine

WC3 is a group A bovine rotavirus isolated from a calf with diarrhea in Chester County, Pennsylvania in 1981. The virus was propagated for use as a vaccine at the twelfth cell culture passage level in either CV1 monkey kidney cells [6] or in tertiary rhesus monkey kidney cells [2]. The surface protein phenotype has been shown to be VP4 = P7[5] and VP7 = G6.

After initial studies had indicated that WC3 vaccine was well tolerated in adults and children, it was administered without detectable adverse effects to infants in doses as high as 10^8 plaque-forming units (pfu). It was well tolerated at a dose of 10^7 pfu in infants as young as one month of age. Further immunogenicity and efficacy studies of WC3 rotavirus (and subsequently of WC3 reassortant rotaviruses) were performed using doses of 1.0 to 3.0×10^7 pfu. Host immune responses were evaluated primarily by assay of serum virus-neutralizing antibody (VNA) measured by a 50% plaque-reduction neutralization test. A three-fold increase in VNA titer was considered to be a positive response to vaccine.

Table 1. Virus-neutralizing antibody (VNA) response of infants to WC3 vaccine

Clinical trial	Age	Number	Number (%) with VNA response to		Number (%) VNA-positive to Wa pre-vaccine
			WC3 (G6)	Wa (G1)	
Philadelphia,	5–11 Mo	21	18 (86)	2 (9.5)	3 (14)
Kanazawa,	12–23 Mo	16	13 (81)	5 (31)	10 (62)
Japan, 1984	2–6 Yr	15	9 (60)	8 (53)	10 (67)
Paris, 1987	5–12 Mo	25	20 (80)	12 (48)	14 (56)
Philadelphia 1985–86	3–12 Mo	49	35 (71)	4 (8.2)	16 (33)
Cincinnati 1988–89	2–12 Mo	103	100 (97)	9 (8.7)	24 (23)

*All infants received one dose of vaccine

The characteristic infant immune response to a single dose of WC3 vaccine is exhibited in Table 1. The most common VNA response is homotypic, with response rates to WC3 ranging from 71 to 97% in subjects less than one year old. Infants less than one year old rarely develop heterotypic VNA to the epidemiologically predominant human serotype G1 (strains Wa or WI79). An increased incidence of G1-specific VNA responses was associated with immunizations of older infants who expressed a higher rate of seropositivity to Wa (G1) rotavirus prior to vaccine administration (Philadelphia-Kanazawa, 1984 [6]). The only cohort of infants less than one year of age to show a substantial (>50%) incidence of G1-specific VNA antibody response to WC3 was a population in Paris (in a Paris orphanage) that exhibited greater than 50% seropositivity to Wa before immunization [16]. Together, these results suggest that WC3 may induce G1 VNA antibody primarily in infants previously primed by G1 rotavirus infection. Specific correlation of prior G1 rotavirus infection with a vigorous G1 VNA antibody response to WC3 vaccine was identified in the Cincinnati clinical trial [25].

WC3 vaccine was not shed in vaccinee stools in quantities detectable by antigen identification (ELISA) tests or by polyacrylamide gel electrophoresis with silver staining performed to detect viral RNA. In the initial Philadelphia/Kanazawa clinical trial, WC3 virus was detected by direct plaque assay in minimal concentrations ($<2.0 \times 10^4$ pfu/gm stool) in 7 of 22 infants (31%) at day 3 to 7 post vaccination [6]. When assessed in subsequent trials, shedding was detected in less than 20% of vaccine recipients.

Efficacy of WC3 vaccine in infants was assessed in a series of four double-blind placebo-controlled efficacy trials (Table 2). Results were evaluated in terms of total episodes of rotavirus-associated gastroenteritis occuring in the (winter) rotavirus season following immunization. Since the most important goal of immunization against rotavirus is prevention of the most severe disease, all

Table 2. Protective efficacy of WC3 bovine rotavirus vaccine

Site	Season	No. doses	Rotavirus disease					
			Total			Clinical score > 8.0		
			Placebo	Vaccine	Protection	Placebo	Vaccine	Protection
Philadelphia	1985–86	1	14/55 (25.5%)	3/49 (6.1%)	76.1%	11/55 (20.0%)	0/49 (0.0%)	100%
Cincinnati	1988–89	1	25/102 (24.5%)	21/103 (20.4%)	16.7%	15/102 (14.7%)	9/103 (8.7%)	40.8%
Bangui	1988–89	2	59/200 (29.5%)	58/214 (27.1%)	< 10%	43/200 (21.5%)	30/214 (14.0%)	34.9%
Shanghai	1989–90	2	37/157 (23.6%)	19/160 (11.9%)	49.6%	3/157 (1.9%)	3/160 (1.9%)	0.0%

disease episodes were graded using a standard scoring system taking into account the magnitude and duration of diarrhea, vomiting, fever, and "behavioral symptoms" [7]. The possible range of severity scores is from 2 to 24. Initially, a score of >8.0 was selected to represent moderate to severe or "clinically significant" disease. In the initial Philadelphia-based efficacy trial, WC3 vaccine-induced protection against all rotavirus disease was 76% and against clinically significant rotavirus disease was 100%. However, in subsequent clinical trials in Cincinnati [2] and Bangui (Central African Republic) [17], WC3 vaccine induced no protection against all rotavirus disease and only modest protection against clinically significant disease. In the Bangui trial, despite 2 doses of vaccine, the serum VNA response to WC3 was only 60% but in the equally non-protective Cincinnati trial a homotypic VNA response rate of 97% to WC3 was observed.

A final WC3 efficacy trial performed in Shanghai was characterized by an approximately 50% protection rate against all rotavirus disease following 2 vaccine doses. Although the scoring system did not detect selective protection against severe disease, WC3 vaccine administration was associated with a selective reduction in days of rotavirus-associated diarrhea (61% reduction) and episodes of fever >38 °C (69% reduction). (Ho Ms, Guo ZP, Xia LD, *et al.* manuscript in prep.).

Because of the inconsistency of protection afforded by WC3 vaccine, this candidate is no longer being pursued for further development.

WC3 reassortant rotavirus vaccine-monovalent

Reassortant rotavirus vaccine candidates were developed to include the genome background of WC3 for the purpose of attenuation while incorporating the gene coding for one of the two surface proteins of human rotavirus for immune specificity to naturally circulating wild-type virus.

Table 3. WC3 (P7[5], G6) reassorted to express surface protein(s) of WI79 (P1A[8], G1).

Vaccine	Phenotype	Clinical Trial	Number of infants	Number (%) with VNA response to:	
				WC3	WI79
A. WI79-9	P7[5], G1	2 doses, phase 1	33	21 (64)	18 (55)
B. WI79-9	P7[5], G1	2 doses, efficacy	38	37 (97)	8 (22)
C. WI79-9 mix with WC3	P7[5], G1 & P7[5], G6	2 doses, phase 1	20	19 (95)	12 (60)
D. WI79-4	P1A[8], G6	2 doses, phase 1	25	19 (76)	7 (28)
E. WI79 (4 + 9)	P1A[8], G1	2 doses, phase 1	27	7 (26)	9 (33)
F. WI79-9 mix with WI79-4	P7[5], G1 & P1A[8], G6	2 doses, phase 1	27	27 (100)	21 (78)
G. WI79-9	P7[5], G1	3 doses, efficacy	43	41 (95)	30 (70)

Initial studies focused upon the gene coding for the human VP7 (G specificity) surface component. This was originally thought to be the "major neutralization antigen" because of the reactivity of sera of parenterally rotavirus-hyperimmunized laboratory animals [18]. Subsequently obtained information indicating that the VP4 (P specificity) surface antigen may be a major immunogen following oral exposure to rotavirus led to study of WC3 reassorted to contain the gene for human rotavirus VP4. Reassortant studies emphasized use of VP7 of G1 type and VP4 of serotype P1A representing the phenotype of human P1A [8], G1 rotaviruses which are most commonly encountered worldwide.

Representative serum VNA antibody responses to P1a and G1 reassortants are listed in Table 3. Like their WC3 parent, when given at a standard dose of 1.0 to 2.0 × 10^7 pfu, these reassortants were very well tolerated clinically. As in the case of WC3, vaccine virus was identified in the stool rarely (< 20% of subjects) and in low concentrations detected only by virus plaque assay.

Immune responses (VNA) to the P7[5], G1 reassortant WI79-9 (Table 3; A, B, C) consistently occurred most frequently to WC3, represented by the VP4 P7[5] surface antigen of the vaccine, and with much less efficiency to WI79, represented by the VP7 G1 surface protein of the vaccine [8, 9]. This observation led to evaluation of a "mirror image" reassortant WI79-4 in which the human virus protein substituted on WC3 was the human VP4 P1A[8] surface protein.

Surprisingly, this reassortant (see Table 3, D) also induced serum VNA response to the WC3 parent more efficiently than it induced VNA to human virus WI79 (Clark *et al.*, unpubl. data).

Incorporation of both human virus surface antigens into the same vaccine was evaluated by use of a reassortant WI79 (4 + 9) in which genes for both human surface proteins VP4 P1A[8] and VP7 G1 were added to an otherwise WC3 rotavirus genome (Table 3, E). WI79 (4 + 9) was compared with a mixture of equal concentrations of the reassortants WI79-9 and WI79-4 (Table 3, F) [11]. Again, surprisingly, the reassortant WI79 (4 + 9) with a totally human strain surface phenotype was a weak human immunogen, inducing VNA to the type 1 parent WI79 in only 33% of subjects. On the contrary, the mixture of two single surface protein reassortants WI79-4 and WI79-9 gave an excellent VNA response to each parental virus (100% to WC3 and 78% to WI79).

We have also demonstrated efficient induction of VNA to both WC3 and human type G1 WI79 virus by administration of 3 doses of WI79-9 reassortant at either one or two months intervals (Clark *et al.* unpubl. data). In the efficacy trial example listed in Table 3G, 95 percent of infants exhibited a VNA response to WC3 and 70% responded with VNA to WI79 virus.

Vaccination with multiple doses of WI79-9 reassortant has consistently led to significant levels of protection against wild-type rotavirus disease (Table 4). This experience has been obtained in seasons when the natural rotavirus challenge was predominantly of serotype G1 as in the case of WC3 vaccine trials. Protection was efficient in a trial, (Philadelphia 1987–1988, 2 doses) [9] in which the VNA response of vaccinees to G1 rotavirus was only 22% (see also Table 3B) as well as in a clinical trial, (Philadelphia 1992–93, 3 doses) in which the incidence of VNA response to G1 rotavirus was 70% (see also Table 3, G) [24]. In the WC3 efficacy trials only one dose (Philadelphia 1985, Cincinnati 1988) or two doses (Bangui, 1988, Shanghai, 1989) of vaccine were administered.

Table 4. Efficacy of monovalent (serotype G1) WC3 reassortant vaccine strain WI79-9

Site	Season	No. Doses	Rotavirus disease (all)			Rotavirus disease (score > 8.0)		
			Placebo	Vaccine	Protection	Placebo	Vaccine	Protection
Philadelphia	1987-88	2	8/39 (20.5%)	0/38 (0%)	100%	n.d.*	n.d.	
Rochester	1991-92	3	3/12 (25.7%)	1/14 (7.1%)	71.6%	n.d.	n.d.	
Rochester	1992-93	3	19/74 (25.7%)	14/152 (9.2%)	64.1%	12/74 (16.2%)	4/152 (2.6%)	83.7%
Philadelphia	1992-93	3	7/41 (17.1%)	2/45 (4.5%)	74.0%	6/41 (14.6%)	0/45 (0.0%)	100%

*n.d. = not determined

Table 5. Effects of protective rotavirus immunization with WI79-9 on total diarrheal disease in one season

	Philadelphia 1987–88	Rochester & Philadelphia 1992–93
No. of Infants[1]	77.0	312.0
No. of vaccine doses[2]	2.0	3.0
Protection (%):		
All rota disease	100.0	64.1
Rota disease score > 8.0	100.0	87.0
Total diarrheal disease episodes	9.0	40.0
Severe diarrheal disease episodes	48.4	65.7
Total diarrhea days	31.1	51.8
Doctor visits for diarrhea	N.C.[3]	48.4

[1] total placebo and vaccine infants
[2] 1 month interval 1987–88 trial, 2 month interval in 1992–93
[3] NC = not calculated

The WI79-9 efficacy trials also demonstrated that, in a developed nation setting, successful immunization against rotavirus has a large impact on the total expression of diarrheal disease, regardless of its etiology (Table 5). In the 1992–1993 efficacy trial, WI79-9 immunization led to a large reduction (40–65%) in total and severe diarrheal disease episodes, total days of diarrhea and doctor's visits for diarrhea. These vaccine-associated protective effects each represent a statistically significant difference from the placebo population.

WC3 reassortant rotavirus vaccine-multivalent

A four-component "quadrivalent" WC3 reassortant vaccine was developed incorporating virus strains proven to be immunogenic in trials of monovalent preparations. This included in approximately equal amounts:

1. WI79-9 = P7[5], G1: $1.0 \times 10^{7.0}$ pfu
2. WI79-4 = P1A[8], G6: $5.0 \times 10^{6.0}$ pfu
3. SC2-9 = P7[5], G2: $1.0 \times 10^{7.0}$ pfu
4. WI78-8 = P7[5], G3: $1.0 \times 10^{7.0}$ pfu

The WI79-9 and WI79-4 were included because this combination had induced the highest levels of VNA to P1A[8] rotavirus strain WI79 in immunogenicity trials. The SC2-9 and WI78-8 reassortants were incorporated to induce VNA specific to G2 and G3 antigens and thereby provide a broad antibody response specific to the three G serotypes most common in the U.S.

In a multicenter double-blind efficacy trial, either the quadrivalent vaccine or placebo was administered to 439 infants [12]. All three doses were given at 6 to

8 week intervals to 417 study participants. On a preliminary analysis, no statistically significant differences were noted between vaccine and placebo recipients with respect to fever, vomiting, or irritability after each dose. Mild diarrhea was reported more often in vaccine recipients than in placebo infants after the first dose only (26.1% diarrhea incidence in vaccine versus 18.2% diarrhea in placebo recipients: $p = 0.045$). None of the diarrhea episodes following the first dose of vaccine exhibited a severity score indicating moderate to severe diarrhea (score > 8.0).

The protection associated with administration of the quadrivalent vaccine is illustrated in Table 6. When the standard of identification of rotavirus diarrhea was only a positive result by ELISA test, the protective rate in 405 evaluable infants was 67.1% ($p = <0.001$). When the standard of identification of rotavirus diarrhea was a positive result by both ELISA and electropherotype analysis, the protective rate in 405 evaluable infants was 72.9% against all rotavirus disease ($p = <0.001$). No selective enhancement of protective effect was detected in evaluation of all diarrheal cases with a clinical score exceeding 8.0. However, rotavirus disease of a severity score exceeding 16.0 (eight episodes) was seen only in the placebo population.

Immune responses associated with administration of the quadrivalent vaccine are being evaluated.

Discussion

Numerous clinical trials have indicated unequivocally that strain WC3 bovine rotavirus given in high titer ($\geqslant 10^{7.0}$ pfu) is well tolerated and is efficient in inducing a homotypic serum VNA response. Four different efficacy trials of WC3 vaccine have yielded protective effects, measured by effect on all rotavirus disease episodes, ranging from none to 76%. Measurable protection against the most severe disease was apparent even in trials where there was no overall reduction in rotavirus disease. Observed protection developed in the absence of detectable human rotavirus type-specific serum antibody; the mechanism is unknown.

Reassortants of WC3 containing the gene for a single human rotavirus surface protein VP7 or VP4 were as well tolerated clinically as the WC3 parent virus. As has been reported in numerous studies of the immunogenicity of human rotavirus VP7 reassortants of strain RRV [3, 20], the human rotavirus VP7 reassortants of WC3 preferentially induced serum VNA neutralizing the donor of the VP4 surface protein, in our case WC3. It was surprising that our reassortant bearing a surface specificity of human virus VP4 and bovine virus VP7 nevertheless also selectively induced VNA to WC3. An inherent lower level of immunogenicity for the human rotavirus was suggested by the fact that a WC3 reassortant with both VP4 and VP7 of human type (P1A[8], G1) expressed on the surface was a particularly inefficient immunogen.

It is surprising that the presence of a bovine surface antigen appears to enhance the immunogenicity of rotavirus in the human, at least in a host-virus system where virus replication appears to be limited. That each of the human

Table 6. Efficacy of quadrivalent WC3 reassortant vaccine: multi-center trial 1993–1994

Population	Prevention of all disease, regardless of severity				Prevention of moderate to severe disease*			
	#cases /subjects in vaccine group	#cases /subjects in placebo group	efficacy (95% CI)	p-value (Fisher's Exact)	#mod/sev case/subjects in vaccine group	#mod/sev cases/subjects in placebo group	efficacy (95% CI)	p-value (Fisher's Exact)
ELISA +, 1st reported incidences	14/199 (7.0%)	44/206 (21.4%)	67/1% (39.9%, 81.9%)	<0.001	10/199 (5.0%)	33/206 (16.0%)	68.6% (36.4%, 84.5%)	<0.001
ELISA +, Electropherotype confirmed, 1st reported incidences	11/199 (5.5%)	42/206 (20.4%)	72.9% (47.3%, 86.0%)	<0.001	9/199 (4.5%)	34/206 (16.5%)	72.6% (42.9%, 86.9%)	<0.001

*Rotavirus cases with a clinical score of nine or higher

VP4 and VP7 human surface proteins on single protein reassortants contribute synergistically to antibody neutralization is suggested by the very high efficiency of the mixture of the human VP4 WI79-4 and human VP7 WI79-9 viruses in inducing neutralizing antibody to the human rotavirus parent P1A[8], G1 WI79.

Vaccines containing a single or multiple WC3 rotavirus reassortants have not shown the variability in protective efficacy rate that was characteristic of WC3 and other animal-origin rotaviruses. In separate clinical trials, protective effects greater than 60% against all rotavirus disease were obtained following multiple doses of serotype G1 reassortant WI79-9, although G1-specific serum VNA levels induced varied from 22% to 70%. A single large multicenter trial of mixed WC3 reassortants expressing respectively G1, G2, G3 and P1A[8] human rotavirus surface proteins gave a protective effect similar to that observed with the monovalent G1 reassortant. The need for G type-specific reassortants homotypic for all extant wild-type rotaviruses cannot be assessed until clinical trials can be performed in a population exposed to predominantly non-type G1 rotaviruses.

It is also important to note that clinical trials with WC3 reassortants performed to date have been restricted to situations characterized by human rotavirus G1 natural challenge in a developed nation milieu. To meet the more difficult challenge of immunizing against rotavirus disease in developing nations, the need for multiple G serotype immunogens and the apparently immunologically synergistic human P immunogen may be more apparent. Until a reliable immune correlate of vaccine-induced protection against rotavirus is identified, answers to these questions will await performance of comparative clinical efficacy trials.

Acknowledgements

We acknowledge the special clinical contributions of research nurses Dorothy Shrager, Diane Lawley and Frances Borian. These studies were supported in part by Merck Research Laboratories, West Point, PA, U.S.A., by Pasteur-Merieux Serum Vaccine, Lyon, France and by Contract NO1-AI-05049 from the National Institute of Allergy and Infectious Disease.

References

1. Acres DS, Radostits OM (1976) The efficacy of a modified live reo-like virus vaccine and an *E. coli* bacteria for prevention of acute neonatal diarrhea of beef calves. Can Vet J 17: 197–212
2. Bernstein DI, Smith VE, Sander DS, Pax KS, Schiff GM, Ward RL (1990) Evaluation of WC3 rotavirus vaccine and correlates of protection in healthy infants. J Infect Dis 162: 1055–1062
3. Bernstein DI, Glass RI, Rodgers G, Davidson BL, Sack DA (1995) Evaluation of rhesus rotavirus monovalent and tetravalent reassortant vaccines in U.S. Children. J Amer Med Assoc 273: 1191–1196
4. Bishop RF, Davidson GP, Holmes H, Ruck BJ (1973) Virus particles in epithelial cells of duodenal mucosa from children with viral gastroenteritis. Lancet 2: 1281–1283

5. Christy C, Madore HP, Treanor JJ, Pray K, Kapikian AZ, Chanock RM, Dolin R (1986) Safety and immunogenicity of live attenuated rhesus monkey rotavirus vaccine. J Infect Dis 154: 1045–1047

6. Clark HF, Furukawa T, Bell LM, Offit PA, Perrella PA, Plotkin SA (1986) Immune response of infants and children to low-passage bovine rotavirus (strain WC3). Amer J Dis Children 140: 350–356

7. Clark HF, Borian EF, Bell LM, Modesto K, Gouvea V, Plotkin SA (1988) Protective effect of WC3 vaccine against rotavirus diarrhea in infants during a predominantly serotype 1 rotavirus season. J Infect Dis 158: 750–587

8. Clark HF, Borian FE, Modesto K, Plotkin SA (1990a) Serotype 1 reassortant of bovine rotavirus WC3, strain WI79-9, induces a polytypic antibody response in infants. Vaccine 8: 327–332

9. Clark HF, Borian FE, Plotkin SA (1990b) Immune protection of infants against rotavirus gastroenteritis by a serotype 1 reassortant of bovine rotavirus WC3. J Infect Dis 161: 1099–1104

10. Clark HF, Offit PA (1994) Rotavirus vaccines. In: Plotkin SA, Mortimer EA (eds) Vaccines. Saunders, Philadelphia, pp 809–822

11. Clark HF, Welsko D, Offit PA (1992) Infant responses to bovine WC3 reassortants containing human rotavirus VP7, VP4, or VP7 & VP4. ICAAC Anaheim, p 343

12. Clark HF, White CJ, Offit PA, Stinson D, Eiden J, Weaver S, Cho I, Shaw A, Krah D, Ellis R + OHBRV Study Group (1995) Preliminary evaluation of safety and efficacy of quadrivalent human-bovine reassortant rotavirus vaccine. Ped Res 37: 172A

13. Delem A, Lobmann M, Zygraich N (1984) A bovine rotavirus developed as a candidate vaccine for use in humans. J Biol Stand 12: 443–445

14. Estes MK, Graham DY, Petrie BL (1985) Antigenic structure of rotaviruses. In: van Regenmortel MHV, Neurath AR (eds) Immunochemistry of viruses. Elsevier, Amsterdam, pp 389–405

15. Estes MK (1996) Rotaviruses and their replication. In: Fields BN et al. (eds) Virology. Lippincott-Raven Publishers, New York, pp 1625–1655

16. Garbag-Chenon A, Fontaine J-L, Lasfargues G, Clark HF, Guyot J, LeMoing G, Hessel L, Bricout F (1989) Reactogenicity and immunogenicity of rotavirus WC3 vaccine in 5–12 month old infants. Res Virol 140: 207–217

17. Georges-Courbot MC, Monges J, Siopathis MR, Roungou JP, Gresenguet G, Bellec L, Bouquety JC, Lanckviet C, Cadoz M, Hessel L, Gouvea G, Clark HF, Georges AJ (1991) Evaluation of the efficacy of a low passage bovine rotavirus (strain WC3) vaccine in children in Central Africa. Res Virol 142: 405–411

18. Hoshino Y, Wyatt RG, Greenberg HB, Flores J, Kapikian AZ (1984) Serotypic similarity and diversity of rotaviruses of mammalian and avian origin as studied by plaque-reduction neutralization. J Infect Dis 149: 694–702

19. Institute of Medicine (1986) The prospects for immunzing against rotavirus. In: New vaccine development. Establishing priorities. Diseases of importance in developing countries, vol II. National Academy Press, Washington, DC, pp 308–318

20. Kobayashi M, Thompson J, Tottlefson SJ, Reed GW, Wright PF (1994) Tetravalent rhesus rotavirus vaccine in young infants. J Infect Dis 170: 1260–1263

21. Mebus CA, Underdahl NR, Rhodes MB, Twiehaus MJ (1969) Calf diarrhea (Scours): reproduced with a virus from a field outbreak. Nebraska Res Bull 233. Nebraska Agricultural Experiment Station, Lincoln, Nebraska

22. Mebus CA, White RG, Stair FL, Rhodes MB, Twiehaus MJ (1972) Neonatal calf diarrhea: results of a field trial using a reo-like virus vaccine. Vet Med Ann Clin 67: 173–174

23. Midthun K, Greenberg HB, Hoshino Y, Kapikian AZ, Wyatt RG, Chanock RM (1985) Reassortant rotaviruses as potential live rotavirus vaccine candidates. J Virol 53: 949–954
24. Treanor JJ, Clark HF, Pichichero M, Christy C, Gouvea V, Shrager D, Palazzo S, Offit PA (1995) Evaluation of the protective efficacy of a serotype 1 bovine-human rotavirus reassortant vaccine in infants. Pediatric Infect Dis J 14: 301–397
25. Ward RL, Sander DS, Schiff GM, Bernstein DI (1990) Effect of vaccination on serotype-specific antibody responses in infants administered WC3 bovine rotavirus before or after a natural rotavirus infection. J Infect Dis 162: 1298–1303

Authors' address: Dr. H. F. Clark, Division of Infectious Diseases, Children's Hospital of Philadelphia, 34th Street and Civic Center Boulevard, Philadelphia, Pennsylvania 19104, U.S.A.

Arch Virol (1996) [Suppl] 12: 199–206

Rotavirus subunit vaccines

M. E. Conner[1,2], S. E. Crawford[2], C. Barone[2], C. O'Neal[2], Y.-J. Zhou[2],
F. Fernandez[3], A. Parwani[3]. L. J. Saif[3], J. Cohen[4], and M. K. Estes[2]

[1] Veterans Administration Medical Center and [2] Baylor College of Medicine,
Houston, Texas, U.S.A.; [3] OARDC/Ohio State University,
Wooster, Ohio, U.S.A.; [4] Unité de Virologie et
d'Immunologie Moleculaire, INRA, Jouy-en-Josas, France

Summary. We evaluated rotavirus subunit vaccines for use in humans and animals. Insect cells were co-infected with combinations of individual baculovirus recombinants expressing human, bovine or simian rotavirus VP2, VP4, VP6 or VP7 to produce virus-like particles (VLPs). To determine whether immunization with VLPs could induce active protective immunity, VLPs were administered parenterally to rabbits, and the immune response and protection from rabbit ALA rotavirus challenge were evaluated. Complete or partial protection was attained, showing that parenteral immunization with VLPs induces active protective immunity. We also examined whether heterotypic immune responses could be induced with a limited number of broadly reactive VP7 proteins or with chimeric particles (multiple VP7 types on individual particles). The feasibility of this approach was determined by immunizing mice with VLPs containing a G3 VP7 or G1 VP7 and chimeric G1/G3 VLPs. Broadly reactive neutralizing antibody was induced by the G1 VLPs. VLPs also have been successfully used to boost lactogenic (colostral and milk) immunity in dairy cows. Taken together, these results show that VLPs can be effective immunogens in rabbits, mice and dairy cattle when administered parenterally, a limited number of VLPs may be sufficient to produce a broadly protective vaccine, and G3 VLPs may serve as an effective subunit vaccine for use in bovines.

Introduction

Rotaviruses are a major cause of diarrhea in young children and animals worldwide with high levels of morbidity in both populations (reviewed in [7] and [9]). Rotavirus disease caused by group A rotaviruses is responsible for approximately one million deaths per year in children, primarily those in developing countries. Development of safe effective rotavirus vaccines for use in children is an international priority [7, 9]. Development of rotavirus vaccines for children have primarily focused on the use of orally administered live virus of animal origin or human-animal viral reassortants. Recent advancements in our understanding of the molecular biology of and immunity to rotavirus have spurred the investigation of rotavirus subunit vaccines. Our approach to developing a potential rotavirus subunit vaccine has been to clone individual rotavirus genes coding for structural proteins and to then co-express these proteins in a baculovirus expression system [10, 14, 17]. The co-expressed proteins self-assemble to form virus-like particles (VLPs). VLPs of different protein composition can be produced by choosing baculovirus recombinants that are co-expressed [10]. This report summarizes our studies of the immunogenicity and protective efficacy of rotavirus VLPs administered parenterally to rabbits, mice and cows [3, 4, 10, 15].

Materials and methods

Viruses and cells

Simian SA11(P[2], G3), lapine ALA (P[14], G3) and human Wa (P1A[8], G1), S2 (P1B[4], G2) and St. Thomas (P2A[6], G4) group A rotavirus strains were cultivated in MA104 cells as previously described [6]. Rotavirus serotype specificity is determined by two neutralization antigens, VP4 (designated by P, for protease sensitive) and a glycoprotein VP7 (designated by G, for glycoprotein). Many P types are defined only by genotypic analysis; P genotypes are designated by P followed by a number in brackets [12]. All viruses used in ELISA and plaque reduction netralization assays were plaque purified. Inactivation of SA11 infectivity was accomplished by formalin inactivation [2].

Production of VLPs

VLPs of differing composition were produced by co-infecting *Spodoptera frugiperda* 9 (Sf9) cells with 5 PFU/cell of different combinations of baculovirus recombinants [10]. The recombinants used coded for bovine VP2, simian P[2] VP4, simian VP6, simian G3 VP7 or human HRV8697 G1 VP7. VLPs were purified by CsCl density gradient purification and then characterized for purity, integrity, concentration, endotoxin levels and for composition by biochemical analysis and electron microscopy (EM) (Figs. 1 and 2) [10]. The VP7 serotype specificity of the serotype G1 and G3 VLPs was confirmed by a monoclonal antibody based ELISA for serotypes G1, G2, G3 and G4 and by immunoelectron microscopy with gold-labeled monoclonal antibody probes [10, 11].

Vaccination and virus challenges

All animals were vaccinated twice intramuscularly (IM) with SA11 rotavirus or VLPs. Twenty rotavirus-free rabbits were vaccinated once or twice at approximately 1 month intervals with either live or inactivated P[2], G3 SA11 [2] or twice with either 10 or 20 µg of P[2], G3 VP2/4/6/7 VLPs in either Freund's complete/incomplete adjuvant or aluminum phosphate [3, 4]. Rabbits were challenged with 10^5 PFU of P[14], G3 lapine ALA virus administered orally approximately one month following the second vaccine dose. Sera and fecal antibody samples were collected pre and post vaccination and post challenge at 1–2 week intervals for one to two months. ALA virus shedding was assessed in fecal samples collected 0–14 days post challenge (dpc). Five rotavirus seronegative CD-1 mice (Charles

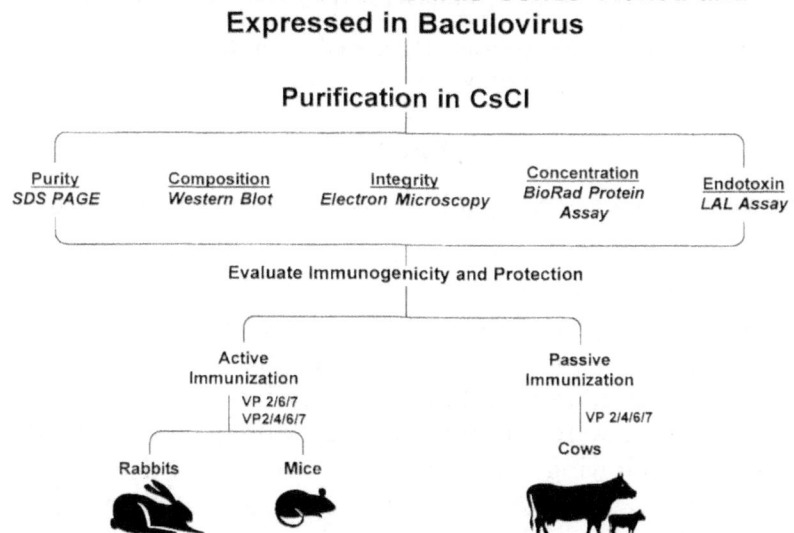

Individual Structural Rotavirus Genes Cloned and Expressed in Baculovirus

Purification in CsCl

| Purity SDS PAGE | Composition Western Blot | Integrity Electron Microscopy | Concentration BioRad Protein Assay | Endotoxin LAL Assay |

Evaluate Immunogenicity and Protection

Active Immunization
VP 2/6/7
VP2/4/6/7

Passive Immunization
VP 2/4/6/7
Cows

Rabbits   Mice

Fig. 1. Overview of methods used to prepare and evaluate rotavirus VLPs as subunit vaccine

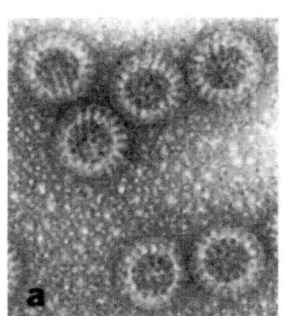 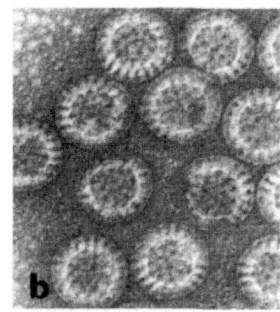

Fig. 2. Electron micrographs of rotavirus VLPs self-assembled after co-expression of baculovirus recombinants of rotavirus structural genes in Sf 9 insect cells. (a) VP2/6/7, (b) VP2/4/6/7 VLPs. Magnification × 142 350

River, Portage, Michigan) were vaccinated twice IM 14 days apart with 2 μg of either VP2/6/7 or VP2/4/6/7 G1 or G3 VLPs. Sera were collected at 2 week intervals for one to two months post vaccination. Fecal antibody samples were collected from individual mice at 61 days post vaccination (dpv) (47 days after the second vaccination) and the mice then were vaccinated a third time in an identical manner as the previous two immunizations. Fecal antibody samples were collected 1 and 2 weeks after the third vaccination. Starting approximately 2 months prior to calving, pregnant seropositive Holstein cows were vaccinated once IM and 14 days later were vaccinated once intramammarily (IMM) by infusion of vaccine into all four teat canals [15]. The cows were vaccinated with either 100 or 250 μg of VP2/4/6/7 VLPs, or mock vaccinated with buffer in incomplete Freund's adjuvant. Sera were collected at the time of each vaccination, sera and colostrum collected on the day of calving, and sera and milk collected 7 days after calving, as previously described [21].

Antibody assays

Neutralizing antibody titers were measured by plaque reduction, as described previously [6, 8, 13, 21]. IgG, IgM, IgA (Total Ig) and isotype specific antibody titers were measured, as previously described [2, 8, 10, 21].

Antigen assays

In rabbits, protection from challenge was determined by the level of virus antigen excretion, measured by ELISA and plaque infectivity assays as described previously [2, 6, 8].

Results

Parenteral administration of SA11 rotavirus and VLPs in rabbits

In previously published studies, we showed that two doses of live or inactivated SA11 rotavirus administered parenterally in complete/incomplete Freund's or aluminum phosphate adjuvant induced rotavirus-specific IgG in the intestine of all rabbits and that all were totally protected (no virus or antigen excretion) from live ALA virus challenge (Table 1) [2]. To determine if VLPs also would induce protection, 10 rabbits were vaccinated twice with either 10 or 20 μg of G3 VP2/4/6/7 VLPs. Following vaccination, all rabbits had detectable anti-SA11 rotavirus IgG in fecal antibody samples. Following challenge with G3 homologous ALA virus, rabbits were totally or partially protected. Partial protection was defined as decreased duration or amount (lower infectivity titer or ELISA optical density) of virus shedding compared to mock or unvaccinated controls challenged with ALA virus.

Parenteral administration of G1 and G3 VP2/6/7 and VP/4/6/7 VLPs in mice

We previously reported that 1–10 μg of P[2], G3 VP2/4/6/7 VLPs were immunogenic in mice, inducing high levels of total Ig and neutralizing antibody to SA11, and that chimeric VLPs that displayed both G1 and G3 VP7 on their surfaces induced neutralizing antibody to both G1 and G3 viruses [10]. Preliminary information indicated that G1 VLPs might induce heterotypic neutralizing antibody. To determine if VP4 is necessary for induction of neutralizing antibody and to determine if G1 VLPs induced broadly cross reactive neutralizing antibody, mice were vaccinated with either G1 or G3 VP2/6/7 or VP2/4/6/7 VLPs [5, 11]. Sera collected 0 and 42 dpv from vaccinated mice were tested in neutralization assays against G1, G2 and G4 human rotaviruses (Wa, S2 and St. Thomas, respectively) and G3 simian SA11 rotavirus. Fecal samples collected after the second or third vaccination had detectable neutralizing antibody to both G1 and G3 rotaviruses and total Ig anti-rotavirus antibody in the intestine. Testing for antirotavirus IgA antibody is in progress. Mice vaccinated with G3 VLPs had neutralizing antibody to G3 SA11 virus but with the exception of one mouse, not to G1, G2 or G4 human rotavirus strains. Protection from challenge was not assessed.

Vaccination to increase levels of passive lactogenic immunity in dairy cows

Holstein cows with antibody to group A rotavirus were vaccinated with P[2], G3 VP2/4/6/7 VLPs or were mock-vaccinated to determine if increased levels of antibody to group A rotavirus could be induced in colostrum and milk. As

Table 1. Immunogencity and protective efficacy of rotavirus VLP subunit and virus vaccines administered parenterally

Immunogen	Route	Dosage	No. of Doses	Animal	Ab Response Following Vaccination				Protection
					Serologic	Mucosal (Intest. or Colost.)			
					NAb	NAb	IgG		
SA11[a]	IM	1×10^7 pfu	1	Rabbit	Yes	No	No		No
SA11[a]	IM	1×10^7 pfu	2	Rabbit	Yes	Yes	Yes		Yes
VP2/4/6/7	IM	10 or 20 µg	2	Rabbit	Yes	No	Yes		Yes[b]
VP2/6/7	IM	2 µg	2	Mouse	Yes	Yes	Yes		ND[c]
VP2/4/6/7	IM	2 µg	2	Mouse	Yes	Yes	Yes		ND
VP2/4/6/7	IM/IMm[d]	100 or 250 µg	2	Cow	Yes	Yes	Yes		ND

[a] Data from previously published studies [2]. Rabbits vaccinated with live or formalin-inactivated SA11 virus
[b] Complete or partial protection. Partial protection defined as significantly decreased virus shedding compared to mock vaccinated controls
[c] Not done
[d] Cows vaccinated IM 7 days before drying off and intramammarily (IMm) 7 days after drying off which is approximately two months prior to calving

measured by ELISA and neutralization compared to mock vaccinated controls, immunization with VP2/4/6/7 VLPs resulted in significant elevations in serum, colostrum and milk antibody titers [15]. Colostrum collected and pooled from immunized cows currently is being tested by feeding to calves to determine whether elevated levels of rotavirus specific lactogenic antibody result in increased protection of calves from virus challenge [20].

Discussion

VLPs may provide a safe and efficacious alternative to live rotavirus vaccines. Our results show the potential of parenterally administered VLPs as a rotavirus subunit vaccine for both human and veterinary use. VLPs administered parenterally to rabbits, mice and cows were highly immunogenic, inducing or increasing neutralizing and isotype specific or total Ig antibodies in serum and at mucosal sites (intestine in rabbits and mice, mammary gland in cows). Vaccination of rabbits with P[2], G3 VP2/4/6/7 VLPs resulted in detection of anti-rotavirus IgG, but not IgA antibody in the intestine. IgG was associated with protection; all vaccinated rabbits showed total or partial protection from lapine P[14], G3 ALA virus challenge. VLPs induced lower titers of IgG than did native virus, resulting in slightly decreased levels of protection. Induction of IgG by VLPs reinforces previous results with live and inactivated SA11 virus, in that induction of protective immunity is not the result of undetectable levels of virus replication. For rotavirus or other mucosal pathogens, safe subunit vaccines may be efficacious when administered parenterally. It is important to note that protection from infection is a more stringent measure of protective efficacy than protection from disease. In field trials of live oral rotavirus vaccines in children, protection from severe disease, not protection from infection, is the measure of protective efficacy [7, 9]. At this juncture, neither the rabbit nor the mouse model can evaluate active immune protection from disease.

The source of the IgG in the intestine has not been determined but it is clear that, when present, anti-rotavirus IgG can mediate protection from virus infection in the intestine. In mice, parenteral vaccination with VLPs also resulted in detection of neutralizing, total Ig and IgG anti-rotavirus antibody in the intestine, showing that induction of IgG following parenteral vaccination is not a species-specific phenomenon. Induction of both IgG and IgA rotavirus specific antibody secreting cells in the lamina propria has been reported in mice vaccinated parenterally with rotavirus [1]. Further studies will be required to determine the source and mechanisms of the IgG induction. It is doubtful that cell mediated immunity, which was not evaluated in the rabbits, played any role in protection from reinfection. Recent findings using gene-knockout and immunodeficient mice infected and challenged with homologous murine rotavirus, indicate that cytotoxic T cells are not important in protection from reinfection, although they do play a role mediating virus clearance [16, 18].

Production of chimeric VLPs that displayed both G1 and G3 VP7s on their surface has previously been reported [10]. Immunization of mice with such chimeric particles resulted in induction of neutralizing antibody to both the G1 and G3 VP7. Further comparison of these results with immunization of mice with VP2/6/7 or VP2/4/6/7 VLPs that displayed only a G1 or G3 VP7 indicated that the G1 VLPs induced homotypic and heterotypic serum and intestinal neutralizing antibody to G1 Wa and G3 SA11 viruses [5, 11]. These data indicate the feasibility of producing a broadly reactive subunit vaccine with a limited number of types of VLPs. The protective efficacy of VLPs lacking VP4 and G1 VLPs to heterotypic challenge is being evaluated in animals.

Rotaviruses are a major cause of diarrhea in young calves, but because very young calves are at the greatest risk of severe diarrhea and death, they have insufficient time to achieve active protective immunity [9, 19, 20]. Rotaviruses administered parenterally have previously been shown to induce levels of antibodies in the colostrum of vaccinated cows sufficient to protect their calves from rotavirus disease [21]. P[2], G3 VP2/4/6/7 VLPs induced significantly increased titers of antibody to rotavirus in serum, colostrum and milk [15]. Protection studies using the colostrum from VLP vaccinated cows are currently in progress.

Rotavirus VLPs show promise as a subunit vaccine based on inducing protective immunity in rabbits and immunogenicity studies in mice and cows. A rotavirus subunit vaccine has potential advantages over live virus vaccines including safety and stability. Further testing of subunit vaccines and alternative routes of immunization for rotavirus vaccines is warranted.

Acknowledgements

The excellent help provided by Rhonda Carter, Kathleen Reed, Suzette Groner, Sheryl Henderson, Silvia Krasuk, Ramon Simon, Luis Flores and Tonja Gray is gratefully acknowledged. This work was supported in part by Veterans Administration Merit Review, Public Health Service grant AI 24998 from the National Institute of Allergy and Infectious Disease, Texas Applied Technology Program Grant 004949-029, and NRICGP Competitive grant No. 93-37204-9201 from the CSRS/USDA.

References

1. Coffin SE, Klinek M, Offit PA (1995) Induction of virus-specific antibody production by lamina propria lymphocytes following intramuscular inoculation with rotavirus. J Infect Dis 71: 874–878
2. Conner ME, Crawford SE, Barone C, Estes MK (1993) Rotavirus vaccine administered parenterally induces protective immunity. J Virol 67: 6633–6641
3. Conner ME, Crawford SE, Barone C, Estes MK (1994) Rotavirus or virus-like particles administered parenterally induce active immunity. In: Vaccines 94. Proceedings of Modern approaches to new vaccines including prevention of AIDS, Cold Spring Harbor 1993. Cold Spring Harbor Press, Cold Spring Harbor, NY, pp 351–355
4. Conner ME, Crawford SE, Barone C, Cohen J, Estes MK (1996) Subunit rotavirus vaccine administered parenterally induces protective immunity (manuscript in preparation)
5. Conner ME, Crawford SE, Barone C, Zhou YJ, Estes MK (1995) Induction of heterotypic neutralizing antibodies by serotype G1 VLPs. Abstract presented at the Fifth

International Symposium on Double-Stranded RNA Viruses, Djerba, Tunisia, March 19–23, 1995

6. Conner ME, Estes MK, Graham DY (1988) Rabbit model of rotavirus infection. J Virol 62: 1625–1631
7. Conner ME, Estes MK, Offit PA, Clark HF, Franco M, Feng N, Greenberg HB (1996) Development of a Mucosal Rotavirus Vaccine. In: Kiyono H, Ogra PL, McGhee JR (eds) Mucosal Vaccines. Academic Press, San Diego
8. Conner ME, Gilger MA, Estes MK, Graham DY (1991) Serologic and mucosal immune response to rotavirus infection in the rabbit model. J Virol 65: 2562–2571
9. Conner ME, Matson DO, Estes MK (1994) Rotavirus Vaccines and Vaccination Potential. In: Ramig RF (ed) Rotaviruses. Springer, Berlin, pp 286–337 (Curr Top Microbiol Immunol, vol 185)
10. Crawford SE, Labbé M, Cohen J, Burroughs MH, Zhou YJ, Estes MK (1994) Characterization of virus-like particles produced by the expression of rotavirus capsid proteins in insect cells. J Virol 68: 5945–5952
11. Crawford SE, Barone C, O'Neal C, Conner ME, Cohen J, Estes MK (1996) Induction of heterotypic neutralizing antibodies by rotavirus-like particles containing a G1 VP7 (manuscript in preparation)
12. Estes MK (1996) Rotaviruses and their replication. In: Fields BN, Knipe DM, Howley PM (ed) Fields Virology, 3rd edn. Lippincott-Raven, Philadelphia, pp 1625–1655
13. Estes MK, Graham DY (1980) Identification of rotavirus of different origins by the plaque reduction test. Am J Vet Res 41: 151–152
14. Estes MK, Crawford SE, Penaranda ME, Petrie BL, Burns JW, Chan W-K, Ericson B, Smith, GE, Summers MD (1987) Synthesis and immunogenicity of the rotavirus major capsid antigen using a baculovirus expression system. J Virol 61: 1488–1494
15. Fernandez FM, Todhunter D, Smith KL, Parwani AV, Estes MK, Crawford SE, Conner ME, Saif LJ (1994) Isotype specific antibody responses to rotavirus and virus proteins in cows inoculated with subunit vaccines composed of recombinant SA11 rotavirus core-like particles (CLP) or virus-like particles (VLP). Vaccine (in press)
16. Franco MA, Greenberg HB (1995) Role of B cells and cytotoxic T lymphocytes in clearance of and immunity to rotavirus infection in mice. J Virol 69: 7800–7806
17. Labbé M, Charpilienne A, Crawford SE, Estes MK, Cohen J (1991) Expression of rotavirus VP2 produces empty core-like particles. J Virol 65: 2946–2952
18. McNeal MM, Barone KS, Rae MN, Ward RL (1995) Effector functions of antibody and CD8[+] cells in resolution of rotavirus infection and protection against reinfection in mice. Virology 214: 387–397
19. Saif LJ, Jackwood JJ (1990) Enteric virus vaccines: theoretical considerations, current status and future approaches. In: Saif LJ, Theil KW (eds) Viral Diarrhea of Man and Animals. CRC Press, Boca Raton, pp 313–329
20. Saif LJ, Redman DR, Smith KL, Theil KW (1983) Passive immunity to bovine rotavirus in newborn calves fed colostrum supplements from immunized or nonimmunized cows. Infect Immun 421: 1118–1131
21. Saif LJ, Smith KL, Landemeir BJ, Bohl EH, Theil KW, Todhunter DA (1984) Immune response of pregnant cows to bovine rotavirus immunization. Am J Vet Res 45: 49–58

Authors' address: Dr. Margaret E. Conner, Division of Molecular Virology, Baylor College of Medicine, Houston, TX 77030, U.S.A.

Arch Virol (1996) [Suppl] 12: 207–215

DNA vaccines against rotavirus infections

J. E. Herrmann[1], S. C. Chen[1], E. F. Fynan[1,3], J. C. Santoro[2], H. B. Greenberg[4], and H. L. Robinson[2]

[1] Division of Infectious Diseases and Immunology, and [2] Department of Pathology, University of Massachusetts Medical School, Worcester, Massachusetts, U.S.A.
[3] Department of Biology, Worcester State College, Worcester, Massachusetts, U.S.A.
[4] Division of Gastroenterology, Stanford University School of Medicine, Stanford, California, U.S.A.

Summary. Plasmid DNA vaccines encoding for murine rotaviral proteins VP4, VP6, and VP7 were tested in adult BALB/c mice for their ability to induce immune responses and provide protection against rotavirus challenge. Serum antibodies were measured by virus neutralization and by ELISA. Cellular immunity was assessed by measuring cytotoxic T cell (CTL) responses. The vaccines were administered by inoculation into cells of the epidermis with an Accell gene gun (Auragen, Inc., Middleton, WI, USA). Each of the three vaccines elicited rotavirus-specific serum antibodies as measured by ELISA. Virus neutralizing antibodies were detected in mice receiving DNA vaccines encoding for VP4 and VP7, but not in those which received the plasmid encoding for VP6. Vaccines encoding for VP4, VP6, or VP7 generated virus-specific CTL responses in recipient mice. Efficacy of the vaccines was determined by challenge with homotypic rotaviruses. Each of the three vaccines was effective in protecting mice against infection after rotavirus ($100\ \text{ID}_{50}$) challenge. Significant reductions ($p < 0.0002$, analysis of variance) in viral excretion measured over a 9 day period were seen in mice receiving the DNA vaccines compared with mice that received control plasmids.

Introduction

Rotavirus infections remain the most important cause of severe acute diarrhea in children worldwide, causing as many as 1 million deaths per year, and because it is thought that diarrhea in developing countries is not likely to be reduced by improved sanitation and water supplies, control measures will require effective vaccines [1]. Rotaviruses are also important causes of pediatric hospitalizations in developed countries. The immune mechanisms (cellular and humoral) for controlling and preventing rotavirus infection and illness are still not well understood. This is true despite extensive investigations concerning immune responses to various rotavirus proteins, experience with live attenuated oral rotavirus vaccines, and seroepidemiological studies. Although progress has been made in the development of live oral vaccines, improved vaccines are still needed, particularly in developing countries where the need is the greatest but where the vaccines have been the least effective [1, 4, 13, 18]. The use of killed rotavirus vaccines and subunit vaccines have been suggested [1], but these types of vaccines do not provide endogenously synthesized proteins and generally do not elicit cytotoxic T cell responses (CTL), responses which are important in controlling infection. A new approach to developing subunit vaccines is the use of direct inoculation of DNA encoding for specific viral proteins. This allows for the expression of immunizing proteins by host cells which take up inoculated DNA, and does not require the use of purified proteins or viral vectors, both of which have limitations. Expression of the immunizing proteins in the host cells results in the presentation of normally processed proteins to the immune system, which is important for raising immune responses against the native forms of proteins [21]. Expression of the immunogen in host cells also results in the immunogen having access to class I major histocompatibility complex presentation, which is necessary for eliciting $CD8^+$ responses. The application of DNA vaccines to viral diseases is still relatively new, but inoculation of DNA has been shown to generate immune responses and protective immunity for influenza A virus infection in mice [9, 19] and ferrets [21], for LCM virus in mice [23], and for rabies virus in mice [22].

Because the parenteral route of inoculation has been demonstrated to generate immunity to challenge rotavirus in both mice and rabbits [3, 15], the use of non-living vaccines such as the DNA vaccines we are proposing hold promise for immunizing against rotavirus infections. The DNA vaccines can be administered by several routes, including intramuscular, intravenous, or intradermal inoculations, or by gene-gun delivery of DNA-coated particles into the epidermis. The latter has proven to be the most efficient method [19] in terms of the amount of DNA required and the degree of immune responses obtained. Thus, we selected the gene gun method for use in studies on immune responses and protective immunity in the adult mouse model. We developed and tested DNA vaccines for rotavirus infections with cloned cDNA encoding for murine rotaviral proteins VP4, VP6 and VP7. VP4 and VP7 appear to be the major proteins which generate humoral and cellular protective immunity [3, 7]. The common rotavirus antigen, VP6, has not been shown to elicit protective immunity, but may be involved in cellular immunity [5]. This report summarizes the major findings we have obtained in development of rotavirus DNA vaccines.

Materials and methods

Preparation of vaccine DNAs

Plasmids encoding for rotaviral proteins were prepared by insertion of cDNAs into the pCMV intron A (IA) expression vector provided to us by Dr. J. Mullins, University of Washington (plasmid JW4303, unpublished). This vector contains sequences from the cytomegalovirus immediate early promotor to drive transcription and sequences from bovine growth hormone genes to provide polyadenylation signals. cDNAs encoding for rotavirus proteins were inserted by blunt end or cohesive end ligation into *Hind* III and *Bam* HI sites. Newly constructed plasmids were tested for correct orientation by restriction endonuclease digestion. Expression of specific rotavirus proteins in transfected COS-1 (SV40-transformed African green monkey kidney) cells was confirmed by indirect immuno-fluorescent staining with monoclonal antibodies to rotavirus VP4, VP6 or VP7. Control plasmids were the plasmids without the rotaviral cDNAs inserts.

Gene gun delivery of DNA

DNA affixed to gold beads of 0.95 μm diameter were coated on plastic tubes and accelerated into the epidermis by use of a gene gun, the Accell instrument (Auragen, Inc., Middleton, WI). This instrument uses a helium pulse as the force to propel the beads into the epidermis. The methods we use to coat the gold beads with DNA and to prepare mice for immunizations have been given in detail [9].

Virus

Epizootic diarrhea of infant mice (EDIM) rotavirus strain EW (P10 [16], G3) was used for virus challenge and for preparation of cDNAs encoding for VP4, VP6 and VP7. A culture-adapted EDIM virus strain EW was used for neutralization and cytotoxic T cell studies. The mice used for vaccine studies were obtained from rotavirus-free colonies (Charles River Laboratories, Portage, MI) at 6 to 8 weeks of age and were housed in plastic microisolaters before and after immunization.

Cytotoxic T cell (CTL) responses

The procedures we used to determine memory CTL activity were based on those we described for influenza virus [12] and by Offit *et al*, for rotaviruses [16]. Briefly, effector cells were prepared from spleens from rotavirus-infected mice. The spleens were minced, the cells suspended, and the cells stimulated by the addition of normal splenocytes infected with EDIM virus. Stimulator cells were prepared by mincing spleens from normal, uninfected BALB/c mice and infecting them with EDIM virus, and were added to the effector cells. The cell mixtures were incubated for 5 days at 37 °C, washed by centrifugation, and added to P815 (H-2^d) target cells. The target cells were infected with EDIM virus or were uninfected. Medium containing ^{51}Cr as NaCrO$_4$ was added to the effector cells in microtiter plates. Other wells containing target cells plus detergent (Renex 30) determine the maximum releasable ^{51}Cr. After incubation for 4 hours at 37 °C, release of ^{51}Cr was measured with a gamma counter and immune lysis determined.

Rotavirus antigen and antibody tests

For monitoring viral antigen shedding in mouse feces, we used a monoclonal antibody-based enzyme-linked immunosorbent assay (ELISA) in microtiter plates [11]. For evaluating serum antibody responses, an indirect ELISA in microtiter plates was used with EDIM virus-coated wells, and neutralizing antibodies were assayed by measuring reduction of EDIM plaque-forming units in MA-104 cells. Intestinal IgA antibodies to EDIM virus were determined by use of IgA-specific peroxidase-labeled antiglobulin in an indirect ELISA [18]. Five percent (wt/vol) stool suspensions in 0.01 M phosphate buffered saline, pH 7.1, were used for assay of intestinal IgA.

Results and discussion

Immune responses in DNA-vaccinated mice

DNA encoding for either VP4, VP6 or VP7 of EDIM virus was inoculated into BALB/c (H-2^d) mice by gene gun delivery of DNA-coated gold beads into the epidermis. The dose given was 0.4 µg of DNA per mouse. Two inoculations were given at 4-week intervals. Mice were tested for serum antibody levels prior to inoculation and again at four weeks after the second inoculation. Serum antibody levels were determined by ELISA against EDIM virus for mice receiving VP4, VP6, or VP7 DNA, or control DNA (plasmid vector without inserts for rotavirus proteins). For comparison with the DNA inoculated mice, mice were given live EDIM virus by oral gavage 4 to 5 weeks prior to testing. The results of the serological studies showed that ELISA antibody developed to EDIM virus in mice receiving live virus or plasmids containing genes for the rotaviral proteins (Fig. 1). Neutralizing antibodies were detectable in mice receiving live virus and in those inoculated with DNAs encoding for VP4 or VP7, but not in mice receiving DNAs encoding for VP6. Both neutralizing antibody and ELISA titers were highest in mice receiving live virus. Antibody titers remained below detectable levels (1:50 or 1:100) in mice inoculated with the control DNA. The specificity of the serum antibodies for the rotavirus protein expected to be generated by the DNA vaccine given was determined by immuno-fluorescence staining of baculovirus-expressed rotavirus proteins in SF9 cells. It was determined that mice inoculated with VP4 or VP6 DNA vaccines generated antibodies specific for those proteins only. Mice receiving VP7 DNA vaccine did not give responses to either VP4 or VP6; studies to determine reactivity with VP7 are in progress.

Cellular immunity was determined by measuring memory CTL activity in mice that were inoculated with DNA vaccines or that were given EDIM virus, as was done for the antibody studies above. Memory CTL activity was measured after *in vitro* stimulation as described above. It was found that EDIM virus specific CTL activity was generated by mice inoculated with each of the DNA vaccines a E:T ratios of both 60:1 and 30:1 (Table 1). The magnitude of the

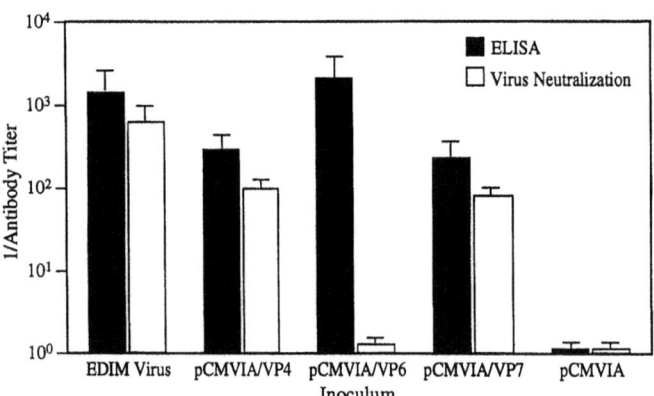

Fig. 1. Serum antibody responses to EDIM virus in mice given DNA vaccines encoding for rotaviral proteins VP4, VP6 or VP7 (six mice per group) or control DNA (pCMVIA, 18 mice), or to live EDIM virus (18 mice). The results are given as mean titers

Table 1. Virus-specific secondary CTL responses in mice receiving rotavirus DNA vaccines

Effector cell: Target cell	Range of % specific lysis*				
	EDIM virus infected (n = 4)	DNA vaccine specific for			
		VP4 (n = 3)	VP6 (n = 4)	VP7 (n = 3)	Plasmid control (n = 3)
60:1	29–95	16–44	15–79	28–64	0–4
30:1	15–46	9–21	8–32	30–34	0–4

*CTL assays were done with spleen cells obtained from mice after their second DNA immunization (18 weeks post-immunization for VP7 vaccine, 4 weeks for the others), followed by virus-restimulation in vitro for 5 days

responses were not as high at E:T ratios of 60:1 as that obtained by infection with EDIM virus, but there were still effective responses. There was minimal lysis of uninfected target cells by the effectors (data not shown).

Protective immunity in adult mice

We used the model developed by Ward et al. for BALB/c mice [20]. In this model, the endpoint is infection rather than illness, because illness is generally limited to infant mice aged 15 days or younger. The adult mouse (6 weeks or older) becomes infected and sheds virus in its feces for approximately one week post infection. Virus shedding in feces was measured by ELISA. The ability of each of the DNA vaccines to provide protective immunity against virus challenge was tested in this model. The mice were challenged orally with $100\,ID_{50}$ of homotypic EDIM rotavirus. The results are shown in Fig. 2A, B and C. Mice (6 per group) which had received the control plasmid showed no protection against challenge virus, whereas mice which had received the plasmid encoding for either VP4, VP6 or VP7 (two inoculations, four weeks apart) showed protection similar to that obtained in mice which had been orally inoculated with EDIM virus 4 weeks earlier.

Intestinal IgA antibody responses in rotavirus challenged mice

Previous studies with EDIM virus given orally to mice have shown that protection against rotavirus challenge was associated with rotavirus-specific fecal IgA responses [8], and we assume that intestinal IgA antibodies may be important in protecting mice immunized with rotavirus DNA vaccines. In preliminary studies, we have seen a marked increase in fecal IgA antibody response in EDIM virus-challenged mice that had been immunized with EDIM virus or with the plasmid encoding for VP6, compared with mice receiving the

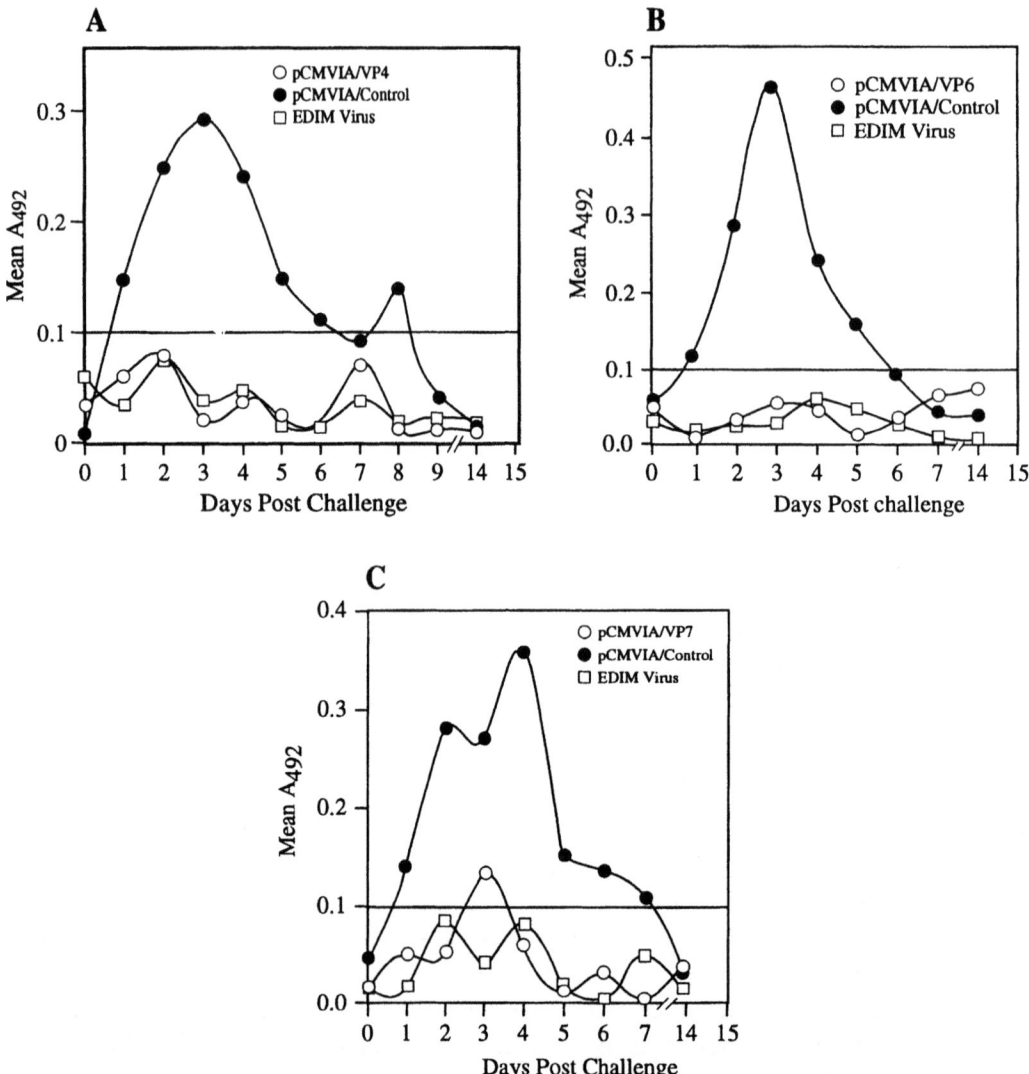

212   J. E. Herrmann et al.

Fig. 2. Protection against EDIM virus challenge in immunized BALB/c mice. Symbols: (○), mice inoculated with pCMVIA/VP4 (Fig. 2A), pCMVIA/VP6 (Fig. 2B), or pCMVIA/VP7 (Fig. 2C) (●) mice inoculated with control plasmid pCMVIA: (□), mice which had been infected with EDIM virus one month prior to challenge. Virus shedding in feces was determined by an ELISA for detecting rotavirus antigen and the results reported as A_{492} values. A positive test is one in which the value is ⩽0.1. There were significant differences (p < 0.0002, analysis of variance with repeated measures) in viral shedding between the mice receiving the plasmids encoding VP4, VP6 or VP7 and the control plasmids for days 1 through 6 post challenge

plasmid control (Fig. 3). Increases were also seen in mice receiving the VP4 or VP7 vaccines, but to a lesser degree.

Our results demonstrate that plasmid DNAs encoding for murine rotaviral proteins VP4, VP6, or VP7 were effective in eliciting both antibody and CTL

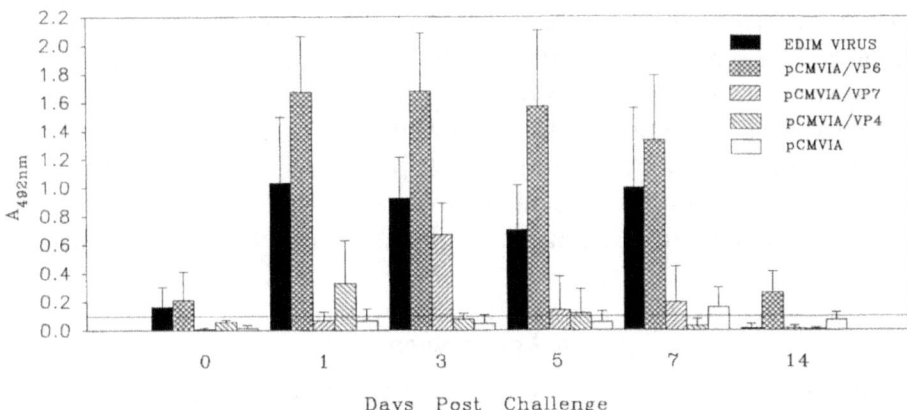

Fig. 3. Rotavirus specific fecal IgA, determined by ELISA, in immunized mice challenged with EDIM virus. Results are expressed as mean A_{492} values, 6 mice per group. A_{492} values > 0.1 are considered positive for IgA

responses, and resulted in protective immunity in an adult mouse model of rotavirus infection. It is interesting to note that genes encoding for murine or simian (SA-11) rotavirus VP7 inserted in vaccinia or adenovirus vectors did not protect adult mice against challenge rotavirus ([6], H. B. Greenberg, unpubl. data), although pups from dams immunized with adenovirus VP7 vectors were protected [2]. Thus, the manner in which plasmid DNAs express proteins may more closely mimic that obtained in natural infections, or may be unique.

The results obtained with the plasmid encoding for VP6 were particularly interesting in that this is the first direct demonstration of VP6 inducing protective immunity. VP6 encoded in an adenovirus or vaccinia virus vector did not protect mice, ([6, 10], H. B. Greenberg, unpubl. data) and monoclonal antibodies to VP6 given orally to mice were not protective [17]. We know that neutralizing antibodies are not produced (Fig. 1), so that the mechanism of protection likely involves cellular immunity, or possibly IgA-mediated intracellular neutralization of incomplete virus, or both. IgA-mediated intracellular virus neutralization has been demonstrated for Sendai virus [14], and recent data [10] suggest that IgA mediated intracellular neutralization may occur with rotaviruses. This was based on the finding that in a murine model, "backpack tumor" transplantation of hybridoma cell lines secreting IgA anti-VP6 antibodies could prevent murine rotavirus infection, whereas the antibodies were not active when presented directly to the luminal side of the intestinal tract. Thus, based on these findings and our finding of enhanced IgA responses in VP6 DNA vaccinated mice after virus challenge, it is likely that IgA-mediated intracellular neutralization may be involved in the protective immunity seen with the VP6 DNA vaccine.

The use of DNA vaccines is a new approach to immunization which offers potential solutions to the difficulties encountered, especially in developing nations, with current live, orally administered rotavirus vaccines. Because our DNA vaccines are not given orally, there should be no interference by maternal antibodies in breast milk, with naturally occurring rotavirus infections, or with other enteric virus infections. The method of delivery, by a portable hand-held gene gun, is one that can be readily developed for widespread application. The use of plasmids encoding for specific rotavirus proteins offers a new approach to investigation of rotavirus immune responses and other factors involved in immunity as well.

Acknowledgements

This work was supported in part by Grant VRD/V27/181/48 from the World Health Organization and by a VA Merit Review Grant.

References

1. Bishop RF (1993) Development of candidate rotavirus vaccines. Vaccine 11: 247–254
2. Both GW, Lockett LJ, Janardhana V, Edwards SJ, Bellamy AR, Graham FL, Prevec L, Andrew ME (1993) Protective immunity to rotavirus-induced diarrhea is passively transferred to newborn mice from naive dams vaccinated with a single dose of a recombinant adenovirus expressing rotavirus VP7sc. Virology 193: 940–950
3. Conner ME, Crawford SE, Barone C, Estes MK (1993) Rotavirus vaccine administered parenterally induces protective immunity. J Virol 67: 6633–6641
4. Conner ME, Matson DO, Estes MK (1994) Rotavirus vaccines and vaccination potential. In: Ramig RF (ed) Rotaviruses. Current Topics in Microbiology and Immunology 185: 285–371
5. Dharakul T, Labbe M, Cohen J, Bellamy AR, Street JE, Mackow ER, Fiore L, Rott L, Greenberg HB (1991) Immunization with baculovirus-expressed recombinant rotavirus proteins VP1, VP4, VP6 and VP7 induce CD8$^+$ T lymphocytes that mediate clearance of chronic rotavirus infection in SCID mice. J Virol 65: 5928–5932
6. Dormitzer P, Burns JW, Greenberg HB, Prevec L, Graham F, Boyle DB, Both G (1993) Vaccination of adult mice with recombinant viruses expressing a protective rotavirus VP7sc. Abstr. IXth Intl. Cong. Virology, W21-2, p 43
7. Estes MR, Cohen J (1989) Rotavirus gene structure and function. Microbiol Rev 53: 410–449
8. Feng N, Burns JW, Bracy L, Greenberg HB (1994) Comparison of mucosal and systemic humoral immune responses and subsequent protection in mice orally inoculated with a homologous or a heterologous rotavirus. J Virol 68: 7766–7733
9. Fynan EF, Webster RG, Fuller DH, Haynes JR, Santoro JC, Robinson HL (1993) DNA vaccines: protective immunizations by parenteral, mucosal, and gene-gun inoculations. Proc Natl Acad Sci USA 90: 11478–11482
10. Burns JW, Pajouh MS, Krishnaney AA, Greenberg HB (1996) Protective effect of rotavirus VP6-specific IgA monoclonal antibodies that lack neutralizing activity. Science 272: 104–107
11. Herrmann JE, Blacklow NR, Perron DM, Cukor G, Krause PJ, Hyams JS, Barrett HJ, Ogra PL (1985) Enzyme immunoassay with monoclonal antibodies for the detection of rotavirus in stool specimens. J Infect Dis 152: 831–832
12. Herrmann JE, West K, Bruns M, Ennis FA (1990) Effect of rimantadine on cytotoxic T lymphocyte responses and immunity to reinfection in mice infected with influenza A virus. J Infect Dis 161: 180–184

13. Kapikian AZ, Flores J, Midthun K, Hoshino Y, Green KY, Gorziglia M, Nishikawa K, Chanock RM, Potash L, Perez-Schael I (1989) Strategies for the development of a rotavirus vaccine against infantile diarrhea with an update on clinical trials of rotavirus vaccine. Adv Exp Med Biol 257: 67–89

14. Mazenec MB, Kaetzel CS, Lamm ME, Fletcher D, Nedrud JG (1992) Intracellular neutralization of virus by immunoglobulin A antibodies. Proc Natl Acad Sci USA 89: 6901–6905

15. McNeal MM, Sheridan JR, Ward RL (1992) Active protection against rotavirus infection of mice following intraperitoneal immunization. Virology 191: 150–157

16. Offit PA, Dudzik KI (1988) Rotavirus-specific cytotoxic T lymphocytes cross-react with target cells infected with different rotavirus serotypes. J Virol 62: 127–131

17. Riepenhoff-Talty M, Dharakul R, Kowalski E, Sterman D, Ogra PL (1987) Rotavirus infection in mice: pathogenesis and immunity. Adv Exp Med Biol 216B: 1015–1023

18. Saif LJ, Jackwood DJ (1990) Enteric Virus Vaccines: Theoretical considerations, current status, and future approaches. In: Saif LJ, Theil KW (eds) Viral Diarrheal of Man and Animals. CRC Press, Boca Raton, pp 313–329

19. Ulmer JB, Donnelly JJ, Parker SE, Rhodes GH, Felgner PI, Dwarki VJ, Gromkowski SH, Deck RR, DeWitt CM, Friedman A, Hawe LA, Leander KR, Martinez D, Perry HC, Shiver JW, Montgomery DL, Liu MA (1993) Heterologous protection against influenza by injection of DNA encoding a viral protein. Science 259: 1745–1749

20. Ward RL, McNeal MM, Sheridan JF (1990) Development of an adult mouse model for studies on protection against rotavirus. J Virol 64: 5070–5075

21. Webster RG, Fynan EF, Santoro JC, Robinson H (1994) Protection of ferrets against influenza challenge with a DNA vaccine to the haemagglutinin. Vaccine 12: 1495–1498

22. Xiang ZO, Spitalnik S, Tran M, Wunner W, Cheng J, Ertl HCL (1994) Vaccination with a plasmid vector carrying the rabies virus glycoprotein gene induces protective immunity against rabies virus. Virology 199: 132–140

23. Zarozinski CC, Fynan EF, Selin LK, Robinson HL, Welsh RM (1995) Protective CTL-dependent immunity and enhanced immunopathology in mice immunized by particle bombardment with DNA encoding an internal protein. J Immunol 154: 4010–4017

Authors' address: Dr. J. E. Herrmann, University of Massachusetts Medical Center, Division of Infectious Diseases and Immunology, 55 Lake Avenue North, Worcester, MA 01655, U.S.A.

Arch Virol (1996) [Suppl] 12: 217–223

Prophylaxis of rotavirus gastroenteritis using immunoglobulin

T. Ebina

Division of Immunology, Research Institute Miyagi Cancer Center,
Natori, Miyagi, Japan

Summary. Oral inoculation of the human group A rotavirus MO strain (G serotype 3) into 5-day-old BALB/c mice causes gastroenteritis characterized by diarrhea. Using this small animal model, passive protection of suckling mice against human rotavirus infection was achieved with the use of immunoglobulin (IgY) from the yolks of eggs of rotavirus-immunized hens. When IgY against the rotavirus strain homotypic with the challenge virus (MO strain) was administered to mice, complete protection was achieved. After immunizing 8-month old pregnant Holstein cows with human rotavirus MO strain, colostrum containing neutralizing antibody to four different G serotypes of human rotavirus, designated Rota colostrum, was obtained. Rota colostrum completely protected suckling mice against rotavirus infection, and purified IgG obtained from Rota colostrum protected against infection with the homologous virus. After randomly grouping 20 infants from a baby care center, 10 infants received 20 ml of Rota colostrum for 2 weeks and 10 control infants received none. Rotavirus-associated diarrhea developed in 7 of the 10 infants in the control group. None of the three infants in the group daily receiving the Rota colostrum had such symptoms, and one of three infants in the group receiving treatment, every other day developed rotavirus-induced diarrhea. Oral administration of Rota colostrum seems to be an effective and safe means of preventing diarrhea caused by human rotavirus infection. Recently, the immunized cows were boosted by reinjection of 4 serotypes of human rotavirus into a superficial cervical lymph node two weeks after delivery, resulting in mass production of cow's milk containing a high-titered antibody to human rotavirus. Therefore, the hyperimmune cow's milk is a candidate for a "physiologically functional food" in Japan.

Introduction

Group A rotaviruses were first recognized in 1973 as causes of infectious gastroenteritis in infants and are now known to be the leading cause of severe diarrhea in infants and young children throughout the world. Vaccine development is in progress, but difficulty has emerged in obtaining sufficiently broad protection against disease caused by the several human rotavirus VP7 (G) serotypes. The principal failure of vaccine candidates in field trials has been due to the limited immune response elicited in mucosal regions where rotavirus infections occur. The most important protection factor against the development of rotavirus diarrhea is the presence of specific antibody in the lumen of the small intestine. In order to analyze the precise mechanism by which rotavirus-induced gastroenteritis can be prevented with immunoglobulins, we needed a small animal model of experimental infection initiated by a human rotavirus. We have succeeded in inducing typical gastroenteritis in suckling BALB/c mice infected with a human rotavirus (MO strain). The MO strain was isolated and adapted to cell culture growth in our laboratory in 1982 [8].

Five day old suckling BALB/c mice were orally inoculated with a single dose of 10^6 fluorescent cell focus forming units, hereafter abbreviated FCFU, of group A human rotavirus serotype 3, MO strain, in 50 µl of minimum essential medium (MEM). Infant mice were inspected daily for past diarrhea as revealed by anal smears and for existing diarrhea after gentle palpation of the abdomen. Successful rotavirus infections in suckling mice were characterized by the development of diarrhea 24 hrs postinoculation. Forty eight hrs after infection, 66/74 (about 90%) of animals inoculated with MO virus developed diarrhea. The infections usually resolved by the 4th to 5th day postinoculation without mortality. Viral shedding in fecal samples was detected by analysis of samples using electron microscopy at 24 and 48 hours after infection. Mice inoculated with the MO virus and MEM-inoculated control mice were sacrificed 24 hrs after inoculation. A loss of microvilli of enterocytes, degenerated enterocytes with intracytoplasmic vacuoles and an interstitial edema of villi were evident in the small intestines of MO-infected mice. Indirect immunofluorescent staining with antirotavirus guinea pig serum and FITC-conjugated anti-guinea pig IgG of the jejunum from a suckling BALB/c mouse was performed. In a rotavirus-infected intestine, 24 hrs after oral administration of the MO strain, the loss of microvilli and intracytoplasmic staining of rotavirus antigens in epithelial cells were noted [3, 4].

Results and discussion

This paper describes use of this animal model to evaluate passive protection of suckling mice using immunoglobulin Y (IgY) from egg yolk and milk from hyperimmune cows. A study in infants also is described.

Passive protection of suckling mice against human rotavirus infection using IgY

Passive protection of suckling mice against human rotavirus infection was achieved with the use of IgY from the yolks of eggs of rotavirus-immunized hens. Five-month-old Diya Cross B34 strain of specific-pathogen-free white leghorn hens were immunized intramuscularly four times with 10^7 FCFU of the Wa (serotype G1) or MO (serotype G3) strain of human rotavirus with complete Freund's adjuvant at intervals of one or two weeks. The water-soluble protein fraction from egg yolk collected every week was passed through filter paper, and assayed for neutralizing antibody titer against human rotavirus. After three immunizations with the Wa strain, neutralizing antibody (NT) titers in water-soluble protein from egg yolks rose to more than 11 000. After a booster injection, NT titers rose to a maximum of 450 000. We also isolated IgY from water-soluble protein fractions. Purified IgY fractions from water-soluble protein were finally filtered through a membrane filter and freeze-dried in powder form. The molecular weight of IgY is slightly larger than that of serum IgG; except for this, IgY is comparable to serum IgG.

The effect of egg-derived anti-MO IgY was tested in MO rotavirus-infected mice. Three hours before they were infected with 10^6 FCFU of the MO strain, 50 µl of anti-MO IgY solution was orally administered to 5-day-old BALB/c mice. Anti-MO IgY completely prevented diarrhea in mice infected with the MO strain. On the other hand, immunoglobulins isolated from egg yolks of unimmunized hens failed to prevent diarrhea. Anti-MO IgY prevented rotavirus gastroenteritis in this mouse model in a dose-dependent manner. Anti-Wa IgY isolated from the yolks of eggs from hens immunized with the heterotypic Wa strain of human rotavirus was capable of preventing MO-induced gastroenteritis; however, anti-Wa IgY solution diluted 10-fold did not prevent gastroenteritis. To determine the relative contribution of homotypic and heterotypic neutralizing activity in protecting suckling mice against MO virus challenge, we titrated the *in vivo* protective activity of a homotypic (MO) and heterotypic (Wa) IgY in suckling mice. Anti-Wa IgY protected mice against MO-induced gastroenteritis at a dilution of 1:5, whereas anti-MO IgY protected against MO-induced gastroenteritis at a dilution of 1:1 000. That is, protection against disease was closely correlated with the *in vitro* neutralizing activity of IgY against the MO virus [3, 7].

Passive protection of suckling mice using hyperimmune cow's milk

Eight-month pregnant Holstein cows were immunized subcutaneously with 10^7 FCFU of the Wa, KUN (serotype G2) or MO strain three times with complete Freund's adjuvant at 10 day intervals [1, 2, 5]. Colostrum collected on the first three days (5 L each) was immediately cooled and stored. Butter fat was removed by a cream separator, and bacteria and red blood cells were removed by three centrifugations at 10 000 g for 30 min. each. Colostral nonfat milk was tested for bacterial count and the absence of *E. coli* before being approved for drinking; approved milk was lyophilized in 20-ml units and stored until administration. MO-immune colostrum (lot 3–16) contained the highest NT titers and unimmunized colostrum contained no NT antibody against any of four human rotaviruses. The effect of Rota colostrum in MO rotavirus-infected mice was observed. One hour before being infected with 10^6 FCFU of the MO strain, 50 µl of colostrum was orally administered to 5-day-old BALB/c mice. Rota colostrum completely prevented diarrhea in mice infected with the MO strain. On the other hand, normal colostrum failed to prevent diarrhea [5].

220   T. Ebina

Passive protection of infants given hyperimmune cow's milk

The effect of Rota colostrum on human rotavirus infections was next investigated. Six infants in a baby care center were orally given 20 ml of Rota colostrum (lot 3–16), (group A) every morning [3] or (group B) every other morning [3]. Four other infants (group C) were given the colostrum when they first showed signs of gastrointestinal upset. Ten control infants did not receive Rota colostrum. Sporadic cases of acute gastroenteritis were observed in the baby care center. No other viruses except rotaviruses in stool samples were detected by electron microscopy. Eventually, rotavirus associated diarrhea developed in 7/10 infants in the control group. All four infants group C who received Rota colostrum after symptoms of gastrointestinal upset appeared developed diarrhea. One of three infants in group B and none of the three infants in group A developed rotavirus induced diarrhea. These results suggest that Rota colostrum prevented the outbreak of diarrhea. Our present work confirms the efficacy of a 2-week course of oral administration of Rota colostrum before an outbreak of rotavirus infection in a baby care center. We also isolated immunoglobulins from Rota colostrum. After casein had been removed by acid precipitation, colostral whey was applied to a protein G-Sepharose affinity column. The IgG fraction prevented rotavirus-induced gastroenteritis in our mouse model in a dose-dependent manner. Moreover, the main fraction of immunoglobulin of colostrum, IgG_1, also prevented rotavirus-induced gastroenteritis in a dose-dependent manner. Therefore, oral administration of Rota colostrum seems to be an effective and safe means of preventing diarrhea caused by human rotavirus infection.

Development of a method to produce large quantities
of hyperimmune cow's milk

In order to produce a large quantity of antibodies for immunization, previously immunized cows were re-immunized after they delivered, resulting in a high titer of neutralizing antibody in normal milk [6]. Table 1 shows an immunization and sampling procedure. Two months prior to delivery, Holstein cows received three intramuscular doses each of four serotypes of human rotaviruses. Immediately after delivery, colostrum was collected. Two weeks after delivery, the Holsteins were re-immunized with the four human rotaviruses in the superficial cervical lymph node, and on day 33 after delivery, 20 liters of normal milk was collected. A large quantity of nonfat milk was collected aseptically, without heating, by continuous centrifugation at $20\,000 \times g$ on day 33 after delivery using a newly developed continuous centrifuge CC-17 manufactured by Hitachi Koki, Co., Ltd., Hitachinaka, Japan. The flow rate of the sample was 10 L/hour. The nonfat milk was then placed in vials (each vial contained 20 ml of milk) and was freeze dried. The titers of neutralizing antibody against the human rotavirus-MO strain in CC-3 (colostrum collected on day 3 after delivery), CC-5 (colostrum collected on day 5 after delivery), and CC-33 (milk collected on day 33 after delivery), were 5 393, 5 132, and 2 652, respectively (Table 2). Even the milk from these immuniz-

Table 1. Immunization and sampling procedure

Date of procedure		Route of injection
July 14:	complete Freund's adjuvant (2 ml) + Wa, KUN, MO and ST-3 (serotype G4) strains of human group A rotavirus (0.5 ml each)	i.m.
Aug. 11:	incomplete Freund's adjuvant (2 ml) + 4 strains of human rotavirus (2 ml)	i.m.
Sept. 8:	IFA (2 ml) + 4 strains of HRV (2 ml)	i.m.
Sept. 25: (day 0)	delivery	
Sept. 28:	colostrum (CC-3) sampling	
Sept. 30:	colostrum (CC-5) sampling	
Oct. 8:	IFA (2 ml) + 4 strains of HRV (2 ml)	intra lymph node booster
Oct. 26:	milk (CC-33) sampling	

Table 2. Neutralizing antibody titers against human rotavirus MO strain in cow colostrum and milk

Sample	NT titer*
Colostrum CC-3	5393
Colostrum CC-5	5132
Milk CC-33	2652

*Titers are expressed as the reciprocal of sample dilution showing a 50% reduction in FCFU per visual field

ed animals contained a high titer of neutralizing antibody. When 50 µl of each sample was administered orally to mice one hour prior to infection, the onset of diarrhea was completely prevented (Table 3). In other words, when previously immunized cows were re-immunized in the lymph node after delivery, a high titer of neutralizing antibody was obtained in subsequent milkings. Approximately 20 L/day of high-titer milk can be collected for 100 days after delivery from each re-immunized cow. Consequently, because large scale production of the antibody is possible, the antibody can be developed for use as a "physiologically functional food" in Japan rather than as a passive vaccine. The definition of a "physiologically functional food" is: 1) the manufacturing process is clearly defined, 2) the functional food contains a "functional agent" that possesses a special property, and the chemical structure of the agent is known, 3) its configuration and quantity in the food are known, 4) the mechanism of action of

Table 3. Preventive effect of cow colostrum and milk against MO-induced diarrhea

Sample	Litter	No. of mice	No. with diarrhea (days after infection)				Diarrhea (affected/total)	% with diarrhea
			1	2	3	4		
Control MEM	A	7	4	1	3	0	10/15	66.7
	B	8	4	5	6	0		
Colostrum CC-3	C	5	0	0	0	0	0/13	0
	D	8	0	0	0	0		
Colostrum CC-5	E	7	0	0	1	0	1/12	8.3
	F	5	0	0	0	0		
Milk CC-33	G	4	0	0	0	0	0/11	0
	H	7	0	0	0	0		

One hour before being challenged with 10^6 FCFU of the human rotavirus MO strain, 6-day-old BALB/c suckling mice were orally inoculated with 50 µl of sample

the agent is known, 5) the agent is stable in the food product, and 6) the product is accepted as a food. Therefore, both colostrum and milk from human rotavirus-immunized cows contain a functional agent, immunoglobulin, and are thus suitable for use as "physiologically functional food" that can prevent the onset of human rotavirus-induced diarrhea. Future applications of this product appear promising.

Acknowledgements

I express my thanks to Dr. Minoru Ohta, Kawatabi Farm, Faculty of Agriculture, Tohoku University, and to Dr. Yoshihiro Kanamaru of the Department of Food Science, Gifu University, for their generous collaboration and to Ms. Eiko Ohkubo for her editorial assistance. This work was supported in part by a grant from the Sendai Institute of Microbiology and from the Ministry of Education, Science, and Culture of Japan.

References

1. Ebina T, Sato A, Umezu K, Ishida N, Ohyama S, Oizumi A, Aikawa K, Katagiri S, Katsushima N, Imai A, Kitaoka S, Suzuki H, Konno T (1983) Prevention of rotavirus infection by cow colostrum containing antibody against human rotavirus. Lancet 2: 1029–1030
2. Ebina T, Sato A, Umezu K, Ishida N, Ohyama S, Oizumi A, Aikawa K, Katagiri S, Katsushima N, Imai A, Kitaoka S, Suzuki H, Konno T (1985) Prevention of rotavirus infection by oral administration of cow colostrum containing antihuman rotavirus antibody. Med Microbiol Immunol 174: 177–185
3. Ebina T, Tsukada K, Umezu K, Nose M, Tsuda K, Hatta H, Kim M, Yamamoto T (1990) Gastroenteritis in suckling mice caused by human rotavirus can be prevented with egg yolk immunoglobulin (IgY) and treated with a protein-bound polysaccharide preparation (PSK). Microbiol Immunol 34: 617–629

4. Ebina T, Tsukada K (1991) Protease inhibitors prevent the development of human rotavirus-induced diarrhea in suckling mice. Microbiol Immunol 35: 583–588

5. Ebina T, Ohta M, Kanamaru Y, Osumi YY, Baba K (1992) Passive immunizations of suckling mice and infants with bovine colostrum containing antibodies to human rotavirus. J Med Virol 38: 117–123

6. Ebina T, Ohta M, Uchiwa H, Murakami U (1994) Production of anti-hair keratin antibody and its characteristics: Utilization of cow colostrum and milk. Anim Sci Technol 65: 580–590

7. Hatta H, Tsuda K, Akachi S, Kim M, Yamamoto T, Ebina T (1993) Oral passive immunization effect of anti-human rotavirus IgY and its behavior against proteolytic enzymes. Biosci Biotech Biochem 57: 1077–1081

8. Kutsuzawa T, Konno T, Suzuki H, Kapikian AZ, Ebina T, Ishida N (1982) Isolation of human rotavirus subgroups 1 and 2 in cell culture. J Clin Microbiol 16: 727–730

Author's address: Dr. T. Ebina, Division of Immunology, Research Institute Miyagi Cancer Center, 47-1 Nodayama, Medeshima-shiode, Natori Miyagi 981-12, Japan

Arch Virol (1996) [Suppl] 12: 225–235

Historical background and classification of caliciviruses and astroviruses

W. D. Cubitt

Department of Virology, Camelia Botnar Laboratories,
Great Ormond Street Hospital, London, U.K.

Summary. Infections caused by caliciviruses, *i.e.*, vesicular exanthema virus of swine were recognised as a major cause of economic loss in the 1930s. However, it was not until the application of electronmicroscopy in the 1970s that caliciviruses and astroviruses were recognised and proven to be a cause of diarrhoea and vomiting. The following review briefly describes the steps which have led to the development of diagnostic tests and enabled the characterization of several members of the *Caliciviridae* and *Astroviridae*. In the past five years this has culminated in the sequencing of their genomes and the expression of viral proteins. This in turn has led to the development of improved diagnostic tests *e.g.*, RT-PCR and enzyme immunoassays, and may pave the way towards producing effective vaccines in the future.

Historical background of the caliciviruses and astroviruses

The 1970s – The era of electron microscopy

The realization that electronmicroscopy (EM) could be used to examine negatively stained preparations of fecal material led to the discovery of many viruses and virus-like particles in samples from patients with diarrhoea. The first of these agents, Norwalk virus (NV) [20], was first visualized in a stool sample from a volunteer challenged with virus from an outbreak of winter vomiting disease in Norwalk, Ohio, USA in 1968 which had affected teachers and schoolchildren [1]. The technique of immune electron microscopy (IEM) was used to concentrate the virus particles, and as a result, the true morphology was obscured resulting in its description as a 27 nm diameter small round virus. This, together with its biophysical and biochemical characteristics resulted in NV being classified as a candidate parvovirus. Over the next few years, electronmicroscopists in the United Kingdom described the morphologically distinct astroviruses [2] [28], named because of their surface morphology which resembles a star (Fig. 1a), and caliciviruses (Fig. 1b) named after their characteristic cups (calices) [29]. Further studies, particularly in the USA, Japan, and Canada revealed other small round viruses which were given arbitrary names based on

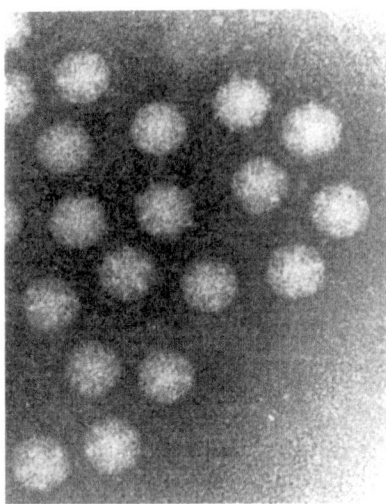

Fig. 1a. Fecal emulsion directly stained with 2% KPTA, pH 6.4 containing astrovirus particles, 27–28 nm diameter. Note that only a few particles display the surface star

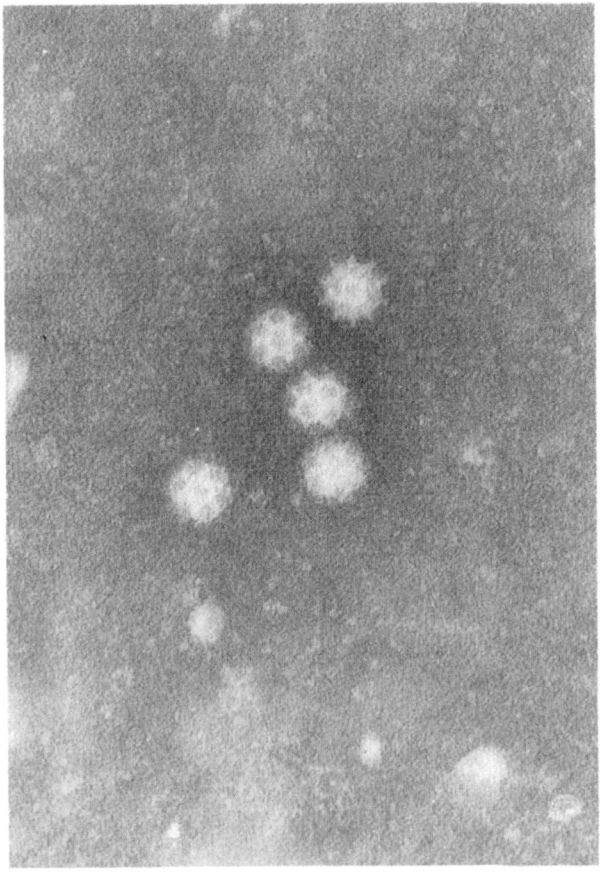

Fig. 1b. Fecal emulsion directly stained with 2% KPTA, pH 6.4 containing calicivirus particles displaying typical surface morphology. Four hollows arranged in a cross = 2 fold axis of symmetry; Star of david = 3 fold axis of symmetry; 10 spiked sphere = 5 fold axis of symmetry. Diameter of particles = 28–32 nm. Reproduced from Cubitt WD (1994) Caliciviruses. In: Kapikian AZ (ed) Viral Infections of the Gastrointestinal tract, pp 551, with permission Marcel Dekker, New York

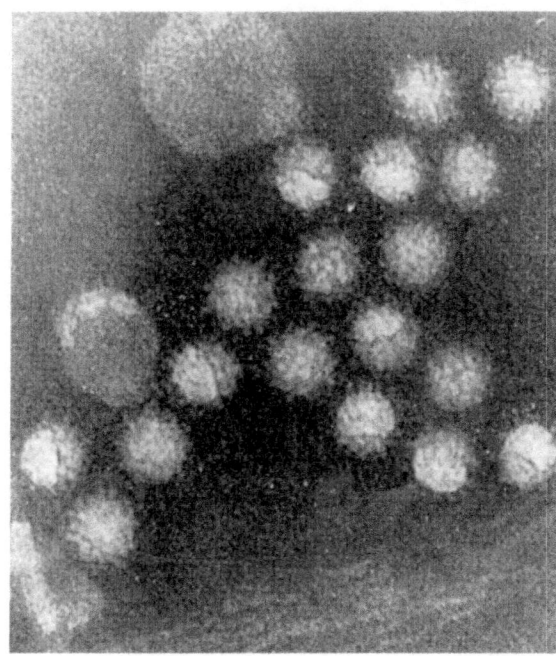

Fig. 1c. Fecal emulsion directly stained with 2% KPTA, pH 6.4 containing SRSVs resembling Norwalk virus. There is no clear evidence of cups on the surface and the particles have a "fuzzy", amorphous appearance. Diameter of particles = 30–35 nm

their place of origin *e.g.*, Marin County agent, Otofuke agent, or their appearance *e.g.*, minireovirus, fuzzy-wuzzy, small round structured virus (SRSV), (Fig. 1c). These arbitrary names created some confusion as to their classification which led Caul and Appleton [3] to propose an interim scheme for differentiating small round viruses (SRSVs) based on their appearance, size and buoyant density (Fig. 2). This morphological classification works well provided the viruses are displaying their characteristic morphology. However, in many preparations this is not the case, *e.g.*, only a small percentage of astroviruses display the characteristic surface star. Factors such as freezing and thawing of samples, degradation of particles by proteolytic enzymes, presence of coproantibodies and incorrect choice of stain may also influence particle morphology and make classification difficult. The electron micrograph (Fig. 3) of a fecal preparation from a patient with diarrhea illustrates such a situation. Using the scheme of Caul and Appleton (Fig. 2), one could say that the patient is infected with three viruses; a small round structured virus, a calicivirus and a small round featureless virus. On the other hand, one could argue that the micrograph is showing the various stages of degradation of a calicivirus.

Clinical importance of calicivirus and astroviruses

The use of EM soon established that caliciviruses, SRSVs and astroviruses were associated with cases of diarrhea and vomiting in many parts of the world where these facilities were available. Although each virus has been shown to cause

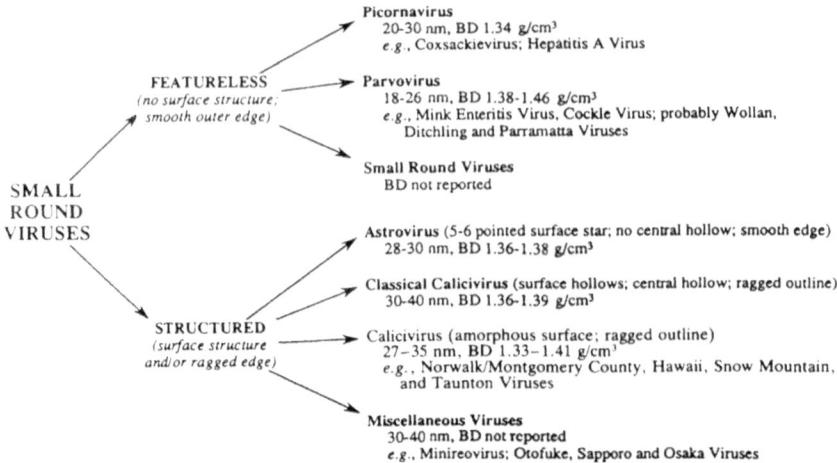

Fig. 2. Classification scheme for small round enteric viruses devised by Caul and Appleton (reference [3]) and modified by Kapikian AZ (1994) In: Kapikian AZ (ed) Virus Infections of the Gastrointestinal tract, chapt 14, pp 477, with permission Marcel Dekker, New York

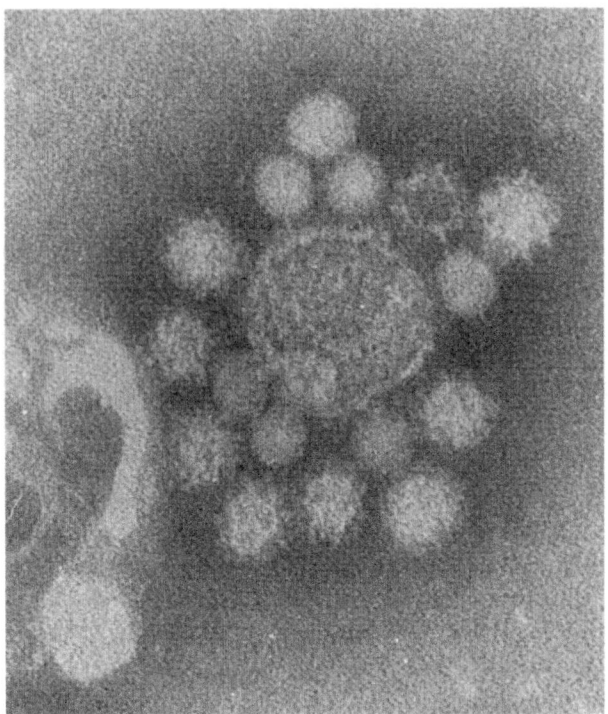

Fig. 3. Fecal sample stained with 2% KPTA, pH 6.4 illustrating various stages of degradation of HuCV. Note particles resembling those in Fig. 1c with an amorphous structure, a particle with distinctive hollows on the surface, a shell showing surface spikes and smaller smooth particles with a diameter of 26–28 nm

illness in all age groups, caliciviruses and astroviruses were most frequently associated with cases in infants and the elderly, whereas SRSVS were more commonly linked to outbreaks in adults and the elderly [20]. Further evidence that astroviruses and SRSVs cause diarrhea was provided by human volunteer studies conducted in the UK [24], USA [40] and Australia [11]. Other recorded features of illness included nausea, pyrexia, abdominal pain, aching limbs and headaches.

Evidence of antigenic variation

Cross challenge experiments in volunteers and IEM provided proof that there were at least three strains of SRSV circulating in the US; Norwalk, (NV); Hawaii, (HV) and Snow Mountain (SMA) agents [40]. Further studies using IEM showed that their were four strains in the UK [27] and nine were reported in Japan. Similar studies showed evidence of several strains of morphologically typical caliciviruses (UK1–3) and Sapporo/Japan [5] and eight strains of astroviruses, HAstV1–8 [23].

The 1980s – Development of diagnostic assays

Numerous attempts to infect animals and to propagate SRSVs and caliciviruses in cell culture met with little success. Therefore development of assays was limited to a few research laboratories which had access to material derived from adult volunteers or from well-defined outbreaks. However, radioimmune assays (RIA) and enzyme immune assays (EIA) were developed for NV, HV, SMA and human calicivirus/Sapporo (refs cited in 4) which were used successfully to provide epidemiological data and to measure antibody responses. As a result, it was clearly established that many of these viruses are prevalent in countries throughout the world and are a common cause of diarrhea.

A similar situation existed with astroviruses, although Lee and Kurtz had succesfully propagated them in primary human embryonic cells, and later demonstrated that they could be serially passaged in a continuous monkey kidney cell line, LLCMK$_2$. Virus particles could be demonstrated in the cytoplasm by EM and detected by immunofluorescence [26]. Success in growing the virus enabled antisera to be raised in rabbits and these sera were used to demonstrate that there were five strains of astrovirus. Using virus obtained from Oxford, Herrmann et al. [13] raised a group-specific monoclonal antibody which led to the development of an EIA [14]. This EIA enabled Marin County agent to be identified as astrovirus type 5 rather than a Norwalk-like agent [12] and has in recent years provided some evidence that astroviruses may be a more important casue of diarrhea than had been previously recognized [10].

Identification of the viral structural proteins

The classification of SRSVs remained a matter of debate although some investigators noted that certain particles appeared to have surface indentations similar to those of caliciviruses. Further evidence that these might be caliciviruses was provided when it was shown that NV, SMA and SRSV 9 contained a single major structural protein with an apparent molecular weight (60–70 kd) similar to HuCV (refs cited in [4]) and other members of the family. Similar studies of astroviruses revealed two to four structural proteins MW 32–36 kd leading to the suggestion that they might be members of the *Picornaviridae* [22].

The 1990s – Molecular characterization and expression of capsid proteins in baculovirus

The successful application of molecular techniques for sequencing viral genomes has resulted in major advances in our understanding of caliciviruses and astroviruses in the past three years. In 1992 groups in the USA and UK published the genomic sequences and predicted genome organization of two SRSVs, Norwalk [17] and Southampton [25] viruses (Fig. 4). This is established that the genome consists of three open reading frames. ORF 1 codes for the non structural proteins and ORF 2 for the capsid protein; the function of ORF 3 at the 3′ end still remains uncertain. On the basis of this additional evidence, it became apparent that these SRSVs had a genomic organization similar to the *Caliciviridae* a classification which has been accepted by the ICTV [6]. Once the sequences were known, many groups around the world applied the reverse transcription polymerase chain reaction (RT-PCR) using primers directed towards what was thought to be a highly conserved region within the RNA dependent RNA polymerase region of ORF1 to amplify and detect the genomic RNA of many SRSVs. Application of this technique met with varying degrees of success [9, 33, 38], and as a result of cloning and sequencing the amplified

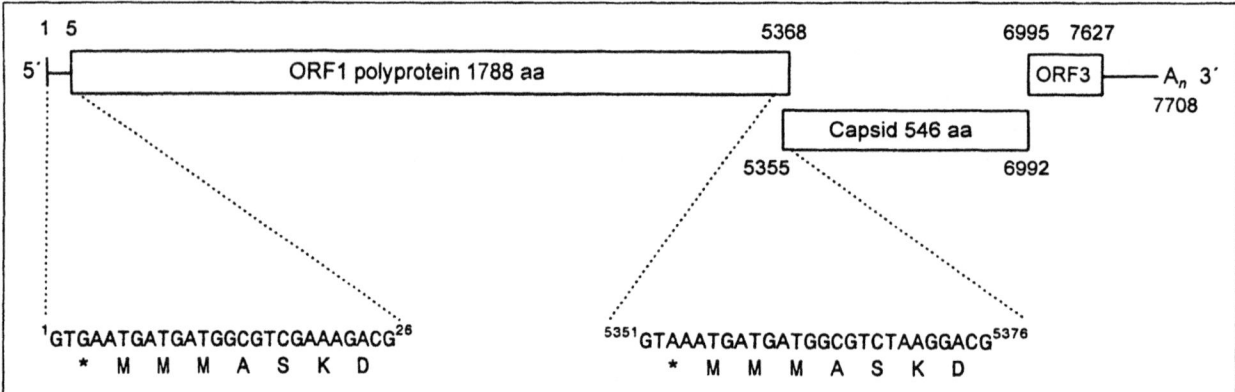

Genome structure and organization of Southampton virus, a small round-structured virus (Genbank Accession no. L07418). Open boxes show computer-predicted open reading frames (ORFs) and their translation products, numbers refer to nucleotide positions. ORF3 is a small basic protein of unknown function with a predicted size of 211 amino acids. The genomic cDNA is 7708 nucleotides, excluding the polyadenylate tail. The conserved nucleotide motifs at the 5′ genomic terminus and at the *ORF1–ORF2* intergenic region are shown below the genome, with their predicted amino acid products below (in the one-letter code). The two sequence motifs have only three base substitutions in the first 26 nucleotides. These base substitutions do not alter the amino acid coded nor the terminator codon immediately adjacent to the first in-frame methionine.

Fig. 4. Reproduced from Lambden PR and Clarke I (1995) Genome organization in the Caliciviridae. Trends in Microbiology 3: 261–265, with permission of Elsevier Publications, Cambridge, UK

products it soon became apparent that SRSVs could be divided into at least two "genogroups" based on their nucleotide and amino acid homologies. Genogroup I includes the prototype Norwalk virus, HuCV/Norwalk/8FIIa/68/US and Southampton/91/UK; representatives of genogroup II include Hawaii/71/US; Snow Mountain/76/US, minireovirus = Toronto/24/77/Can and some morphologically typical caliciviruses. A third genogroup based on data on the polymerase region is now known to include morphologically typical HuCVs represented by the prototype japanese strain HuCV/Sapporo/82/Japan which had previously been shown to differ antigenically from some HuCV strains found in the UK [5]. In 1995, this third group of viruses was shown to have a unique genomic organization differing from that of the other two genogroups of human caliciviruses and that of other members of the *Caliciviridae* (27a) but with sufficient similarities to suggest they are a seperate genus within the family. Although RT-PCR has proven to be very successful, there are still many samples containing virus particles (SRSV and HuCV) that are readily detectable by EM which fail to react with the primers currently in use. A more detailed review of recent studies on Caliciviruses is present by Jiang *et al.* in this volume.

Similar studies conducted on astroviruses resulted in sequencing of the complete genome of types 1 [15] and 2, [39], (Fig. 5). The genomic organization is unlike any other family of animal viruses and resembles the plant family *Luteoviridae*. The genome is 7 kb long and contains three sequential ORFs, (Fig. 5). The two ORFs closest to the 5' end are linked by a ribosomal frame shifting

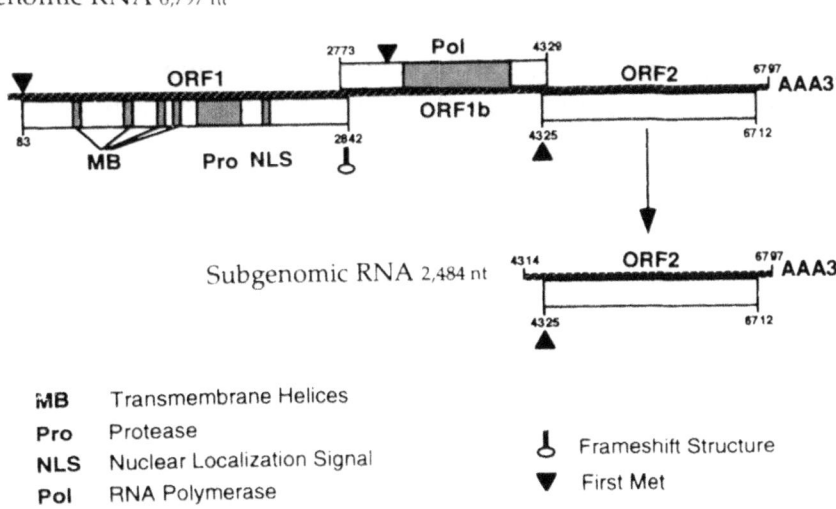

Fig. 5. The arrangements of the genome, subgenomic RNA and deduced coding information for human astrovirus are shown. ORF 1b encoding a putative polymerase is in a different reading frame to that of ORF1a. Reproduced from Monroe *et al.* (1995), reference [31] with permission of Springer-Verlag, Wien, Austria

motif and contain sequence motifs indicative of non structural proteins, and which likely function as a serine protease and RNA-dependent RNA polymerase. A nuclear localization sequence is also located in ORF1. The 3′ ORF encodes the structural proteins. On the basis of the unique characteristics of the astroviruses they have been classified in a new family, the *Astroviridae* [31]. RT-PCR has been applied to the detection of astroviruses using primers directed to the capsid region. This method has been found to be more sensitive than EM or EIA and cloning of the amplified products have shown that there is sequence variation between at least 6 of the 7 known serotypes [32]. A more detailed account of the molecular biology of astroviruses is presented by Carter in this volume.

Expression of structural proteins in baculovirus

Once the sequence of NV was known, Jiang *et al.* succeeded in expressing the capsid protein in baculovirus and found that the protein self-assembles into complete capsids, [19]. Recently a genogroup II virus, (Mexico virus) has been expressed in a similar way [16].

Structure of caliciviruses

The application of electron cryomicroscopy to study the structure of NV [35] has shown that the virus is compromised of 90 dimers of capsid assembled in T = 3 icosahedral symmetry A more detailed description is presented by Prasad *et al.* in this volume. Their findings support the view that caliciviruses may degenerate to produce several distinct forms including the smooth structureless particles of 22–30 nm shown in Fig. 3, which represents the shell surrounding the RNA.

Development of EIAs

The availability for the first time of large amounts of pure antigen enabled the production of reference antisera in rabbits and guinea pigs, which in turn led to the development of sensitive and specific EIAs to measure isotype-specific antibody responses, (IgA, IgG and IgM) [8, 37] and to detect virus in fecal samples [18]. These accomplishments seemed to herald the approach of useful and readily available diagnostic assays, but surveys throughout the world provided surprisingly little evidence of the presence of SRSVs antigenically related to NV [18]. In contrast, seroepidemiological surveys suggested infection was widespread [34]. The reason for the discrepancy may be that many infections are subclinical; alternatively other strains of SRSV evoke a response detectable by serological assays using the NV antigen. Support for this view comes from studies with sera obtained from volunteers challenged with SMA and HV [37] in that rNV particles were found to be capable of detecting heterologous seroresponses. In contrast, the rNV antigen assay was found to be type-specific (7a).

The rapid advances that have occurred in our knowledge of the molecular biology of astroviruses and caliciviruses during the past three years should enable more sensitive and specific diagnostic tests to become readily available. This will allow more detailed surveys to be conducted to establish whether or not these viruses cause a wider spectrum of disease than presently recognized. Studies in the veterinary field provide evidence that such a situation may exist [36].

References

1. Adler JL, Zickl R (1969) Winter vomiting disease. J Infect Dis 119: 668–673
2. Appleton H, Higgins PG (1975) Viruses and gastroenteritis in infants. Lancet 1: 1297
3. Caul EO, Appleton H (1982) The electronmicroscopical and physical characteritics of small round human fecal viruses: an interim scheme of classification. J Med Virol 9: 257–265
4. Cubitt WD (1989) Diagnosis, occurrence and clinical significance of the human "candidate" caliciviruses. Progr Med Virol 36: 103–119
5. Cubitt WD, Blacklow NR, Herrmann JE, Nowak NA, Nakata S, Chiba S (1974) Antigenic relationships between human caliciviruses and Norwalk virus. J Inf Dis 129: 709–714
6. Cubitt WD, Bradley D, Carter MJ, Chiba S, Estes MK, Saif L, Schaffer F, Smith AW, Studdert M, Thiel HJ (1995) Caliciviridae. In: Murphy FA, Fauquet CM, Bishop DHL, Ghabrial SA, Jarvis AW, Martelli GP, Mayo MA, Summers MD (eds) Virus Taxonomy. 6th report of the International Comittee on the Taxonomy of Viruses. Arch Virol Suppl 10. Springer, Wien New York, pp 359–363
7. Dolin R, Blacklow NR, Dupont H, Buscho RF, Thornhill TW, Kapikian Z, Chanock RM (1971) Transmission of acute infectious non-bacterial gastroenteritis by oral transmission of stool filtrates. J Infect Dis 123: 307–312
7a Graham DY, Jiang X, Tanaka T, Opekun A, Madore P, Estes MK (1994) Norwalk virus infection of volunteers: new insight based on improved assays. J Infect Dis 17: 34–43
8. Gray JJ, Cunliffe C, Ball J, Graham DY, Desselberger U, Estes MK (1993) Detection of immunoglobulin IgM, IgA and IgG Norwalk specific antibodies by indirect enzyme immunosorbent assay with baculovirus expressed Norwalk virus capsid antigen in adult volunteers challenged with Norwalk virus. J Clin Microbiol 32: 3059–3063
9. Green J, Norcott JP, Lewis D, Arnold C, Brown DWG (1994) Norwalk-like viruses. Demonstration of genomic diversity of polymerase chain reaction. J Clin Microbiol 31: 3007–3012
10. Greenberg H, Matsui SM (1992) Astroviruses and caliciviruses; emerging enteric pathogens. Infect Agents Dis 1: 71–91
11. Grohmann GS, Murphy AM, Christopher PJ (1981) Norwalk virus gastroenteritis in volunteers consuming depurated oysters. AJEBAK 59: 219–228
12. Herrmann JE, Cubitt WD, Hudson RW, Perron-Henry DM, Oshiro LS, Blacklow NR (1990) Immunological characterization of Marin County strain of astrovirus. Arch Virol 110: 213–220
13. Herrmann JE, Hudson RW, Perron-Henry DM, Kurtz JB, Blacklow NR (1988) Antigenic characterization of cell cultivated astrovirus serotypes and development of astrovirus specific antibodies. J Infect Dis 158: 182–185
14. Herrmann JE, Nowak NA, Perron-Henry DM, Hudson RW, Cubitt WD, Blacklow NR (1990) Diagnosis of astrovirus gastroenteritis by antigen detection with monoclonal antibodies. J Infect Dis 161: 226–229

15. Jiang BS, Monroe SS, Koonin EZ, Stine SE, Glass RI (1993) RNA sequence of astrovirus: distinctive genomic organization and a putative retrovirus-like ribosomal frameshifting signal that directs the viral replicase synthesis. Proc Natl Acad Sci USA 90: 10539–10543

16. Jiang X, Matson DO, Ruiz-Pallacios GM, Hu J, Treanor J, Pickering LK (1995) Expression, self assembly, and antigenicity of a snow mountain-agent-like calicivirus capsid protein. J Clin Microbiol 33: 1452–1455

17. Jiang X, Wang M, Estes MK (1993) Sequence and genomic organization of Norwalk virus. Virol 195: 51–61

18. Jiang X, Wang M, Estes MK (1995) Characterization of SRSVs in different geographic locations by EIA and RT-PCR. Arch Virol 140: 363–374

19. Jiang X, Wang M, Graham DY, Estes MK (1992) Expression, self-assembly, and antigenicity of the Norwalk virus capsid protein. J Virol 66: 6527–6532

20. Kapikian AZ, Wyatt RG, Dolin R, Thornhill TS, Kalica AR, Chanock RM (1972) Visualization by immune electron microscopy of a 27 nm particle associated with infectious non-bacterial gastroenteritis. J Virol 5: 1075–1080

21. Kurtz JB, Cubitt WD (1989) Astroviruses and caliciviruses. In: Farthing MTG, Keutsch GT (eds) Enteric Infection. Chapman Hall, pp 207–215

22. Kurtz JB, Lee TW (1987) Astroviruses: human and animal. In: Ciba Symposium 128. Wiley & Sons, pp 92–101

23. Kurtz JB, Lee TW (1994) Prevalence of human astrovirus serotypes in the Oxford region 1976–1992 with evidence of two new serotypes. Epidem Inf. 112: 187–193

24. Kurtz JB, Lee TW, Craig JW, Reed SE (1979) Astrovirus infection in volunteers. J Med Virol 3: 221–230

25. Lambden PR, Caul EO, Ashley CR, Clarke IN (1993) Sequence and genome organisation of a human small round structured (Norwalk-like) virus. Science 259: 516–519

25a Lambden PR, Clarke IN (1995) Genomic organization in the Caliciviridae. Trends in Microbiol 3: 261–265

26. Lee TW, Kurtz JB (1981) Serial propagation of astrovirus in tissue culture with the aid of trypsin. J Gen Virol 57: 421–424

27. Lewis D (1991) Norwalk agent and other small round structured viruses in the UK. J Infect 23: 220–222

27a Liu BL, Clarke IN, Caul EO, Lambden PR (1995) Human enteric caliciviruses have a unique genome structure distinct from Norwalk-like viruses. Arch Virol 140: 1345–1356

28. Madeley CR, Cosgrove BP (1975) 28 nm particles in faeces in infantile gastroenteritis. Lancet 2: 451–452

29. Madeley CR, Cosgrove BP (1976) Caliciviruses in man. Lancet i: 199

30. Matson DO, Zhong WM, Nakata S, Numata K, Jiang X, Pickering LK, Chiba S, Estes MK (1995) Molecular characterization of a human candidate calicivirus with sequence relationships closer to animal caliciviruses than other human caliciviruses. J Med Virol 45: 215–222

31. Monroe SS, Carter MJ, Herrmann JE, Kurtz JB, Matsui M (1995) *Astroviridae*. In: Murphy FA, Fauquet CM, Bishop DHL, Gabrial SA, Jarvis AW, Martelli GP, Mayo MA, Summers MD (eds) Virus Taxonomy. Sixth Report of the International Committee on Taxonomy of Viruses. Arch Virol Suppl 10. Springer, Wien New York, pp 363–367

32. Noel JS, Lee TW, Kurtz JB, Glass RI, Monroe SS (1995) Typing of human astroviruses from clinical isolates by enzyme immunoassay and nucleotide sequencing. J Clin Microbiol 33: 797–801

33. Norcott JP, Green J, Lewis D, Estes MK, Barlow KL, Brown DWG (1994) Genomic diversity of small round structured viruses in the UK. J Med Virol 44: 280–286
34. Parker SP, Cubitt WD, Jiang X, Estes MK (1994) Seroprevalence studies using a recombinant Norwalk-virus protein enzyme immunoassay. J Med Virol 41: 179–184
35. Prasad BV, Rothnagel R, Jiang X, Estes MK (1994) Three dimensional structure of baculovirus expressed Norwalk virus capsids. J Virol 68: 5117–5125
36. Schaffer FL (1979) Caliciviruses. In: Fraenkel-Conrat H, Wagner RR (eds) Comprehensive Virology 14. Plenum Publishing, New York, pp 249–284
37. Treanor JJ, Jiang X, Madore P, Estes MK (1993) Subclass specific serum antibody responses to recombinant Norwalk virus capsid antigen (rNV) in adults infected with Norwalk, Snow Mountain or Hawaii virus. J Clin Microbiol 31: 1630–1634
38. Wang J, Jiang X, Madore P, Gray J, Desselberger U, Ando T, Seto Y, Oishi I, Lew F, Green KY, Estes MK (1994) Sequence diversity of small round structured viruses in the Norwalk virus group. J Virol 44: 280–286
39. Willcocks MM, Carter MJ (1994) The complete sequence of a human astrovirus. J Gen Virol 75: 1785–1788
40. Wyatt RG, Dolin R, Blacklow NR, Dupont H, Buscho RF, Thornhill TW, Kapikian AZ, Chanock RM (1974) Comparison of three agents of acute infectious non bacterial gastroenteritis by cross challenge experiments in human volunteers. J Infect Dis 129: 307–312

Author's address: Dr. W. D. Cubitt, Department of Virology, Camelia Botnar Laboratories, Great Ormond Street Hospital, London, U.K.

Arch Virol (1996) [Suppl] 12: 237–242

© Springer-Verlag 1996

Structure of Norwalk virus

B. V. V. Prasad[1,*], **M. E. Hardy**[2], **X. Jiang**[2,3], and **M. K. Estes**[2]

[1] Verna and Marrs McLean Department of Biochemistry,
Baylor College of Medicine, Houston, Texas, U.S.A.
[2] Division of Molecular Virology, Baylor College of Medicine,
Houston, Texas, U.S.A. [3] Center for Pediatric Research, Eastern Virginia
Medical School, Norfolk, Virginia, U.S.A.

Summary. Norwalk virus is the major cause of epidemic viral gastroenteritis of humans. Attempts to grow this virus in laboratory cell lines have been unsuccessful. However, the Norwalk virus capsid protein, when expressed in insect cells infected with a recombinant baculovirus, spontaneously assembles into virus-like particles. We have determined the 3-dimensional structure of baculovirus-expressed Norwalk virus using electron cryomicroscopy and computer image reconstruction to a resolution of ~22 Å. These particles, having a diameter of 380 Å exhibit T = 3 icosahedral symmetry. The 3-dimensional structure is composed of 90 dimers of the 58 000 molecular weight (58 K) capsid protein, each of which forms an arch-like capsomere. The structure of the protein subunit is modular with three distinct domains. The distal globular domain that appears bilobed is connected to the lower shell domain by a central stem-like domain. We also have been able to grow crystals of the baculovirus-expressed Norwalk virus particles suitable for high resolution X-ray crystallography.

Introduction

Norwalk virus, a member of the family *Caliciviridae*, genus *Calicivirus*, causes epidemic acute gastroenteritis in humans [12, 14]. It is estimated that about 42% of outbreaks of acute, epidemic non-bacterial gastroenteritis in the United States are caused by Norwalk and Norwalk-like viruses, which include Hawaii, Snow Mountain and Montgomery County viruses [2]. Norwalk virus contains a genome of single-stranded positive-sense RNA of about 7.6 kb [12]. Three open reading frames (ORFs) have been identified in the genome [12]. The first and longest ORF (ORF1) encodes a polyprotein precurser to non-structural proteins. The second ORF encodes the capsid protein (apparent molecular weight of 58 000 [58K]; Ref [11]) and the third ORF codes for a protein whose functional properties are unknown.

Norwalk virus and other members of the *Caliciviridae* are unique among the animal viruses. The icosahedral capsids of these viruses are composed of a single structural protein. In Norwalk virus, 180 molecules of the structural protein form the protein shell that encapsidates the viral genome. While common among plant viruses, virus capsids made of a single structural protein are unusual among animal viruses. Therefore, all the functional entities required for structural integrity, immunogenicity, and infectivity are encoded in one structural protein. Thus, while caliciviruses possess all the characteristics of animal viruses, they have the structural simplicity of plant viruses, and are excellent model systems for studying the molecular basis of viral assembly, genome encapsidation, immunogenicity and pathogenesis. Norwalk virus is a candidate of choice for such studies because of its clinical importance and because its genome has been cloned, allowing expression and genetic manipulations of the structural and non-structural proteins.

We have undertaken 3-dimensional structural studies of the members of the *Caliciviridae* using electron cryomicroscopy techniques and X-ray crystallography to understand structure-function relationships in these viruses [17, 18]. We report here the progress we have made in understanding the structural details of Norwalk virus, the prototype human calicivirus. We have carried out structural studies on the baculovirus-expressed Norwalk capsids, in lieu of native virus, which currently cannot be obtained in large quantities. When expressed using the baculovirus system, the Norwalk virus capsid protein spontaneously assembles into virus-like particles [11]. These particles can be purified in large quantities, making them amenable to high resolution structural analysis.

Materials and methods

Baculovirus-expressed recombinant Norwalk virus (rNV) particles were produced and purified from insect cells infected with a recombinant baculovirus that contains the 3′ end of the viral genome as previously described [11, 16]. Electron cryomicroscopy and the 3-dimensional structure determination of the baculovirus-expressed Norwalk virus capsids were carried out as described by Prasad *et al.* [17].

Results and discussion

A surface representation of the three-dimensional structure along the icosahedral 3-fold axes is shown in Fig. 1a. The structure has an overall diameter of 380 Å, and exhibits T = 3 icosahedral symmetry.

The T = 3 icosahedral organization is a common feature in several ssRNA viruses. The list includes viruses of plant origin, such as tomato bushy stunt virus (TBSV), southern bean mosaic virus (SBMV), turnip crinkle virus (TCV), cowpea mosaic virus [6, 21], and insect viruses such as black beetle virus [10] and flock house virus [5]. Picornaviruses, one of the important classes of animal ssRNA viruses, which includes rhinoviruses [20], polioviruses [8], mengo virus [15], and foot-and-mouth disease viruses [1], exhibit T = 1 icosahedral symmetry. If we ignore the distinction between VP1, VP2 and VP3, which make up the picornavirus capsid structure with 60 copies each, we could describe the structure of picornaviruses as having a pseudo T = 3 structure. In its architecture, rNV particles share more similarities with TBSV and TCV than with other T = 3 viruses. Unlike other known structures of T = 3 viruses, the TBSV and TCV capsid proteins form prominent projections in their structures, as is seen with rNV. These projections form the main feature. In other T = 3 structures, these projections are not as obvious.

The main feature of the 3-dimensional structure of rNV particles is the presence of 90 arch-like capsomeres located at all the local and strict 2-fold axes of the T = 3 icosahedral lattice. These capsomeres are arranged in such a way that there are prominent hollows at all the icosahedral 5- and 3-fold axes (Fig. 1a). The arches begin at a radius of ~ 145 Å and extend to a radius of about 190 Å. A rectangular platform of a dimension ~ 50 × 70 Å is located at the top of the arch. The hole between the two sides of the arch is about 20 Å in diameter. These arches surround the hollows which are about 40 Å deep and 90 Å wide. Each virus particle has 32 such hollows, 12 of which are on the icosahedral 5-fold axes with the remaining 20 hollows being at the icosahedral 3-fold axes. The hollows at the 5-fold axes have a small hump at the center with a mote around it, while the hollows at the icosahedral 3-fold positions are rather flat.

The 90 capsomeres can be classified into two types based on their location with respect to the icosahedral axes of symmetry (Fig. 1a). The first type of capsomere, designated as type A, are the capsomeres surrounding the icosahedral 5-fold axis. These are located at the local 2-fold axes, midway between the icosahedral 5- and 3-fold positions. The second type, designated as type B, are located at all the strict 2-fold axes of the icosahedral structure (midway between any two neighboring 5-fold or 3-fold axes).

Each side of the arch represents a single molecule of the capsid protein. Dimeric clustering of the capsid proteins is a common feature in the T = 3 virus structures. The protein subunit appears to have a modular structure consisting of three domains: the distal globular domain, called P1, which appears bilobed; the central stem domain, called P2, and a shell domain called S (Fig. 1b). These domains are so named following the convention used in describing the capsid protein structure in TBSV, TCV and other T = 3 structures [6, 21]. The S domains of these subunits merge together to form the continuous mass density that is found between the radii 115 and 145 Å (Fig. 1c). The arches (P1 + P2) emanate from this contiguous shell. No significant density is found inside the radius of 115 Å.

We have examined the sequences of the Norwalk and Norwalk-like viruses to identify any similarity between this group of viruses and other T = 3 ssRNA viruses. While the primary sequences of the capsid protein of Norwalk and other Norwalk-like viruses, along with the two known calicivirus sequences show identifiable homology between themselves, these sequences show no homology with the capsid protein sequences of other viruses. However, similarities in the architectural features between the Norwalk capsid and the other T = 3 viruses prompt us to speculate that the S domain may fold into an 8-stranded beta-barrel domain.

It is conceivable that the organization of the domain structures along the length of the polypeptide chain in Norwalk virus is similar to that found in TBSV

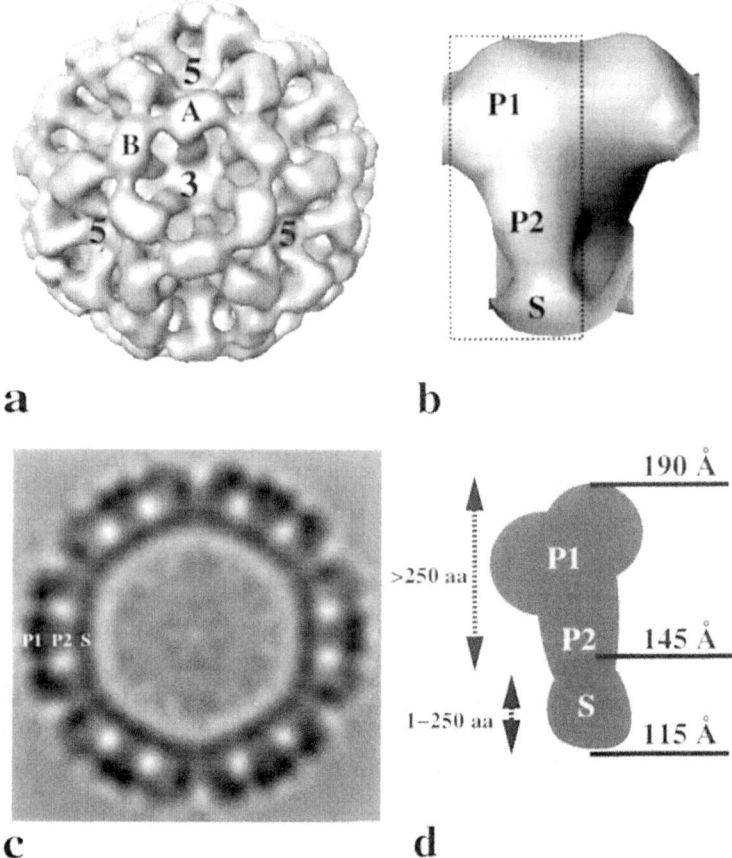

Fig. 1. a) Surface representation of the 3-dimensional structure of the baculovirus-expressed Norwalk virus capsid viewed along the icosahedral 3-fold axes. The icosahedral 5- and 3-fold symmetry axes, two types of capsomeres, A and B, are indicated. Type A capsomeres surround the icosahedral 5-fold axes. Type B capsomeres are located at the icosahedral 2-fold axes. Each virion has 60 type A and 30 type B capsomeres. Application of the icosahedral 5-, 3-, and 2-fold symmetry operators to one set of A and B capsomeres generates these 90 capsomeres; **b)** An isolated view of the type B capsomere extracted from the 3-dimensional density map. The arch-like structure of the capsomere is clearly seen. One side of the arch which represents a protein subunit is enclosed inside the dashed rectangle. The three domains: P1, P2, and S in the protein subunit are denoted; **c)** A central section along the icosahedral 3-fold axis. The S domains associate to form a contiguous shell of thickness $\sim 30\,\text{Å}$ from which the arches (P1 + P2) emanate. Dark regions correspond to the protein density and the lighter regions represent solvent density; **d)** A cartoon representation of the proposed domain organization along the primary sequence of the capsid protein

and TCV [7, 9]. In the TBSV and TCV structures, the internal R domain is formed by the N-terminal 80 residues, the S domain by the middle 167 residues and the P domain by the C-terminal 114 residues. When we examined the sequences of Norwalk and Norwalk-like viruses, the most conserved region was found to be between amino acid residues 30 and 250. The C-terminal regions exhibit significant variation [4]. It is quite possible that the P1 and P2 domains in the Norwalk structure are formed by the residues beyond 250 and the N-terminal 250 residues may constitute the S domain. Our preliminary analysis using secondary structure prediction algorithms and the 3-D profile search methods proposed by Bowie *et al.*, indicates that residues 30 to 250 may fold into an 8-stranded beta-barrel motif (A. Liu, S. Adams and B.V.V. Prasad, unpubl. results). The proposed domain organization is shown in Fig. 1d.

The only way to confirm the proposed domain organization is to determine the atomic resolution structure by X-ray crystallography. Towards this goal, we have recently crystallized rNV capsids suitable for X-ray crystallographic analysis. These crystals, grown using hanging drop methods, diffract to higher than a 3.2 Å resolution. Data collection has been carried out on crystals at room temperature and also on a frozen crystal at liquid nitrogen temperature ($-170\,°C$) using the synchrotron radiation source at CHESS (Cornell High Energy Synchrotron Radiation Source). Analysis of the oscillation photographs has indicated that the unit cell dimensions are $606\,\text{Å} \times 606\,\text{Å} \times 467\,\text{Å}$ with a tetragonal space group [19].

Conclusions

Structural analysis of the baculovirus-expressed Norwalk virus capsids by electron cryomicroscopy and computer processing at 22 Å resolution has provided the architectural details of the capsid. The architecture of the Norwalk virus capsid is similar to that found in other animal caliciviruses, such as primate calicivirus [18] and feline calicivirus (S. Pracht, M. Liu and B. V. V. Prasad, unpubl. results). Architectural similarities between caliciviruses, including Norwalk virus and other $T = 3$ ssRNA viruses, and picornaviruses have led us to suggest a possible domain organization of the primary sequences of caliciviruses. An exciting aspect of our studies is the progress we have made towards determining the atomic resolution structure of the Norwalk virus capsid. Knowledge of the atomic resolution structure of the capsid will provide insight into the nature of chemical interactions that govern the assembly, uncoating, receptor recognition and antigen neutralization properties of Norwalk virus. Establishing structure-function relationships will provide a rational basis for formulating antiviral strategies for this human virus and for other caliciviruses that cause acute or persistent infections.

Acknowledgements

Our work is supported in part by grants from the NIH (AI 36040, AI 30448, AI38036, T32DK-07664), National Center for Research Resources (RR 02250), NSFBIR-9413229, W. M. Keck Foundation and R. Welch Foundation.

References

1. Acharya R, Fry E, Stuart D, Fox G, Rowlands D, Brown F (1989) The three-dimensional structure of foot-and-mouth disease virus at 2.9 Å resolution. Nature (London) 337: 709–716

2. Blacklow NR, Greenberg HB (1991) Viral gastroenteritis. New Engl J Med 325: 252–253
3. Bowie JU, Luthy R, Eisenberg D (1991) A method to identify protein sequences that fold into a known three-dimensional structure. Science 253: 614–616
4. Estes MK, Hardy ME (1995) Norwalk virus and other enteric calicivirus. In: Blaser MJ, Smith PD, Ravdin JI, Greenberg HB, Guerrant RL (eds) Infections of the Gastrintestinal Tract. Raven Press, New York
5. Fisher AJ, Johnson JE (1992) Ordered duplex RNA controls the capsid architecture of an icosahedral animal viruses. Nature (London) 361: 176–179
6. Harrison SC, Skehel JJ, Wiley DC (1996) Virus structure. In: Fields BN, Knipe DM, Howley PM, Chanock RM, Melnick JL, Monath TP, Roizman B, Straus SE (eds) Virology, 3rd edn, vol 1. Lippincott-Raven Publishers, Philadelphia, pp 59–99
7. Harrison SC, Olson A, Schutt CE, Winkler FK, Bricogne G (1978) Tomato bushy stunt virus at 2.9 Å resolution. Nature (London) 276: 368–373
8. Hogle JM, Chow M, Filman DJ (1985) Three-dimensional structure of poliovirus at 2.9 Å resolution. Science 229: 1358–1365
9. Hogle JM, Maeda A, Harrison SC (1986) The structure and assembly of turnip crinkle virus I: X-ray crystallographic analysis at 3.2 Å. J Mol Biol 191: 625–638
10. Hosur MV, Schmidt T, Tucker RC, Johnson JE, Gallaghar TM, Selling BH, Rueckert RR (1987) Structure of an insect virus at 3.0 Å resolution. Proteins 2: 167–176
11. Jiang X, Wang M, Graham DY, Estes MK (1992) Expression, self-assembly and antigenicity of the Norwalk virus capsid protein. J Virol 66: 6527–6532
12. Jiang X, Wang M, Wang K, Estes MK (1993) Sequence and genomic organization of Norwalk virus. Virology 195: 51–61
13. Jiang X, Graham DY, Wang K, Estes MK (1990) Norwalk virus genome: cloning and characterization. Science 250: 1580–1583
14. Kapikian AZ, Estes MK, Chanock RM (1996) Norwalk group of viruses. In: Fields BN, Knipe DM, Howley PM, Chanock RM, Melnick JL, Monath TP, Roizman B, Straus SE (eds) Virology, 3rd edn, vol 1. Lippincott-Raven Publishers, Philadelphia, pp 783–810
15. Luo M, Vriend G, Kamer G, Minor I, Arnold E, Rossmann MG, Boege U, Scraba DG, Duke GM, Palmenberg AC (1987) The structure of Mengo virus at atomic resolution. Science 235: 182–191
16. Hardy ME, White LJ, Ball JM, Estes MK (1995) Specific proteolytic cleavage of recombinant Norwalk virus capsid protein. J Virol 69: 1693–1698
17. Prasad BVV, Rothnagel R, Jiang X, Estes MK (1994) Three-dimensional structure of baculovirus-expressed Norwalk virus capsids. J Virol 68: 5117–5125
18. Prasad BVV, Matson DO, Smith AW (1994) Three-dimensional structure of Calicivirus. J Mol Biol 240: 256–264
19. Prasad BVV, Hardy ME, McKenna R, Schmidt T, Vyas M, Rossmann MG, Estes MK (1995) X-ray crystallographic studies on baculovirus-expressed Norwalk virus capsids. 14th Annual Meeting of American Society for Virology, University of Texas, Austin, 8–12 July 1995
20. Rossmann MG, Arnold E, Erickson JW, Frankenberger EA, Griffith JP, Hecht HJ, Johnson JR, Kamer G, Luo M, Mosser AG (1985) Structure of a human common cold virus and functional relationship to other picornaviruses. Nature (London) 317: 145–133
21. Rossmann MG, Johnson JE (1989) Icosahedral RNA virus structure. Ann Rev Biochem 58: 533–573

Authors' address: Dr. B. V. V. Prasad, Verna and Marrs McLean Department of Biochemistry, Baylor College of Medicine, Houston, TX 77030, U.S.A.

Arch Virol (1996) [Suppl] 12: 243–249

Recombinant Norwalk virus-like particles as an oral vaccine

J. M. Ball[1], **M. K. Estes**[1], **M. E. Hardy**[1], **M. E. Conner**[1], **A. R. Opekun**[2], and **D. Y. Graham**[1,2]

[1] Division of Molecular Virology, Baylor College of Medicine, Houston, Texas, U.S.A. [2] Department of Medicine, Baylor College of Medicine, Houston, Texas, U.S.A.

Summary. Viruses that infect cells in the gastrointestinal tract are well suited for examining the immune response to oral delivery of antigen and for exploring the advantages and pitfalls of oral vaccines. Norwalk virus (NV) (family *Caliciviridae*, genus *Calicivirus*) causes acute gastroenteritis in all age groups. The NV capsid is composed of 180 copies of a single 58 000 molecular weight protein which spontaneously forms virus-like particles (VLPs) that can be purified in extremely high yields (22 mg per 300 ml culture) when produced using the baculovirus expression system. We are testing the potential of these recombinant NV (rNV) particles for use as an oral vaccine by administering them to mice and volunteers. Mice were orally inoculated four times with rNV particles in concentrations ranging from 5 to 500 µg in the absence of adjuvant or from 5 to 200 µg with 10 µg of cholera toxin. Serum IgG and fecal IgA immune responses were monitored. rNV particles were found to be immunogenic when orally given to mice with or without adjuvant. These particles also were safe and immunogenic when orally given to volunteers. These studies show that rNV particles are an excellent model to test the oral delivery of mucosal immunogens in general, and that rNV particles are ideal candidates for vaccine development in particular.

Introduction

Norwalk virus (NV), a member of the family *Caliciviridae* (genus *Calicivirus*), causes acute epidemic gastroenteritis in humans [9]. Extensive studies of the molecular biology and pathogenesis of this important pathogen have been hindered by the lack of either a cell culture system or an animal model that supports virus replication. The only source of NV virions is stool from infected individuals, and levels of virus excreted by these individuals are extremely low. In addition, most of the virion antigen detected in stool is soluble protein which has been shown to be unfolded capsid protein specifically cleaved by an intestinal protease [5].

The NV genome, like that of other caliciviruses, has three open reading frames, one of which (ORF 2) encodes a single capsid protein with an apparent

244 J. M. Ball et al.

molecular weight of 58 000 (58 K [7, 8]). Expression of the second and third
ORFs of the NV genome in insect cells infected with a recombinant baculovirus
results in the spontaneous assembly of virus-like particles (VLPs) composed of 90
dimers of the 58 K protein arranged in a T = 3 symmetry [7, 17]. Recombinant
NV (rNV) particles are morphologically and antigenically similar to native
particles as demonstrated by ELISA and EM using convalescent serum from
infected volunteers and by the observation that they are highly immunogenic
when inoculated parenterally with adjuvant into experimental animals [4, 7].

The rNV particles have several unique properties that could be advantageous
for a mucosal immunogen: (i) NV capsids contain 180 copies of a single 58 K
protein which spontaneously fold into VLPs lacking nucleic acid; (ii) rNV
particles are easily produced and purified in large quantities (22 mg from 9×10^8
cells); (iii) NV VLPs are stable at low pH (such as the pH of the stomach), when
lyophilized, and after long term storage in water or phosphate buffered saline (PBS);
and (iv) rNV particles are particulate and therefore might be targeted to Peyer's
patches. Consequently, we have explored the potential for using rNV particles as
an oral immunogen in mice and in volunteers with pre-existing antibody.

Materials and methods

Vaccine preparation and characterization

The rNV vaccine was prepared in *Spodoptera frugiperda* (Sf9) cells infected at a multiplicity
of infection of 1 with a recombinant baculovirus that expresses the NV capsid protein,
essentially as previously described [5, 7] (Fig. 1). Following 7 days of culture, intact cells and

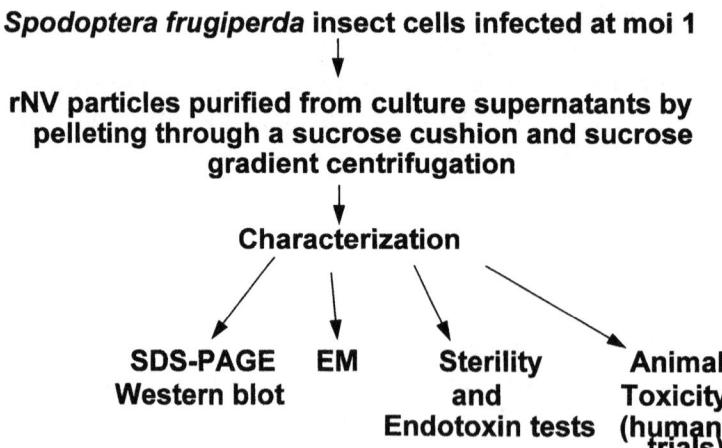

Fig. 1. Schematic of the preparation and purification of the recombinant Norwalk virus
(rNV) vaccine grown in insect cells in a baculovirus-expression system. Particles were
concentrated from the medium of infected insect cells by ultracentrifugation and purified by
banding on a sucrose gradient. Purified particles were evaluated by SDS-PAGE, Western
blot, and EM. Vaccine preparations given to volunteers also were evaluated in animals for
toxicity. Protein concentrations were determined by the Pierce BCA assay (Rockford, IL)

debris were pelleted and discarded, and the rNV particles in the supernatant fluid were pelleted through a 35% sucrose cushion. The pelleted particles were suspended in sterile Milli-Q water (Milli-Q Water System, Millipore), layered onto a sucrose step gradient (20%–65%), and centrifuged (90000 × g). Peak fractions were pooled, pelleted at 360000 × g, and suspended in sterile Milli-Q water.

The particles were characterized by SDS-PAGE, Western blot, and negative-stain electron microscopy (Fig. 2). Sterility of the preparation was confirmed by endotoxin measurement (LAL assay, Ass. of Cape Cod, Inc., Woods Hole, MA), and bacteriologic cultures. All vaccine preparations given to volunteers were tested for toxicity by oral and intraperitoneal administration to mice and guinea pigs.

rNV particles also were expressed in transgenic tobacco leaves (t-rNV) and potato tubers (p-rNV); these particles were morphologically similar to the baculovirus-expressed VLPs [11].

Oral delivery of the rNV vaccine

The experimental design of the oral administration of the rNV particles is outlined in Fig. 3. Mice were given 50 μl of rNV of various concentrations in the presence or absence of 10 μg of cholera toxin (CT). Alternatively, tobacco leaf extracts containing t-rNV particles were orally administered to CD-1 mice by gavage in the presence or absence of CT. Potato tubers (p-rNV) were peeled, sliced, and fed to mice [11]. Pre-inoculation serum and fecal samples were acquired before administering the first dose of vaccine. Each mouse received four doses of vaccine which was administered on days 1, 2, 11, and 28 by oral gavage or feeding. Control mice received PBS.

Informed consent was obtained from adult volunteers prior to the administration of two doses of rNV in 100 ml of sterile Milli-Q water at a 3 week interval. Pre-inoculation serum and fecal samples were acquired one week prior to administration of the first dose; serum samples were collected two weeks after each dose was administered; and stools were obtained for two consecutive days after each dose.

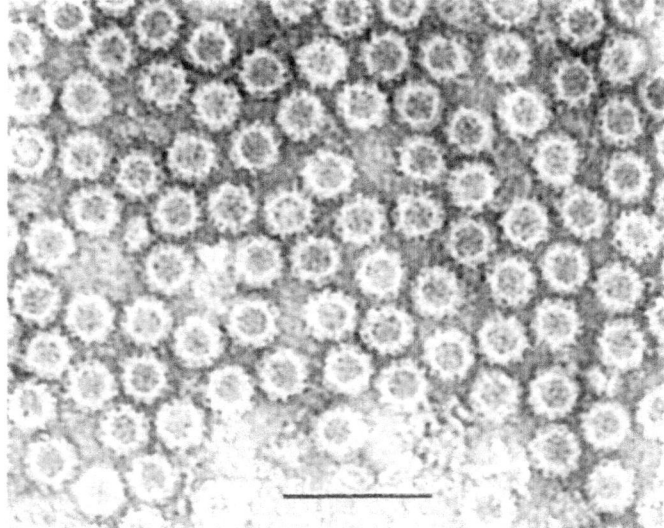

Fig. 2. Electron micrograph of the rNV vaccine. Particle integrity of the purified vaccine preparation was confirmed by negative stain (1% ammonium molybdate) electron microscopy. The integrity, high concentration, and purity of the rNV particles can be seen. Bar = 100 nm

Immune assays

Serum rNV-specific total antibody titers and fecal IgA levels were monitored by ELISA [3]. Briefly, rNV particles were absorbed to polyvinyl chloride 96-well plates, and blocked overnight at 4 °C with 5% Carnation dry milk in PBS (Blotto). Serial dilutions of the sera or fecal extract were added to the wells and plates were incubated for 2 hr at 37 °C. Following extensive washing, peroxidase-conjugated, species-specific antibody was added and incubated for 1 hr at 37 °C followed by washing again. The reaction was developed with TMB substrate (Kirkegaard and Perry Labs, Gaithersburg, MD) and stopped by the addition of 1 M phosphoric acid.

Total fecal IgA was determined by a modification of the procedure by Conner *et al* [2]. Species-specific anti-IgA antibody was coated onto 96-well microtiter plates overnight at room temperature. The plates were blocked for two hours, fecal extracts were added and incubated at 37 °C. Bound IgA was detected by the addition of peroxidase-conjugated anti-IgA and TMB substrate.

Results

rNV particles

The rNV vaccine was not bacterially contaminated (two weeks of culture at 37 °C), and endotoxin levels were consistently below 0.02 endotoxin units per mg of vaccine. SDS-PAGE and Western blot analyses revealed a major 58 K band and a minor band at approximately 30 K (data not shown). EM analysis showed numerous VLPs that appeared structurally similar to native NV virions (Fig. 2). No toxicity was observed in mice and guinea pigs.

Orally delivered rNV particles induce an immune response in mice

To determine the immunogenicity of rNV VLPs, baculovirus-expressed rNV particles were administered to CD-1 mice by oral gavage, and serum total and fecal IgA antibody responses were monitored (Table 1). Seventy-one percent of the mice that received less than 200 μg of VLPs demonstrated a $\geqslant$ 4-fold rise in antibody titer, and 100% of the mice that received $\geqslant$ 200 μg had a positive immune response. In the presence of CT, a larger percentage of animals (98%) responded to doses of less than 200 μg of rNV. Control mice did not have antibody to rNV (data not shown). These data indicate that rNV VLPs are immunogenic when administered orally to mice and have potential for use as an oral vaccine.

Both tobacco-expressed rNV particles given orally or rNV expressed in potato tubers and fed to mice stimulated NV-specific serum antibodies [11]. The tobacco leaf extracts also stimulated NV-specific intestinal IgA. These data support the feasibility of an edible vaccine.

Orally delivered rNV VLPs are safe and immunogenic in volunteers

The safety and immunogenicity of rNV VLPs were evaluated in volunteers with relatively low pre-existing antibody titers to NV in a Phase I human trial. The vaccine was administered in sterile water in the absence of adjuvant. A $\geqslant$ 4-fold

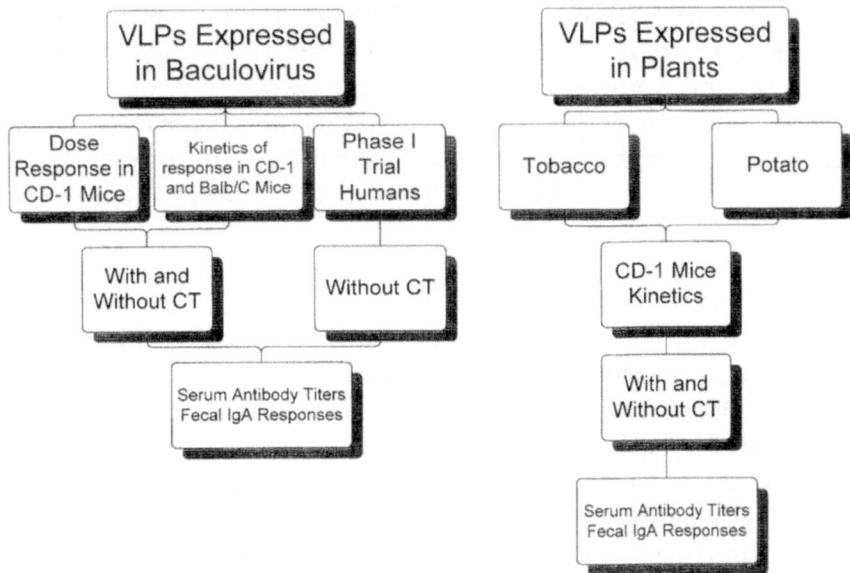

Fig. 3. Schematic of the oral immunization of mice and volunteers with rNV particles. The VLPs were expressed in a baculovirus expression system or in transgenic plants. Mice were orally administered the rNV vaccine in the presence or absence of CT. The dose response and the kinetics of the immune response in mice were studied. Volunteers were orally given rNV particles in water. Vaccine immunogenicity was monitored by evaluation of serum and fecal antibody responses

Table 1. Immunogenicity of rNV administered orally to mice and humans

Immunogen	Given To	Route	Dose (µg)	Adjuvant	% Positive Responders	
					Serum Antibody	Fecal IgA
rNV from baculovirus	mice	oral gavage	< 200 ≥ 200	None	71 100	21 40
rNV from baculovirus	mice	oral gavage	< 200 ≥ 200	CT[a]	98 100	80 88
rNV from tobacco	mice	oral gavage	50	None	89	62
rNV from tobacco	mice	oral gavage	50	CT	90	56
rNV from potato	mice	oral as food	50	None	40	10
rNV from potato	mice	oral as food	45	CT	70	0
rNV from baculovirus	humans	oral in water	< 200 > 200	None	60 100	in progress

[a]Cholera toxin

increase in antibody titer was elicited following a single oral dose (Table 1). Following the administration of a second dose, antibody titers did not increase. No adverse reactions were observed or reported by the volunteers irrespective of the dose of rNV particles given. When the vaccine dose exceeded 200 μg, all volunteers had a significant rise ($\geq$ 4-fold) in antibody titer, whereas only 60% of the volunteers had a rise in antibody titer when the vaccine dose was less than 200 μg.

Discussion

Many pathogens initiate infection at a mucosal surface but most vaccines are administered by a parenteral route. Several studies have demonstrated the advantages of oral administration of antigen, including ease of delivery, reduced side effects, and the production of secretory IgA (sIgA) at mucosal surfaces [12, 13, 14]. Although most oral vaccines are attenuated live bacteria or viruses, a number of adjuvants and encapsulation formulations recently have been introduced to enhance the uptake of antigen by Peyer's patches [1, 10, 15, 16]. Adjuvants have been designed to capture an antigenic protein into a particle that has the potential to enhance antigen uptake as well as stimulate the immune response. We report here the oral delivery of a non-replicating recombinant viral particle that is immunogenic, appears to be safe, and does not require the addition of an adjuvant to stimulate an immune response.

Recombinant NV particles were orally delivered to mice and were shown to stimulate a significant increase in antibody titer in most animals. We also have shown that administration of rNV to humans was safe and immunogenic based on observed increases in rNV-specific antibodies. These data suggest the potential use of rNV particles as an oral immunogen.

These promising results have stimulated testing of other formulations of oral rNV vaccines. One example that has been tested is the concept of edible vaccines. Expression of rNV in tobacco and potato has been successful [6, 11], and oral delivery of tobacco-expressed rNV particles stimulated a serum and mucosal immune response in most mice in the absence or presence of CT. Potato tubers containing expressed rNV also were immunogenic when fed to mice. These data support the feasibility of an edible vaccine.

The ability to stimulate serum and intestinal antibody with these non-replicating rNV particles given without adjuvant initially seemed surprising because experience with poliovirus vaccine indicated mucosal responses require oral immunization with live attenuated virus [14]. However, these data are exciting relative to the possibilities of using rNV particles as simple vaccines to prevent calicivirus diseases or as carriers to immunize with heterologous, protective epitopes of other human or animal pathogens. Our hypothesis is that the rNV particles bind to a specific intestinal receptor and then are endocytosed, processed and presented to the mucosal immune system. Alternatively, uptake by the Peyer's patch is possible because the recombinant particles are particulate. In either case, these VLPs appear to contain all the properties required of an effective oral immunogen.

Acknowledgements

This work was supported in part by an Applied Technology Program grant from the Texas Higher Education Coordinating Board, and from Public Health Service Awards AI 36519, T 32 DK 07664, RR 00350, and RR 00188 from the National Institutes of Health.

References

1. Alving CR (1991) Liposomes as carriers of antigens and adjuvants. J Immunol Methods 140: 1–14
2. Conner ME, Crawford SE, Barone C, Estes ME (1993) Rotavirus vaccine administered parenterally induces protective immunity. J Virol 67: 6633–6641
3. Graham DY, Jiang X, Tanaka T, Opekun AR, Madore HP, Estes MK (1994) Norwalk virus infection of volunteers: new insights based on improved assays. J Infect Dis 170: 34–43
4. Green KY, Lew JF, Jiang X, Kapikian AZ, Estes MK (1993) Comparison of the reactivities of baculovirus-expressed recombinant Norwalk virus capsid antigen with those of the native Norwalk virus antigen in serologic assays and some epidemiologic observations. J Clin Micro 31: 2185–2191
5. Hardy ME, White LJ, Ball JM, Estes MK (1995) Specific proteolytic cleavage of recombinant Norwalk virus capsid protein. J Virol 69: 1693–1698
6. Haq TA, Mason HS, Clements JD, Arntzen CJ (1995) Oral immunization with a recombinant bacterial antigen produced in transgenic plants. Science 268: 714–716
7. Jiang X, Wang M, Graham DY, Estes MK (1992) Expression, self-assembly, and antigenicity of the Norwalk virus capsid protein. J Virol 66: 6527–6532
8. Jiang X, Wang M, Wang M, Estes MK (1993) Sequence and genomic organization of Norwalk virus. Virology 195: 51–61
9. Kapikian AZ, Estes MK, Chanock RM (1996) Norwalk group of viruses. In: Fields BN, Knipe DM, Howley PM, Chanock RM, Melnick JL, Monath TP, Roizman B, Straus SE (eds) Virology. Raven Press, New York, pp 671–693
10. Kreuter J (1988) Possibilities of using nanoparticles as carriers for drugs and vaccines. J Microencap 5: 115–127
11. Mason HS, Shi J, Ball JM, Estes MK, Arntzen CJ (1996) Expression of Norwalk virus capsid protein in transgenic tobacco and potato and its oral immunogenicity in mice. PNAS 93: 5335–5340
12. Mestecky J, McGhee JR (1987) Immunoglobulin A: molecular and cellular interactions involved in IgA biosynthesis and immune response. Adv Immunol 40: 153–245
13. Mestecky J, Russel MW, Jackson J, Brown TA (1986) The human IgA system: A reassessment. Clin Immunol Immunopath 40: 105–114
14. Ogra PL, Karzan DT, Righthand F, MacGillivary M (1968) Immunoglobulin response and secretions after immunization with live and inactivated polio vaccine and natural infection. N Engl J Med 279: 893–899
15. O'Hagan DT, Palin KJ, Davis SS (1989) Poly(butyl-2-cyanoacrylate) particles as adjuvants for oral immunization. Vaccine 7: 213–216
16. Stieneker F, Kreuter J, Lower J (1991) High antibody titers in mice with polymethylmethacrylate nanoparticles as adjuvant for HIV vaccines. AIDS 5: 431–435
17. Prasad BVV, Rothnagel R, Jiang X, Estes MK (1994) Three-dimensional structure of baculovirus-expressed Norwalk virus capsids. J Virol 68: 5117–5125

Authors' address: Dr. Judith M. Ball, Division of Molecular Virology, Baylor College of Medicine, One Baylor Plaza, Houston, TX 77030, U.S.A.

Arch Virol (1996) [Suppl] 12: 251–262

© Springer-Verlag 1996

Genetic and antigenic diversity of human caliciviruses (HuCVs) using RT-PCR and new EIAs

X. Jiang[1], D. O. Matson[1], W. D. Cubitt[2], and M. K. Estes[3]

[1] Center for Pediatric Research, Children's Hospital of The King's Daughters,
Eastern Virginia Medical School, Norfolk, U.S.A.
[2] Department of Virology, Institute of Child Health, London, U.K.
[3] Division of Molecular Virology, Baylor College of Medicine, Houston, U.S.A.

Summary. RT-PCR using primers from conserved regions of calicivirus genomes, followed by sequencing, permits characterization of genetic variation within the family. EIAs based on baculovirus-expressed viral capsid proteins and hyperimmune antisera against the capsid proteins were developed to detect HuCV antigens and antibodies. Serologic surveys using recombinant Norwalk virus (rNV) and recombinant Mexico virus (rMX, a SMA-like virus) EIAs showed that infections by HuCVs are common and that children acquire antibodies to HuCVs at an early age in both developed and developing countries. Three HuCV genogroups have been described that are represented by Norwalk virus (NV), Snow Mountain agent (SMA), and Sapporo virus, although recently accumulated sequences of HuCV strains indicate these genogroups can be further divided. These genogroups also correspond to distinct antigenic groups based on the results of the recombinant EIAs. The three genogroups co-circulate and have a worldwide distribution, although the SMA genogroup seems to be predominant currently. Application of these new assays for further characterization of the genetic and antigenic properties of HuCVs remains an important task for HuCV research.

Introduction

Human caliciviruses (HuCVs) include morphologically "typical" and "atypical" small round structural viruses (SRSV) that cause acute gastroenteritis in humans. Atypical SRSVs also were called NV-like or NV-related viruses before the description of the viral genomes. The major public health concern about HuCVs has been their ability to cause large outbreaks of gastroenteritis. Such outbreaks usually have high attack rates and have occurred in schools, restaurants, summer camps, hospitals, nursing homes, and cruise ships. Exposure to a common source of the virus, such as contaminated food or water, usually can be identified. Outbreaks resulting from consumption of uncooked shellfish are common. HuCVs also can be spread by person-to-person transmission. The syndrome of HuCV-associated gastroenteritis includes diarrhea, vomiting, nausea, abdominal cramps, fever, and malaise; diarrhea and vomiting are most common. The inability to passage HuCVs in cell culture or in an animal model severely hampered the progress of HuCV studies and caused the acquisition of data about HuCVs to lag behind that of most other enteric viruses.

Cloning of the NV genome initiated a molecular phase of HuCV research. Although HuCVs still cannot be passaged in cell culture or in an animal model, many virus features have been described at the molecular level. HuCVs contain a single-stranded, positive-sense RNA of 7.7 kb with a poly-A tail [8, 11, 17]. The genomic organization of HuCVs is similar to that of animal caliciviruses. Motifs like those present in picornaviruses, such as the non-structural 2C, 3C, and 3D proteins and the structural polypeptide VP3, have been identified in the HuCV genome [11]. Molecular characterization of HuCVs has led to the development of new diagnostic assays, including RT-PCR, to detect viral RNA, and EIAs based upon baculovirus-expressed capsid proteins, to detect HuCV antigens and antibodies. Use of these new assays in epidemiologic and virologic studies has rapidly expanded our knowledge of HuCVs [2, 10, 12, 13, 15, 24, 33–35]. This article updates the genetic and antigenic characterization of HuCVs with these new assays.

Recombinant NV capsid and derived EIAs

Expression of the NV capsid gene in a baculovirus system generated a single viral capsid protein of 58 K that self-assembled into virus-like particles (VLPs) similar morphologically and antigenically to authentic virions found in human stools [9]. An EIA using the rNV capsid to coat microtiter plates was developed to measure antibodies in patient specimens [5, 7, 9, 31]. The rNV capsid antigen also was used to generate hyperimmune antisera by immunization of laboratory animals. These antisera were used in an EIA to detect viral antigen in stool specimens. Application of the rNV EIAs in volunteer studies showed that the new assays were more sensitive than RIA and generated new information about infection and immunity of volunteers following challenge with NV [5].

Recombinant MX capsid and derived EIAs

MX virus was first detected by RT-PCR in a stool specimen from a Mexican child with diarrhea [13]. The sequence of the amplified products showed that MX virus is genetically close to SMA, and distantly related to the NV and Sapporo genogroups. A 3.3 kb fragment of cDNA from the 3' end of the MX genome was cloned by RT-PCR using primer 36 and an oligo-dT primer [13]. The MX viral capsid protein expressed in the baculovirus system also self-assembled into VLPs that were morphologically similar to but antigenically distinct from rNV VLPs [14]. Two EIAs based on the expressed rMX particles

Table 1. Seroresponses to rMX and rNV among volunteers infected
with Snow Mountain agent, Hawaii agent, or Norwalk virus

Volunteer		Antibody Titer Pre- and Post-challenge to			
		rMX	Fold-Change	rNV	Fold-Change
SMA	86–25	1 600/102 400	64	400/1 600	4
	86–28	1 600/25 600	16	100/400	4
Hawaii	86–33	1 600/25 600	16	400/400	0
	86–38	1 600/102 400	64	400/400	0
NV	523	1 600/6 400	4	1 600/102 400	64
	529	1 600/6 400	4	400/25 600	64
	544	6 400/6 400	0	400/25 600	64
	543	1 600/25 600	16	400/25 600	64
	77–02	6 400/25 600	4	6 400/102 400	16

and hyperimmune antisera against the rMX capsid antigen were developed to
detect MX viral antibody and antigen, respectively [14, 15].

Specificity of the antibody EIAs

The rNV EIAs detected antibody responses in volunteers infected with NV and
low level (2–4 fold increase between pre- and post-infection sera) antibody
responses in volunteers infected with SMA and Hawaii agent (HA; 33). The rMX
antibody EIA detected antibody responses in children infected with MX virus
and in volunteers infected with SMA and HA [14]. A low level (2–4 fold increase
between pre- and post infected sera) of cross-reactivity in serum specimens of
volunteers infected with NV also was detected by the rMX EIA (Table 1). This
cross-reaction indicates the presence of common epitopes on the viral capsid
proteins between the two genogroups. However, in a blocking EIA, rNV VLPs
blocked the binding of antibodies from volunteers infected with NV, but not
those from volunteers infected with SMA and HA, and rMX VLPs blocked the
binding of antibodies from volunteers infected with SMA and HA, but not from
volunteers infected with NV [14]. These data suggest that the antibodies
detected by the two EIAs are predominantly type-specific.

Specificity of the antigen EIAs

The EIAs to detect virus utilizing hyperimmune antisera against the rNV and
rMX capsid antigens appear to be more specific than the EIAs to detect
antibody. The rNV and rHX virus EIAs detected homologous VLPs but did not
detect heterologous VLPs even at a high antigen concentration (2 µg/ml; 15).

The rNV EIA detected prototype NV and closely related strains in the NV genogroup, but not viruses in the SMA and Sapporo genogroups [12]. The rMX EIA detected prototype MX and closely related strains, including SMA and HA, but did not detect NV and Sapporo virus [15]. However, the use of rMX antigen EIA to detect SMA and HA needs to be further evaluated because only 2/8 stool specimens from volunteers infected with SMA and 1/8 stool specimens from volunteers infected with HA were positive by the rMX EIA [15]. The low detection rate may reflect the antigenic diversity between these viruses indicated by the genomic variation; MX virus shares 89% amino acid sequence in the RNA polymerase region with SMA and HA RNA polymerases.

Seroepidemiology of HuCVs using the recombinant antibody EIAs

The rNV and rMX EIAs have been useful in describing the epidemiology of HuCVs. Initial seroprevalence studies using the rNV antibody EIA showed that infection with NV was common in both developed and developing countries [7, 13, 21, 28, 30, 32]. In England, two large-scale surveys were conducted using the rNV antibody EIA. In a multicenter survey [7], 2 382 (73%) of 3 250 tested serum specimens collected in 1991–1992 were positive for antibody to rNV. The lowest (25%) prevalence of antibody was among 6- to 11-month-old infants and the highest (90%) was among persons more than 60 years old. In London [30], there was little evidence of infection during the first two years of life. However, the prevalence of antibody to rNV rose steadily throughout the age range of children attending school, reaching a peak of 70% among children 11 to 16 years old. A similar seroprevalence to NV was found in Japan, in which antibody prevalence to NV remained at a low level throughout childhood and then showed a steep rise during school age and early adulthood [28]. In Mexico [13], 85% of children had antibodies to NV in a cohort study of diarrhea involving 200 children monitored from birth to 2 years of age. This high prevalence was comparable to that found for rotaviruses in the same group of children. A similar high seroprevalence of antibody to rMX also was observed in the Mexican (our unpubl. obs.) and London [32] children described above. In London, a serologic survey of 338 children showed that infection with MX strain occurred earlier in life than did NV infection. Primary infections were common after the age of 6 months, with more than 70% of children having antibody by age 2 years, whereas only 12% of children had antibody to NV at that age [32].

Detection of HuCVs in field samples by the recombinant antigen EIAs

In spite of the high prevalence of antibody to rNV and rMX, few NV-like viruses were detected by the rNV antigen EIA among specimens collected recently from different areas of the world. In one large survey, including 451 stool samples from 26 outbreaks and 175 sporadic cases of acute gastroenteritis from 6 countries, <1% of samples were positive for NV-antigen [12]. All of these positive samples contained calicivirus-like particles by electron microscopy (EM) or immune EM (IEM). We concluded that the low rate detection of NV antigen was due to the

Table 2. SMA-like viruses found in the UK by rMX EIA

| Sample | rMX EIA | | RT-PCR |
	OD_{450} (pre/post)	P/N	
1	0.029/**0.232**[c]	**8.0**	+
2	0.048/**0.228**	**4.8**	+[a]
3	0.064/**0.272**	**4.3**	+[a]
4	0.050/**0.194**	**3.9**	+
5	0.058/**0.223**	**3.8**	+[a]
6	0.065/**0.226**	**3.5**	+
7	0.076/**0.228**	**3.0**	+
8	0.061/**0.145**	**2.4**	+[a]
9	0.013/0.048	3.7[b]	−
10	0.049/0.075	1.5	−
11	0.053/0.077	1.5	−
12	0.051/0.070	1.4	−
13	0.070/0.090	1.3	+
14	0.056/0.074	1.3	−
15	0.049/0.057	1.2	−
16	0.078/0.083	1.1	+
17	0.051/0.055	1.1	+
18	0.059/0.061	1.0	−
19	0.062/0.048	0.8	−
20	0.076/0.056	0.7	−

[a] The RT-PCR products from these viruses were cloned and sequenced
[b] This specimen was considered to be negative due to the low OD_{450} value of the well coated with the rabbit hyperimmune serum, although the P/N ratio was more than 2
[c] Bold type shows positive data

high specificity of the rNV antigen EIA, with most samples containing calici-viruses other than the prototype NV. This has been confirmed by determining the nucleotide sequences of the RT-PCR products of these strains; these sequences showed a wide genetic variability [12].

After the development of the rMX EIAs, we re-tested some of these same NV antigen-negative specimens by the rMX EIA and showed that many specimens contained MX-like viruses. For example, eight of 20 diarrhea stool specimens from children admitted to the Hospital for Sick Children in London and included in the NV studies described above were positive in the rMX EIA; these results were confirmed by RT-PCR [15]. Four of the eight RT-PCR products were sequenced [2] and all four revealed sequences close to SMA and MX strain (Table 2).

The rMX EIA also has been used to screen 1 032 diarrhea stool specimens from children attending day-care centers (DCCs) in Norfolk, Virginia, and enrolled in a weekly surveillance study of diarrhea between November, 1993, and

November, 1994. Twenty-four diarrhea stool specimens from 21 children in five DCCs were positive for MX strain. Twenty-three of the 24 positive specimens were confirmed by RT-PCR. Six of the 23 RT-PCR products were sequenced and shown to contain SMA genogroup viruses [15]. We also used the rMX EIA to investigate four nursing home-related outbreaks of acute gastroenteritis that occurred in Virginia during 1993–1994, and found SMA-like viruses in each by serologic and/or virologic parameters [16]. In addition, SMA-like viruses have been found in the UK [3] and South Africa [34] using the rMX EIAs. In England, we have detected SMA-like viruses in stool specimens collected as early as 1982 and SMA-like viruses have been found continuously since then [3].

Genetic characterization of HuCVs based on RT-PCR and sequencing

The RT-PCR technique developed for detection of NV [10] was used to detect different HuCV strains with primers from the conserved RNA-dependent RNA-polymerase region of the genome [12, 35]. Sequence analysis of the RT-PCR amplified products showed that HuCVs were genetically diverse. Three genogroups of HuCVs exist: including the NV, SMA, and Sapporo genogroups [12]. Similar genetic variation of HuCVs also has been reported by others in different countries [1, 2, 6, 18–20, 25, 26, 34, 35], with most of the reports describing the NV and SMA genogroups. The Sapporo genogroup has been described less frequently, which may reflect a technical problem in that the primers used were mainly designed from the NV sequence, while the Sapporo genogroup is distantly related to NV.

A correlation between antigenic and genetic types has been noticed by comparison of the sequences and the results of the new EIAs (15, Table 3). Viruses detected by the rMX EIA have high (>83%) amino acid (aa) sequence identity with prototype SMA and MX strain, but low aa sequence identity with NV genogroup (57–61%) and Sapporo genogroup (30–33%) HuCVs [15]. None of the rMX EIA-positive specimens was detected by the rNV EIA and vice versa [15]. Sequence variation among rMX EIA positive strains has been noted. For example, only 2/8 of the prototype SMA and 1/8 of the Hawaii strain were detected by the rMX EIA, and the sequences of the RNA polymerase region of these two strains share <89%, while other strains detected by the rMX EIA share >93% nucleotide identity with the prototype MX strain (Table 3, Fig. 1) Therefore, we conclude that the SMA genogroup may be further divided antigenically and genetically.

We also identified additional strains in the Sapporo genogroup by RT-PCR. The prototype Sapporo virus was first identified in 1982 in children with diarrhea in Sapporo, Japan [27]. This virus has a typical calicivirus morphology by EM. In 1986, strains similar to the prototype strain were found in children with diarrhea in Phoenix, Arizona, and in children attending DCCs in Houston, Texas [24]. Sequence analysis showed that the prototype Sapporo and these Sapporo-like viruses shared high levels (>96%) of sequence identity in the

Table 3. Genetic and antigenic relationships of HuCVs determined by type-specific EIAs and RT-PCR

Virus[a]	aa similarity (%) to				EIA (+/tested)		Reference
	NV	SMA	Sapp	MX	rNV	rMX	
8FIIa NV/68/US[b]	100	57	30	58	+ (4/4)	− (0/4)	Jiang et al., 1990
SMA/76/US[b]	57	100	32	89	− (0/8)	+ (2/8)	Madore et al., 1990
HA/71/US	61	91	33	83	− (0/8)	+ (1/8)	Madore et al., 1990
MX 34/89/Mex[b]	58	89	30	100	−	+	Jiang et al., 1995b
SRSV/4S/90/UK	60	91	32	98	−	+	Cubitt et al., 1994
Nfk 1135/94/US[c]	60	90	32	97	−	+	
Nfk 1184/94/US[c]	60	90	32	97	−	+	
Nfk 770R1/94/US[c]	60	90	32	97	−	+	
HuCV/3C/92/UK	60	90	32	97	−	+	Cubitt et al., 1994
HuCV/5C/92/UK	60	90	31	97	−	+	Cubitt et al., 1994
Nfk 5230/94/US[c]	59	89	30	94	−	+	
Nfk 5432/94/US[c]	58	88	30	93	−	+	
Nfk 6610/94/US[c]	60	93	32	88	−	+	
HuCV/12C/92/UK	59	89	31	89	−	+	Cubitt et al., 1994
HuCV/Sapp/82/J[b]	30	31	100	31	−	−	Matson et al., 1995
HuCV UT/86/US	30	32	95	30	−	−	Matson et al., 1995

[a] Individual viruses are named by designated strain/year of isolation/country. Prefix "HuCV" indicates those strains with typical caliciviral morphology, and prefix "SRSV" indicates a strain without typical caliciviral morphology

[b] Prototype NV, SMA, Sapporo virus, and MX strain were used as references in the sequence comparisons. J, Japan

[c] Indicates viruses detected in the day care centers in Norfolk, Virginia

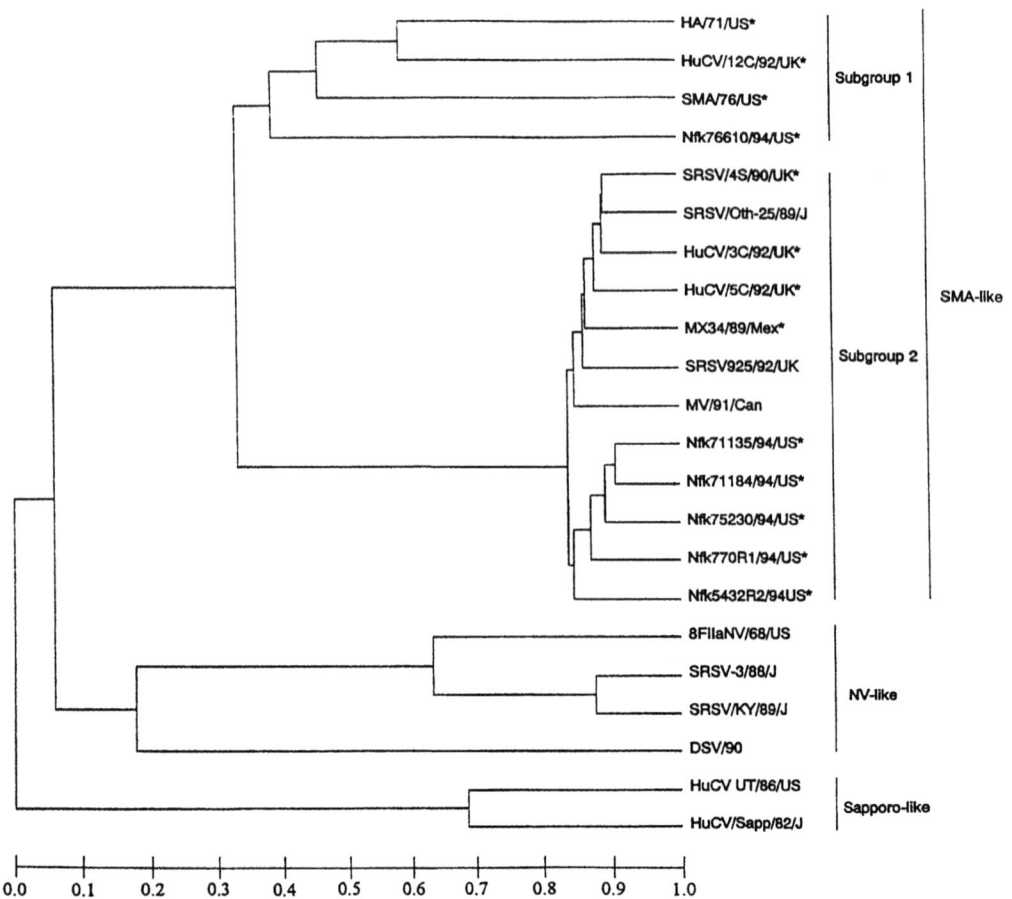

Fig. 1. Dendrogram of the RNA polymerase region of 22 HuCVs. The prefix "HuCV" indicates strains with typical calicivirus morphology, and the prefix "SRSV" indicates strains without typical calicivirus morphology, as described in the footnote of Table 3. The amino acid sequences between primers 36/39 (~386 bases) in the RNA polymerase region of the viral genomes were aligned using the PC/GENE multiple alignment program. The length of each horizontal line indicates the genetic distance between the viruses normalized to 1.0 for the dendrogram. *Indicates viruses positive in the rMX EIA

RNA-dependent RNA polymerase region of the genome. In the past year, new strains of Sapporo-like viruses have been found in children with diarrhea in the US (Matson, unpubl. data), UK (Cubitt, unpublished data) and South Africa (Taylor, unpubl. data) by RT-PCR. Phylogenetic analysis of the sequences showed that the new strains clustered with previously identified Sapporo-like viruses and that this cluster was closer to animal caliciviruses than to other known human caliciviruses (Fig. 2). The finding of a high level of sequence diversity among these strains also suggested that the Sapporo genogroup may be further divided.

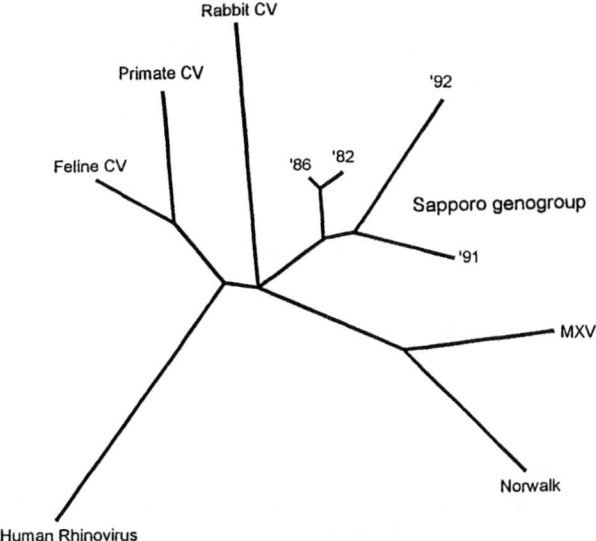

Fig. 2. Phylogenetic relationship among caliciviruses. Four Sapporo-like viruses are denoted by the year they were isolated: HuCV/Sapporo/82/Japan, HuCV/Sapporo/86/US, HuCV/Sapporo/91/US, and HuCV/Sapporo/92/UK. Norwalk and MXV represent the NV and SMA genogroups, respectively. Three animal caliciviruses are noted as feline CV, primate CV and rabbit CV. One picornavirus was used as an outgroup for the analysis. The network represents the maximum likelihood estimate of the phylogeny utilizing the DNAML program of the phylogenetic package Phylip 3.5 c

Conclusions and future directions

In conclusion, although new knowledge on the genetic and antigenic relationships of HuCV has been generated by using RT-PCR and new EIAs, classification of HuCVs remains preliminary. Further application of these new assays and development of new assays to study HuCVs are important. First, accumulation of sequences of strains within and among genogroups and search for new strains are necessary. Extension of sequences to the viral capsid region in addition to the RNA polymerase region may promote understanding of genetic and antigenic relationships among HuCV strains. Second, development of type-specific assays, particularly for type-specific antibody detection, is required. Expression of viral capsid proteins for all antigenic groups of HuCVs and generation of monoclonal antibodies against the expressed antigens may be necessary. Third, limited information has shown that predominating antigenic groups of HuCVs may change over time in a pattern exhibited by other enteric viral pathogens. The NV genogroup, which was found to be predominant in 1970s and 1980s, has been rarely detected recently, whereas the SMA genogroup is commonly detected. Further study to confirm and understand this pattern is necessary. Finally, our knowledge of HuCVs causing gastroenteritis in humans may be extended. The finding that the Sapporo genogroup is genetically closer to animal caliciviruses than to other HuCVs suggest that similar diseases may occur in animals and humans, and that animal-to-human transmission is possible. In general, the molecular characterization of HuCVs has just started. With the continual development and improvement of new assays, the role(s) of HuCVs in human diseases will be better understood.

Acknowledgments

This study was supported by grants from the US Public Health Service (HD-13021-16, AI 28855 and AI 30448) and from the Jeffress Research Grant Foundation (J-303).

References

1. Ando T, Monroe SS, Gentsch JR, Jin Q, Lewis DC, Glass RI (1995) Detection and differentiation of antigenically distinct small round-structured viruses (Norwalk-like viruses) by reverse transcription-PCR and Southern hybridization. J Clin Microbiol 33: 64–71
2. Cubitt WD, Jiang X, Wang J, Estes MK (1994) Sequence similarity of human caliciviruses and small round structured viruses. J Med Virol 43: 252–258
3. Cubitt WD, Jiang X, The use of a recently developed enzyme immunoassay to study the occurrence of human calicivirus (Mexico strain) as a cause of sporadic cases and outbreaks of diarrhea in the United Kingdom, 1982–1994 J Med Virol (in press)
4. Dolin RR, Reichman C, Rocssner KD, Tralker TS, Schooley RT, Gary W, Morens D (1982) Detection by immune electron microscopy of the Snow Mountain agent of acute gastroenteritis. J Infect Dis 146: 184–189
5. Graham DY, Jiang X, Tanaka T, Opekun AR, Madore HP, Estes MK (1994) Norwalk virus infection of volunteers: new insights based on improved assays. J Infect Dis 170: 34–43
6. Green J, Norcott JP, Lewis D, Arnold C, Brown DWG (1993) Norwalk-like viruses: Demonstration of genomic diversity by polymerase chain reaction. J Clin Microbiol 31: 3007–3012
7. Gray JJ, Jiang X, Desselberger U, Estes MK (1993) The prevalence of antibody to Norwalk virus in East Anglia, England: Detection by indirect EIA with baculovirus-expressed Norwalk capsid antigen. J Clin Microbiol 31: 1022–1025
8. Jiang X, Graham DY, Wang K, Estes MK (1990) Norwalk virus genome cloning and characterization. Science 250: 1580–1583
9. Jiang X, Wang M, Graham DY, Estes MK (1992) Expression, self-assembly and antigenicity of the Norwalk virus capsid protein. J Virol 66: 6527–6532
10. Jiang X, Wang J, Graham DY, Estes MK (1992) Detection of Norwalk virus in stool using polymerase chain reaction. J Clin Microbiol 30: 2529–2534
11. Jiang X, Wang M, Wang K, Estes MK (1993) Sequence and genomic organization of Norwalk virus. J Virol 195: 56–61
12. Jiang X, Wang J, Estes MK (1995) Characterization of SRSVs using RT-PCR and a new antigen EIA. Arch Virol 140: 363–374
13. Jiang X, Matson DO, Velazquez FR, Zhong WM, Hu J, Ruiz-Palacios G, Pickering LK (1995) A study of Norwalk-related viruses in Mexican children. J Med Virol 47: 309–316
14. Jiang X, Matson DO, Ruiz-Palacios G, Hu J, Treanor J, Pickering LK (1995) Expression, self-assembly, and antigenicity of a Snow Mountain-like calicivirus capsid protein. J Clin Microbiol 33: 1452–1455

15. Jiang X, Cubitt D, Treanor JJ, Hu J, Dai X, Matson DO, Pickering LK (1995) Development of an EIA to detect MX strain, a human calicivirus in the Snow Mountain agent genegroup. J Gen Virol 76: 2739–2747

16. Jiang X, Turf E, Barrett E, Monroe S, Matson DO, Pickering LK, Outbreaks of gastroenteritis in elderly nursing homes and retirement facilities associated with human caliciviruses (HuCVs) J Med Virol (in press)

17. Lambden PR, Caul EO, Ashley CP, Clarke IN (1993) Sequence and genome organization of a human small round-structured (Norwalk-like) virus. Science 259: 516–519

18. Lew JF, Petric M, Kapakian AZ, Jiang X, Estes MK, Green KY (1994) Identification of minireovirus as a Norwalk-like virus in pediatric patients with gastroenteritis. J Virol 68: 3391–3396

19. Lew JF, Kapikian AZ, Jiang X, Estes MK, Green KY (1994) Molecular characterization and expression of the capsid protein of a Norwalk-like virus found in Desert Shield troops with gastroenteritis. Virology 200: 319–325

20. Lew JF, Kapikian AZ, Valdesuso J, Green KY (1994) Molecular characterization of Hawaii virus and other Norwalk-like viruses: Evidence for genetic polymorphism among human caliciviruses. J Infect Dis 170: 535–542

21. Lew JF, Valdesuso J, Vesikari T, Kapikian AZ, Jiang X, Estes MK, Green KY (1994) Detection of Norwalk virus or Norwalk-like virus infection in Finnish infants and young children. J Infect Dis 169: 1364–1367

22. Lewis DC (1990) Three serotypes of Norwalk-like virus demonstrated by solid-phase immune electron microscopy. J Med Virol 30: 77–81

23. Lewis DC (1991) Norwalk agent and other small-round structural viruses in the U.K. J Infect 23: 220–222

24. Matson DO, Estes MK, Glass RI, Bartlett AV, Penaranda M, Calomeni E, Tanaka T, Nakata S, Chiba S (1989) Human calicivirus-associated diarrhea in children attending day-care centers. J Infect Dis 159: 71–81

25. Matson DO, Zhong WM, Nakata S, Numata K, Jiang X, Pickering LK, Chiba S, Estes MK (1995) Molecular characterization of a human calicivirus with sequence relationships closer to animal caliciviruses than other known human caliciviruses. J Med Virol 45: 215–222

26. Moe CL, Gentsch J, Grohmann G, Ando T, Monroe SS, Jiang X, Wang J, Estes MK, Seto Y, Humphrey C, Stine S, Glass RI (1994) Application of PCR to detection of Norwalk virus in fecal specimens from outbreaks of gastroenteritis. J Clin Microbiol 32: 642–648

27. Nakata S, Chiba S, Terashima H, Sakuma Y, Kogasaka R, Nakao T (1983) Microtier solid-phase radioimmunoassay for detection of human calicivirus in stools. J Clin Microbiol 17: 198–201

28. Numata K, Nakata S, Jiang X, Estes MK, Chiba S (1994) Epidemiological study of Norwalk virus infection in Japan and Southeast Asia by enzyme-linked immunosorbent assays with Norwalk virus capsid protein produced by the baculovirus expression system. J Clin Microbiol 32: 121–126

29. Okada S, Sekine S, Ando T, Hayashi Y, Murao M, Yabuuchi K, Miki T, Ohashi M (1990) Antigenic characterization of small, round-structured viruses by immune electron microscopy. J Clin Microbiol 28: 1244–1248

30. Parker SP, Cubitt WD, Jiang X, Estes MK (1994) Seroprevalence studies using a recombinant Norwalk virus protein enzyme immunoassay. J Med Virol 42: 146–150

31. Parker SP, Cubitt WD, Jiang X, Estes MK (1993) Efficacy of a recombinant Norwalk virus protein enzyme immunoassay for the diagnosis of infections with Norwalk virus and other "candidate" caliciviruses. J Med Virol 41: 179–184

32. Parker SP, Cubitt WD, Jiang X, The application of an enzyme immunoassay using a baculovirus expressed recombinant human calicivirus (Mexico virus) for the study of outbreaks of gastroenteritis and determining its seroprevalence in children in London, UK. J Med Virol (in press)
33. Treanor JJ, Jiang X, Madore P, Estes MK (1993) Subclass specific serum antibody responses to recombinant Norwalk capsid antigen (rNV) in adults infected with Norwalk, Snow Mountain, or Hawaii viruses. J Clin Microbiol 31: 1630–1634
34. Wolfaardt M, Taylor MB, Grabow WOK, Cubit WD, Jiang X, Molecular characterization of small round structured viruses associated with gastroenteritis in South Africa J Med Virol 47: 386–391
35. Wang JX, Jiang X, Madore P, Desselberger U, Gray J, Ando T, Oishi I, Estes MK (1994) Sequence diversity of SRSVs in the Norwalk virus group. Virol 68: 5982–5990
36. Wyatt RG, Dolin R, Blacklow NR, DuPont HL, Buscho RF, Thornhill TS, Kapikian AZ, Chanock RM (1974) Comparison of three agents of acute infectious nonbacterial gastroenteritis by cross-challenge in volunteers. J Infect Dis 129: 709–714

Authors' address: Dr. X. Jiang, Center for Pediatric Research, Children's Hospital of The King's Daughters and Eastern Virginia Medical School, 855 West Brambleton Avenue, Norfolk, Virginia 23510-1001, U.S.A.

Arch Virol (1996) [Suppl] 12: 263–270

The epidemiology of human calicivirus/Sapporo/82/Japan

S. Nakata[1], K. Kogawa[1], K. Numata[1], S. Ukae[1], N. Adachi[1], D. O. Matson[2], M. K. Estes[3], and S. Chiba[1]

[1] Department of Pediatrics, Sapporo Medical University School of Medicine, Sapporo, Japan; [2] Center for Pediatric Research, Eastern Virginia Medical School and Children's Hospital of King's Daughter, Norfolk, U.S.A. [3] Division of Molecular Virology, Baylor College of Medicine, Houston, U.S.A.

Summary. Based on genome analysis of the RNA-dependent RNA polymerase region, it has been proposed that human caliciviruses (HuCV) can be classified into at least three genogroups: genogroup I is represented by Norwalk virus (NV), genogroup II by Snow Mountain agent (SMA) and genogroup III by HuCV/Sapporo/82/Japan (HuCV/Sa/82/J) virus. HuCV/Sa/82/J strain is genetically unique and more closely related to animal caliciviruses than are other known HuCVs, such as NV and SMA.

HuCV/Sa/82/J strain was detected in four outbreaks of HuCV gastroenteritis occurring between 1977 and 1982 in an infant home in Sapporo. The HuCVs detected from these four outbreaks all showed a typical "Star of David" configuration by electron microscopy (EM), and they were identical antigenically and genetically. This strain has also been detected in other prefectures in Japan, as well as in the USA, UK, Saudi Arabia and Kenya. Seroepidemiological studies have shown a worldwide distribution of this virus, including Japan, USA, UK, Southeast Asia, Canada, China and Kenya. This virus has been circulating in Sapporo for at least 19 years (1977–1995). HuCV/Sa/82/J strain is thought to be one of the common causes of viral gastroenteritis worldwide.

The HuCV/Sa/82/J strain has been detected mainly in infants. Age-related prevalence of antibody to this strain also shows that infections commonly occur in children less than 5 years old, although viruses in the NV and SMA genogroups commonly infect adults. The pattern of acquisition of antibodies to strain HuCV/Sa/82/J is similar to that of other common viral infections. HuCV/Sa/82/J strain is unique virologically and clinically among caliciviruses.

264 S. Nakata et al.

Introduction

Human calicivirus Sapporo strain (HuCV/Sa/82/J) is one of the causative viruses of acute gastroenteritis of humans, especially of infants [2, 3, 15, 16]. This virus has a characteristic "Star of David" configuration by EM and contains a single major polypeptide [20]. The epidemiological importance of this virus in infants has been confirmed by many studies in and outside of Japan during the past 15 years [2, 3, 10, 12, 13, 15–18]. Recently, based on genome analysis of the RNA-dependent RNA polymerase region, it has been proposed that human caliciviruses be classified into at least three genogroups. These are genogroup I represented by Norwalk agent (NV), genogroup II by Snow Mountain agent (SMA) and genogroup III by HuCV/Sa/82/J group [5, 11, 14]. The viruses in these three groups also are different from each other antigenically. HuCV/Sa/82/J is one of three genogroups of human caliciviruses and is unique clinically and genetically. In this paper, the historical background of this virus is reviewed and epidemiological studies which used immunological assays or dot blot hybridization assays will be summerized. Recent advances in studies of this virus learned by using new methods, such as PCR and sequencing of RT-PCR products, will also be discussed. HuCV/Sa/82/J related strains are defined as viruses which react with antibody to HuCV/Sa/82/J in enzyme immunoassay (EIA), radioimmunoassay (RIA) or immune EM (IEM).

Four outbreaks of HuCV gastroenteritis in an infant home in Sapporo from 1977 to 1982 and the development of immunoassays

Eleven outbreaks of viral gastroenteritis occurred in an infant home in Sapporo between 1977 and 1982. Five of these outbreaks were caused by group A rotaviruses [1], four by HuCV/Sa/82/J [2, 3, 15, 16], one by Sapporo agent [8], and one by enteric adenovirus type 40 [4]. The viruses detected from the four outbreaks of HuCV/Sa/82/J gastroenteritis all have a typical "Star of David" configuration by EM (Fig. 1). These viruses are antigenically identical to each other by IEM. Antisera to the Sapporo HuCV were prepared by immunizing guinea pigs, mice and rabbits with viruses purified from stool samples collected from two outbreaks in 1981 and 1982 [15, 16]. The RIA [15] and EIA [17] for detection of HuCV/Sa/82/J antigen and a blocking RIA and EIA for detection of antibody to that strain have been developed and used for epidemiological studies in and outside of Japan.

Detection of outbreaks or sporadic cases of HuCV gastroenteritis in Sapporo from 1983 to 1994

Since 1983, many outbreaks of acute gastroenteritis have occurred in the same infant home in Sapporo. Group A rotaviruses have caused outbreaks almost every year, whereas the HuCV/Sa/82/J strain has not been detected during the study period. To determine whether HuCV/Sa/82/J strains are circulating regularly in Sapporo, we examined by EIA diarrheal stool samples from children

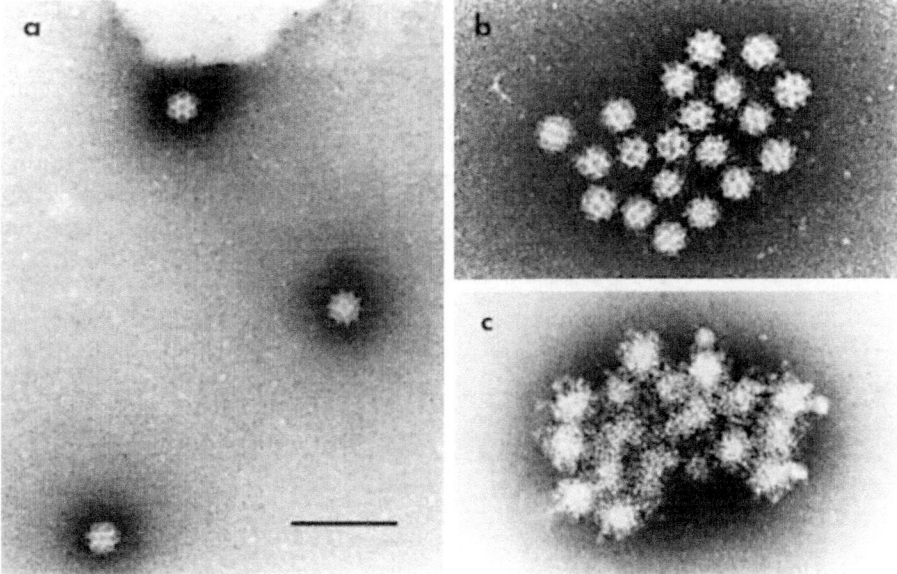

Fig. 1. Electron micrographs of the human calicivirus/Sapporo/82/Japan (HuCV/Sa/82/J) obtained in 1982 in Sapporo. (a) A Star of David configuration is apparent. (b) and (c) Immune EM results showing two different degrees of reactivity of HuCV/Sa/82/J with a guinea pig antiserum to the homologous virus. The samples were stained with phosphotungstic acid. Bar = 100 nm

who visited outpatient clinics in Sapporo. Detection of viruses from children with diarrhea in Sapporo is shown in Fig. 2. Group A rotaviruses were dominant every year, followed by small round structured viruses (SRSVs) including HuCV/Sa/82/J and then enteric adenoviruses. HuCV/Sa/82/J was detected between 1987 and 1994 in Sapporo [10], although the rate was low. These data indicate that strains related to HuCV/Sa/82/J have been circulating in Sapporo for a long time period.

Detection of HuCV/Sa/82/J in Japan

Apart from Sapporo city, strain HuCV/Sa/82/J was also detected in other regions of Japan, including in Ehime prefecture, Aichi prefecture and Tokyo (Fig. 3). In Ehime prefecture, 50 SRSV positive samples collected from 1984 to 1988 were tested for HuCV/Sa/82/J by EIA. Three of 19 samples in 1984 and one of 10 specimens in 1988 were positive, suggesting that HuCV/Sa/82/J related strains were circulating in remote areas from Sapporo (unpubl. data). This strain was found also in Tokyo by IEM using guinea pig immune serum to HuCV/Sa/82/J (pers. comm. with T. Ando).

266 S. Nakata et al.

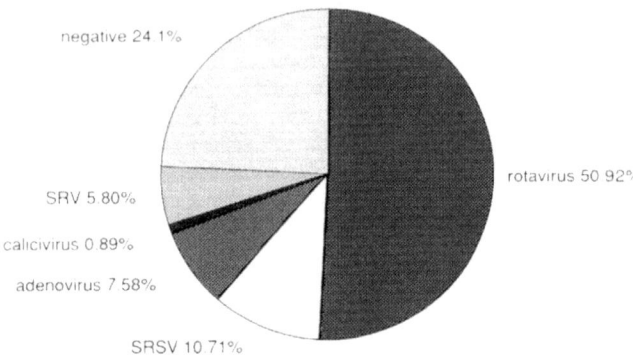

Viruses detected from infants with gastroenteritis
in Sapporo from November 1987 to March 1988

negative 24.1%

SRV 5.80%

calicivirus 0.89%

adenovirus 7.58%

SRSV 10.71%

rotavirus 50.92%

Fig. 2. A distribution of the viruses detected by EM from infants with acute gastroenteritis in Sapporo from November 1987 to March 1988. Group A rotaviruses were dominant (50.92%) followed by SRSVs plus HuCV/Sa/82/J (11.60%) and enteric adenoviruses (7.58%)

Hokkaido

Miyagi

Kyoto   Saitama

Fukuoka   Tokyo

Aichi

Ehime

Fig. 3. Distribution of HuCV/Sa/82/J in Japan. HuCV/Sa/82/J-related viruses were detected in Hokkaido, Ehime, Aichi and Tokyo by EIA. More than 80% of serum samples collected from Hokkiado, Miyagi, Saitama, Kyoto and Fukuoka had antibodies to HuCV/Sa/82/J as determined by a blocking-RIA

Seroepidemiology of HuCV/Sa/82/J in Japan

Seroepidemiological studies of strain HuCV/Sa/82/J were done by a RIA-blocking test [18]. Two hundred forty sera were collected from healthy adults, aged 20 to 49 years, in five prefectures (Hokkaido, Miyagi-ken, Saitama-ken, Kyoto-fu and Fukuoka-ken) of Japan (Fig. 3). More than 80% of serum samples collected in each of the five districts had antibodies to HuCV/Sa/82/J. There was no significant difference in the prevalence of antibody by age group or by district, and the total proportion of seropositive results in Japan was 87% (209/240). These results suggest a broad distribution of this virus as a common infectious agent in many regions of Japan, and that infection occurs early in life.

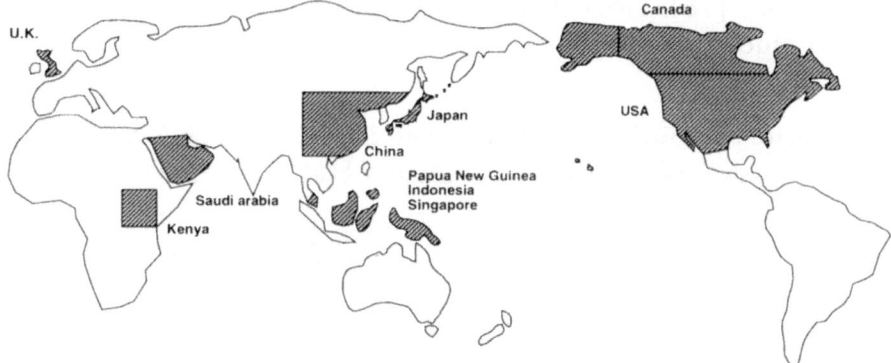

Fig. 4. Distribution of HuCV/Sa/82/J in the world. HuCV/Sa/82/J related viruses were detected in Japan, the USA, UK, Saudi Arabia and Kenya by EIA or RIA. A high prevalence of antibody to this virus was found in serum samples from healthy adults in Canada, USA, Papua New Guinea, Indonesia, Singapore and in each of five immunoglobulin pools from 2000 adults donors from China

Distribution of HuCV/Sa/82/J in the world

HuCV/Sa/82/J has also been detected outside Japan, including Canada, the USA, UK, Saudi Arabia and Kenya. (Fig. 4) Epidemiological studies of HuCV/Sa/82/J in the USA [17] have been done using ELISAs. Only one (0.6%) of 163 stool specimens collected from children hospitalized with diarrhea was positive. However, HuCV/Sa/82/J was detected in 2.9% of diarrheal stools in a day-care center [12]. This incidence was about half that noted for group A rotaviruses but higher than that noted for bacterial pathogens in the original study. These data indicate that many infections with HuCV/Sa/82/J may not require hospitalization but that HuCV/Sa/82/J is an important cause of diarrhea in the day-care center setting. Moreover, a predominantly high rate of asymptomatic infection with this strain was demonstrated in a longitudinal study of diarrhea in another day-care center [13]. Diarrhea stool specimens collected in the UK and Saudi Arabia and containing caliciviruses by EM were positive by RIA. This strain of caliciviruses also was detected in Kenya from children aged 0–6 years old with complaints of diarrhea (S. Nakata, unpubl. data).

The worldwide distribution of HuCV/Sa/82/J strain has also been demonstrated by seroepidemiological studies in which sera or immunoglobulin pools from many countries were tested. A high prevalence of antibody to HuCV/Sa/82/J was found in serum samples from healthy adults in Canada (80%), the USA (Arkansas, Hawaii, Texas, and New York [>90%]) and in 5/5 immunoglobulin pools from 2,000 adult donors in China (100%) [17]. Testing of 150 sera from healthy adults showed that the prevalence of antibody to this virus is 70% in Singapore, 88% in Indonesia, and 94% in Papua New Guinea. The mean prevalence of antibody in adults in the three countries is 84%, similar to the 87% in Japan [18]. These data suggest that infections with HuCV/Sa/82/J strain are common worldwide.

Detection of HuCV by dot blot hybridization with cDNA probe derived from RDRP region of HuCV/Sa/82/J

Although the epidemiological importance of HuCV/Sa/82/J strain has been confirmed by EM, IEM, RIA and ELISA as described above, immunological methods have not been widely used for routine diagnosis because of limitations in the amount of positive control immune sera and antigen availability. Recent success with the cloning of the NV genome [6] resulted in the development of new methods to investigate human caliciviruses and SRSVs. An RT-PCR technique to detect the RNA-dependent RNA polymerase (RDRP) region is one of the most useful methods for detection and analysis of SRSVs [5, 7]. Recently, cDNA for the RDRP region of HuCV/Sa/82/J has been obtained by RT-PCR [14] and subcloned into an AT1000 vector. We have developed a dot blot hybridization assay for detecting HuCV in stool specimens using this cDNA as a probe [9]. This assay is specific for HuCV/Sa/82/J and antigenically related viruses. The sensitivity is similar to that of ELISA.

When clinical samples positive by EM for SRSV were tested by ELISA and dot blot assay, a higher positive rate was obtained with the dot blot assay than by ELISA. Because the RDRP region is more conserved than the capsid protein region, the dot blot assay may detect HuCVs more broadly than the ELISA. Another possibility is that the dot blot assay may detect HuCV covered with fecal antibody which interferes with antigen-antibody reaction in the ELISA [9].

Nucleotide and amino acid similarity among HuCV/Sa/82/J detected between 1979 and 1990

To test the above hypothesis, the RDRP region of HuCV/Sa/82/J-related strains collected from children in Sapporo between 1979 and 1990 was amplified by RT-PCR and the products were sequenced [10]. Nucleotide and amino acid sequences of the PCR products showed a high degree of identity among two strains from two outbreaks, one strain from a sporadic case in Sapporo and the original HuCV/Sa/82/J strain. Although additional samples need to be analysed, these data indicate a correlation between the antigenicity and sequence similarity of the RDRP region among HuCV/Sa/82/J related viruses. Sequence analysis of the samples positive by dot blot but negative by ELISA is now in progress.

Age-related prevalence of antibody to HuCV/Sa/82/J in Sapporo, Japan, in Houston, USA and in Nairobi, Kenya

Several epidemiological studies have been carried out in Japan, the USA and Kenya to determine the age at which antibodies to HuCV/Sa/82/J are acquired. The age-related acquisition of antibody to this strain in Sapporo has been demonstrated by IEM [19]. Current results suggest that infection with HuCV/Sa/82/J mainly occurs in children < 6 years of age. The patterns of acquisition of antibody to HuCV and to group A rotaviruses were compared by ELISA in Houston, Texas [17]. All infants < 3 months of age had antibody to HuCV. The prevalence of antibody declined to 25% in infants aged 4 to 11 months, then steadily increased thereafter, reaching a maximum in pre-school children, 4 to 6 years old. This pattern was similar to that for group A rotaviruses but for the latter the highest prevalence of antibody was found at an earlier age, i.e., by the age of 3 years. In Kenya, although the tested sample numbers are small, the pattern of acquisition of antibody to HuCV/Sa/82/J is similar to those found in Japan and in USA. By the age of 6 years the majority of the population has been infected with HuCV/Sa/82/J (S. Nakata, unpubl. data). Overall, infections by HuCV/Sa/82/J commonly occur in children < 5 years old in Asia, Africa and North America.

References

1. Chiba S, Akihara M, Kogasaka R, Horino K, Nakao T, Urasawa T, Urasawa S, Fukui S (1979) An outbreak of acute gastroenteritis due to rotavirus in an infant home. Tohoku J Exp Med 127: 265–271
2. Chiba S, Sakuma Y, Kogasaka R, Akihara M, Horino K, Nakao T, Fukui S (1979) An outbreak of gastroenteritis associated with calicivirus in an infant home. J Med Virol 4: 249–254
3. Chiba S, Sakuma Y, Kogasaka R, Akihara M, Terashima H, Horino K, Nakao T (1980) Fecal shedding of virus in relation to the days of illness in infantile gastroenteritis due to calicivirus. J Infect Dis 142: 247–249
4. Chiba S, Nakata S, Nakamura I, Taniguchi K, Urasawa S, Fujinaga K, Nakao T (1983) Outbreak of infantile gastroenteritis due to type 40 adenovirus. Lancet 2: 954–957
5. Estes MK, Hardy ME (1995) Norwalk virus and other enteric caliciviruses. In: Blaser MJ, Smith PD, Ravdin JI, Greenberg HB, Guerrant RL (eds) Infections of the Gastrointestinal Tract. Raven Press, New York, pp 1009–1034
6. Jiang X, Graham DY, Wang J, Estes MK (1990) Norwalk virus genome cloning and characterization. Science 250: 1580–1583
7. Jiang X, Wang J, Graham DY, Estes MK (1992) Detection of Norwalk virus in stool by polymerase chain reaction. J Clin Microbiol 30: 2529–2534
8. Kogasaka R, Nakamura S, Chiba S, Sakuma Y, Terashima H, Yokoyama T, Nakao T (1981) The 33- to 39-nm virus like particles, tentatively designated as Sapporo agent, associated with an outbreak of acute gastroenteritis. J Med Virol 8: 187–193
9. Kogawa K, Nakata S, Ukae S, Adachi S, Numata S, Matson DO, Estes MK, Chiba S (1996) Dot Blot Hybridization with a cDNA Probe Derived from the Human Calicivirus Sapporo 1982 Strain. Arch Virol (in press)
10. Kogawa K, Nakata S, Numata K, Ukae S, Adachi N, Estes MK, Chiba S (1994) Nucleotide sequence diversity in RNA polymerase region of human calicivirus antigenically related to Sapporo 1982 strain. The 28th Joint Viral Panels Meeting: The US-Japan Co-operative Medical Science Program. Tokyo, Japan
11. Lew JF, Kapikian AZ, Valdesuso J, Green KY (1994) Molecular characterization of Hawaii virus and other Norwalk-like viruses. Evidence for genetic polymorphism among human caliciviruses. J Infect Dis 170: 535–542

12. Matson DO, Estes MK, Glass RI, Barlett AV, Penaranda M, Calomeni E, Tanaka T, Nakata S, Chiba S (1989) Human calicivirus-associated diarrhea in children attending day care centers. J Infect Dis 159: 71–78
13. Matson DO, Estes MK, Tanaka T, Barlett AV, Pickering LK (1990) Asymptomatic human calicivirus infection in a day-care center. Pediatr Infect Dis J 9: 190–196
14. Matson DO, Zhong W-M, Nakata S, Numata K, Jiang X, Pickering LK, Chiba S, Estes MK (1995) Molecular characterization of a human calicivirus with sequence relationships closer to animal caliciviruses than known human caliciviruses. J Med Virol 45: 215–222
15. Nakata S, Chiba S, Terashima H, Sakuma Y, Kogasaka R, Nakao T (1983) Microtiter solid-phase radioimmunoassay for detection of human calicivirus in stools. J Clin Microbiol 17: 198–201
16. Nakata S, Chiba S, Terashima H, Yokoyama T, Nakao T (1985) Humoral immunity in infants with gastroenteritis caused by human calicivirus. J Infect Dis 152: 274–279
17. Nakata S, Estes MK, Chiba S (1988) Detection of human calicivirus antigen and antibody by enzyme-linked immunosorbent assays. J Clin Microbiol 26: 2001–2005
18. Nakata S, Chiba S, Terashima H, Nakao T (1985) Prevalence of antibody to human calicivirus in Japan and Southeast Asia determined by radioimmunoassay. J Clin Microbiol 22: 519–522
19. Sakuma Y, Chiba S, Kogasaka R, Terashima H, Nakamura S, Horino K, Nakao T (1981) Prevalence of antibody to human calicivirus in general population of northern Japan. J Med Virol 7: 221–225
20. Terashima H, Chiba S, Sakuma Y, Kogasaka S, Nakata S, Minami R, Horino K, Nakao T (1983) The polypeptide of a human calicivirus. Arch Virol 78: 1–7

Author's address: Dr. S. Nakata, Department of Pediatrics, Sapporo Medical University School of Medicine, South-1, West-17, Chuo-ku, Sapporo 060, Japan.

Arch Virol (1996) [Suppl] 12: 271–276

Reverse transcription-polymerase chain reaction detection and sequence analysis of small round-structured viruses in Japan

K. Yamazaki[1], M. Oseto[2], Y. Seto[3], E. Utagawa[4], T. Kimoto[1], Y. Minekawa[1], S. Inouye[4], S. Yamazaki[4], Y. Okuno[1], and I. Oishi[1]

[1] Laboratory of Virology, Osaka Prefectural Institute of Public Health, Osaka, Japan
[2] Ehime Prefectural Institute of Public Health, Ehime, Japan
[3] Osaka Municipal Institute of Environment and Science, Osaka, Japan
[4] National Institute of Health, Tokyo, Japan

Summary. Between 1985 and 1995, mass outbreaks of acute gastroenteritis caused by small round-structured virus (SRSV), occurred in eight prefectures in Japan. Fecal samples from 59 patients ill during these outbreaks were recently examined in our laboratory by electron microscopy (EM) and by reverse transcription-polymerase chain reaction (RT-PCR). For RT-PCR, we prepared two sets of primers, a set corresponding to the polymerase region of open reading frame 1 (ORF-1) and a set corresponding to the capsid region of ORF-2 of Norwalk virus (NV). The SRSV nucleic acid detection rate with these primers was more than double that achieved with EM. Most samples found by EM to contain virus particles were also positive by PCR. When the two sets of primers were used separately, the virus detection rate differed depending on the primer used, suggesting that the viral strains examined were not genetically not homogeneous. We then selected nine strains of the virus, cloned their PCR products and analyzed their base sequences. The base sequences of these strains were compared with those of reference strains including prototype NV and Snow Mountain agent (SMA). This comparison yielded the following findings: (1) SRSVs that cause mass outbreaks of gastroenteritis in Japan are genetically variable; (2) SRSV strains that are genetically similar to SMA and SRSV-OTH25/89/J (OTH25) are dominant in Japan, but strains similar to NV are also present in this country; and (3) a strain (MI1/94) which is genetically identical to Southampton virus (SHV) was detected. Detection of SRSV using sensitive RT-PCR and analysis of the sequences of the amplification products seems to provide a useful means of studying the molecular epidemiology of SRSV.

Introduction

Small round-structured viruses (SRSVs) are now classified as caliciviruses [2, 15], based on their capsids containing a single major structural protein [3, 10], and because of the structural features of the genome [4]. SRSV is a major pathogen causing human acute gastroenteritis. In the field of public health, this virus has recently been attracting close attention because it has been shown that ingestion of shellfish (*e.g.*, oysters) contaminated by this virus serves as a major risk factor for food-borne gastroenteritis. SRSV is difficult to culture. For this reason, detection of these viruses has primarily depended on the direct detection of viral particles in fecal samples by electron microscopy. Since the complete base sequence of Norwalk virus (NV) [7] was recently determined [5], efforts have been made to develop a reverse transcription-polymerase chain reaction (RT-PCR) based assay to detect the nucleic acid of [6, 8, 9, 14].

We collected fecal samples from 59 patients who had developed acute gastroenteritis during mass outbreaks of this disease, probably caused by SRSV, over the past 10 years in Japan. We attempted to identify SRSV in these samples using electron microscopy (EM) and RT-PCR. A subset of the isolates was subjected to sequence analysis of the PCR products. The sequences were compared with those of prototype (NV, Snow Mountain agent [SMA], etc.) viruses with known base sequences, for the purpose of genetically characterizing the viruses prevailing in Japan.

Results and discussion

Table 1 shows the epidemiological data for the patients involved in these outbreaks. With the exception of two samples collected from children in 1993 and 1994, all samples were from adults. Fourteen of these patients had eaten oysters 1 to 2 days before the onset of the disease. All samples, excluding the four collected in 1993 and 1994, had been confirmed to be SRSV positive by EM before selection for this study. Fecal samples, collected during the acute stage of the disease, were suspended in phosphate buffered saline and then concentrated by ultra-centrifugation ($100,000 \times$ **g**) as previously described [12]. After negative

Table 1. Epidemiology of outbreaks of acute gastroenteritis associated with SRSV in 8 prefectures of Japan

Year	No. of cases	No. of specimens	Age group	Prefecture
1985	1	1	Adult	Kanagawa
1989	2	2	Adult	Osaka, Saitama
1990	1	1	Adult	Kanagawa
1991	1	1	Adult	Saitama
1992	2	4	Adult	Osaka, Kanagawa
1993	3	20	Adult, Children	Osaka, Kanagawa
1994	6	27	Adult, Children	Osaka, Chiba, Tochigi, Shiga
1995	3	3	Adult	Osaka, Ehime, Hiroshima
Total	19	59		

staining using 1.5% phosphotungstic acid (PTA, pH 6.5), the samples were examined by EM. Extraction of viral nucleic acids from fecal samples using CTAB and subsequent analysis by RT-PCR was carried out according to the method of Jiang et al. [6]. For PCR, we used a set of primers (36 and 35) from the ORF-1 polymerase region (Pol) of the NV genome developed by Wang et al. [15]. Another primer set (S9 and S10) was designed from the ORF-2 region of the NV genome encoding the capsid (Cap) [16]. The base sequences were as follows: S9 (position of 5′ first nucleotide in NV genome: 5 346) 5′-ATGATGAT-GGCGTCTAAGGAC-3′ and S10 (5800) 5′-ACATCAGCAATCACATGTGG-3′. The molecular sizes of the amplification products were 470 bp and 455 bp, respectively. The PCR products were subjected to agarose gel electrophoresis and then observed under ultraviolet illumination following staining with ethidium bromide. The PCR products were cloned as follows. First, the gel fragments, cut from agarose, were subjected to a cDNA extraction process. The cDNA was treated with Klenow fragment (Toyobo, Japan) to create blunt ends, and was then ligated to the SmaI site of the plasmid pBluescript SK$^+$ (Stratagene, USA) and transformed in E. coli (JM109, Toyobo, Japan). DNA from positive clones was partially purified and its sequence was determined by the dideoxy method, using a combination of an ALF sequencing kit (Pharmacia, Sweden) and an autosequencer. The nucleotide sequences and the predicted amino acid sequences were compared with those of the following reference strains to identify any similarities: NV (EMBL/GenBank: Accession No. M87661), SRSV-KY89/89/J [13, 15] (KY89, L23828), Southampton virus (SHV, L07418), SRSV-OTH25/89/J [13, 15] (OTH25, L23829, L23930) and SMA (L23830).

First, we compared the SRSV detection rate using EM and RT-PCR to examine samples from five epidemic cases (Table 2). When PCR was carried out using both sets of primers, viral nucleic acids were detected in 23 (62%) of the 37 samples. This rate of detection was more than double the rate achieved by EM (27%, 10/37). Of the samples rated by EM as SRSV positive, 97% were found to

Table 2. Comparison of SRSV detection from the stool specimens of the patients with gastroenteritis by EM and RT-PCR methods

Case No.	Code	Year	EM	PCR		
				Total	Pol	Cap
1	TOB	1993	5/11	7/11	7/11	3/11
2	TOC	1993	4/8	5/8	4/8	5/8
3	IZ	1994	1/13	7/13	7/13	5/13
4	H	1994	0/4	3/4	2/4	3/4
5	MI	1994	0/1	1/1	0/1	1/1
No. of Positive/Tested			10/37	23/37	20/37	17/37
(%)			27%	62%	54%	46%

Table 3. Relationships of the sequences of the RNA polymerase region among SRSVs

SRSV strain	Norwalk		KY89		SMA		OTH25	
	NA	AA	NA	AA	NA	AA	NA	AA
TOC/93	64	65	63	65	73	80	71	80
H104/94	65	65	63	65	74	82	73	82
SA/89	62	62	61	62	83	96	80	92

The numbers show the percent similarity of the aligned nucleic acid (NA) sequences and the predicted amino acid (AA) sequences

contain SRSV nucleic acids by RT-PCR. The detection rate, however, differed between use of PCR with the Pol primer set alone (54%, 20/37) and PCR with the Cap primer set alone (46%, 17/37). This difference seemed to reflect differences in the base sequences of the viral strains examined. On the basis of this finding, we selected seven representative strains from each outbreak and cloned the PCR-amplified products. Their nucleotide sequences were then analyzed and compared with the sequences of the reference strains obtained from the database. In the Pol region (Table 3), the nucleic acid (NA) sequence of SA/89 had an 80–83% homology with that of SMA and OTH25, amino acid (AA) sequence homology between these strains was 92–96%. Polymerase homology of SA/89 with NV and KY89 was lower (61–62%). The two strains detected in 1993 and 1994 (TOC/93 and H104/94) had 73–74% homology with SMA, and 64–65% with NV. In the Cap region (Table 4), the homology of SA/89 and SA/91 to OTH25 was higher (93–94% in terms of the NA sequence and 98–99% in terms of the AA sequence). Strain MI1/94, detected by the PCR on outbreak in 1994, could not be distinguished from SHV. These two strains had a homology of 99% in terms of the NA sequence and 100% in terms of the AA sequence.

Table 4. Relationships of the sequences of the capsid region among SRSVs

SRSV strain	Norwalk		Southampton		OTH25	
	NA	AA	NA	AA	NA	AA
TOC/93	59	54	61	50	76	81
IZ10/94	61	56	61	53	76	82
MI1/94	76	84	99	100	62	51
SA/89	60	57	62	53	93	98
SA/91	61	57	61	52	94	99

The numbers show the percent similarity of the aligned nucleic acid (NA) sequences and the predicted amino acid (AA) sequences

At present, various primers are used for RT-PCR of SRSV. The Pol-region primer 36/35 has been used most frequently and initially sometimes at low temperature to detect heterogeneous viral nucleic acid and to obtain new sequences [11, 15]. Although this primer set has extensive reactivity and has been useful, RT-PCR using this primer set alone did not allow satisfactory SRSV detection rates in the present study. The SRSV detection rate may be increased if this primer set is used in combination with other primers. Recently it has also become desirable to prepare primers for detecting human calicivirus genomes that recently have been identified [11]. Most of the samples that were found by EM to contain SRSV were found to contain SRSV when examined by RT-PCR, using either of the two primers. However, the detection rate differed between the two primers. This suggests differences in base sequences among different strains of SRSV. The base sequences determined for representative strains indicate that viruses akin to SMA and OTH25 are now prevalent in Japan, but that viruses similar to NV are also present. It is particularly noteworthy that the genetic properties of MI1/94 resembled those of SHV, and KY89 and NV also resembled each other, although these pairs were isolated in different years and from different districts. This finding seems to be significant in considering the molecular epidemiology of SRSV. A new method for analyzing genotypes of SRSV has recently been developed [1]. We are now typing SRSV using this method. Our preliminary results using strains isolated from epidemic or sporadic cases of gastroenteritis indicate that there are a number of viruses similar to SMV in circulation.

Acknowledgements

The author is indebted to Drs. MK Estes, J Noel, T Ando, SS Monroe and RI Glass for their valuable advice regarding this study. The authors thank Dr. M. Murao who kindly provided samples for this study.

References

1. Ando T, Monroe SS, Gentsch JR, Jin Q, Lewis DC, Glass RI (1995) Detection and differentiation of antigenically distinct small round-structured viruses (Norwalk-like viruses) by reverse transcription-PCR and southern hybridization. J Clin Microbiol 33: 64–71
2. Gree SM, Dingle KE, Lambden PR, Caul EO, Ashley CR, Clarke IN (1994) Human enteric Caliciviridae: a new prevalent small round-structured virus group defined by RNA-dependent RNA polymerase and capsid diversity. J Gen Virol 75: 1883–1888
3. Greenberg HB, Valdesuso JR, Kalika AR, Wyatt RG, McAuliffe VJ, Kapikian AZ, Chanock RM (1981) Proteins of Norwalk virus. J Virol 37: 994–999
4. Jiang X, Graham DY, Wang K, Estes MK (1990) Norwalk virus genome cloning and characterization. Science 250: 1580–1583
5. Jiang X, Wang M, Wang K, Estes MK (1993) Sequence and genomic organization of Norwalk virus. Virology 195: 51–61
6. Jiang X, Wang J, Graham DY, Estes MK (1992) Detection of Norwalk virus in stool by polymerase chain reaction. J Clin Microbiol 30: 2529–2534

7. Kapikian AZ, Wyatt RG, Dolin R, Thornhill TS, Kalica AR, Chanock RM (1972) Visualization by immune electron microscopy of a 27 nm particle associated with acute infectious nonbacterial gastroenteritis. J Virol 10: 1075–1081
8. Leon RD, Matsui SM, Baric RS, Herrmann JE, Blacklow NR, Greenberg HB, Sobsey MD (1992) Detection of Norwalk virus in stool specimens by reverse transcriptase-polymerase chain reaction and nonradioactive oligoprobes. J Clin Microbiol 30: 3151–3157
9. Lew JF, Kapikian AZ, Valdesuso J, Green K (1994) Molecular characterization of Hawaii virus and other Norwalk-like viruses: evidence for genetic polymorphism among human caliciviruses. J Infect Dis 170: 535–542
10. Madore HP, Treanor JJ, Dolin R (1986) Characterization of the Snow Mountain agent of viral gastroenteritis. J Virol 58: 487–492
11. Matson DO, Zhong W-M, Nakata S, Numata K, Jiang X, Pickering LK, Chiba S, Estes MK (1995) Molecular characterization of a human calicivirus with sequence relationships closer to animal caliciviruses than other known human caliciviruses. J Med Virol 45: 215–222
12. Oishi I, Yamazaki K, Minekawa Y, Nishimura H, Kitaura T (1985) Three-year survey of the epidemiology of rotavirus, enteric adenovirus, and some small spherical viruses including "Osaka-agent" associated with infantile diarrhea. Biken J 28: 9–19
13. Oishi I, Yamazaki K, Kimoto T, Minekawa Y (1992) Demonstration of low molecular weight polypeptides associated with small, round-structured viruses by western immunoblot analysis. Microbiol Immunol 36: 1105–1112
14. Utagawa ET, Takeda N, Inouye S, Kasuga K, Yamazaki S (1994) 3′-Terminal sequence of a small round structured virus (SRSV) in Japan. Arch Virol 135: 185–192
15. Wang J, Jiang X, Madore HP, Gray J, Desselberger U, Ando T, Seto Y, Oishi I, Lew JF, Green KY, Estes MK (1994) Sequence diversity of small, round-structured viruses in the Norwalk virus group. J Virol 68: 5982–5990
16. Yamazaki K, Oishi I, Okuno Y, Shibata T (1995) Outbreaks of acute gastroenteritis-genotypic diversity of SRSVs detected by RT-PCR. Clin Virol 23: 251–256 (in Japanese)

Author's address: Dr. K. Yamazaki, Laboratory of Virology, Osaka Prefectural Institute of Public Health, Nakamichi 1-chome, Higashinari-ku, Osaka 537, Japan.

Arch Virol (1996) [Suppl] 12: 277–285

The molecular biology of astroviruses

M. J. Carter and **M. M. Willcocks**

School of Biological Sciences, University of Surrey, UK

Summary. Astroviruses (genus *Astrovirus*) are assigned to a newly established virus family, the *Astroviridae*. The molecular biology of these agents reveals many features unique amongst the non-enveloped animal viruses and resembles that of members of certain plant virus families. In particular, their possession of a serine protease and use of ribosomal frameshifting to express the RNA polymerase are similar to the luteoviruses. Many aspects of the astrovirus replication strategy are still unclear, but replication may involve a nuclear step and non-structural proteins may influence host cell range.

Introduction

Astroviruses were first identified in 1975 in the stools of infants with diarrhoea [12]. Particles are 28 nm across with a smooth outer margin. In their centre, a proportion of particles display a prominent 5- or 6-pointed star-like motif. This is unique among viruses and at present there is no satisfactory model for how this structure might be produced. However, this feature led to the recognition of these viruses as distinct from other enteric viruses and also offered an obvious name; the astroviruses (astron [Greek]; a star).

The occurrence of similar particles in the stools of symptom-free children implied that virus infection would be generally associated with mild illness, and hospitalization figures show that documented astrovirus illnesses account for only 2–3% of hospital admissions for diarrhoea [2]. However, the detection of astroviruses still rests largely on morphological criteria, not necessarily a reliable guide. Few laboratories are equipped to conduct more detailed tests and it is likely that the incidence of astrovirus-induced illness has probably been underestimated. Recent surveys, using more objective techniques, have found a much higher incidence of astroviruses in symptomatic illness [4, 21]. While infection is often mild, this is by no means always so and astroviruses have been implicated in severe illnesses in adults.

Immune electron microscopy indicated that astroviruses could be grouped into serotypes; 5 were initially recognised [8, 10], although this number was subsequently extended to 7 [11] and now to 8 [Kurtz; unpublished]. Data from a small survey in the UK suggest that infection is widespread. Seropositivity to

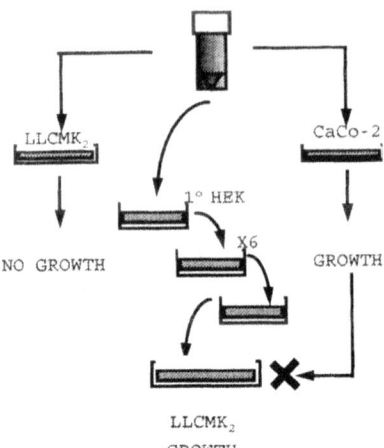

Fig. 1. Diagrammatic representation of the adaptation procedure used to rescue astroviruses in different cell lines. Virus in the stool does not grow in LLCMK$_2$ cells, but can be induced to do so after six passages in primary human embryo kidney cells. Alternatively virus in stool specimens can be grown directly in CaCo-2 cells but does not become adapted to growth in LLCMK$_2$ cells by this process. All techniques require the addition of trypsin to the medium for release of infectious virus to occur

type 1 rises rapidly in childhood, reaching 80% by age 5 [7]. However, infections by other serotypes are not acquired as rapidly and these viruses could cause significant disease in older persons if cross-reactive immunity, induced by a prior infection with type 1 virus, is poor. This may apply particularly to serotype 4, which shows greatest divergence in structural protein sequence from other serotypes.

Astroviruses were first cultivated by adapting them to replicate in cell cultures by serial blind passage in primary human embryo kidney cells with a protease supplement [9]. After six passages, virus released from these cells had acquired the ability to replicate in a continuous monkey kidney cell line, LLCMK$_2$, which cannot support the growth of virus directly from stool specimens (Fig. 1). Each of the five astrovirus serotypes then known was adapted to culture in this way, but all such laboratory-adapted isolates must be host range mutants. In view of the difficulties in performing this adaptation, it was not surprising that the number of such adapted isolates remained few. However, in 1990 astroviruses were cultivated directly from fecal samples in a continuous cell line derived from a human colonic carcinoma (CaCo-2) [20]. Virus grown in CaCo-2 cells does not acquire the ability to grow in LLCMK$_2$ cells (Fig. 1). Other cell lines derived from gut tissue have since been used successfully to cultivate astroviruses, but all require a proteolytic supplement.

Virus replication

Studies of virus replication in both LLCMK$_2$ and CaCo-2 cells have shown that two virus-specific RNAs are induced in infected cells [1, 14]. The first is full-length genomic RNA (6.8 kb), whilst the second is smaller (2.8 kb). Both species are polyadenylated at their 3′ termini. An earlier report identified only a single, full-sized RNA in these cells [18], but this work was conducted late in infection; at this point even cellular RNA had been extensively degraded and only encapsidated RNA had survived. Thus only genomic RNA was observed. This also implies that astroviruses do not encapsidate their subgenomic RNA.

There are few data on astrovirus protein synthesis. Non-structural proteins are made in low amount and are difficult to detect; only the structural protein precursor has been studied in detail. This is a trypsin-labile protein of approximately Mr 90,000 which reacts with antiserum to purified virions [14]. Virus particles contain several proteins but the consensus opinion is that there are two of approximately Mr 30,000 and a smaller moiety (Mr 27,000 in human astrovirus type 1) which varies in size between serotypes [22]. Together these three proteins account for almost all of the putative 90,000 precursor protein, and this can be shown to be processed to smaller molecules, presumably reflecting maturational cleavage [14]. However, at present the positions of cleavage in the polyprotein are not known. N-terminal sequence analysis has suggested a confusing pattern of overlapping proteins [17], but this may derive from partial processing *in vitro* or unsuitability of the protease supplement (trypsin), which may not reflect accurately the host protease normally exploited by the virus during replication in the gut. At present there is no information on the synthesis and maturation of non-structural proteins.

The astrovirus genome

Astroviruses contain a single-stranded RNA genome (approximately 6.8 kb) of positive sense (Fig. 2). The 3′ terminus is polyadenylated and the 5′ end is believed to be linked covalently to a protein, VPg, although direct evidence for this is lacking. Complete genome sequences are known for three human astroviruses, cell-culture adapted serotypes 1 and 2 [5], and CaCo-2 cell-grown serotype 1 [23], and capsid protein sequences have been determined from 7 of the 8 serotypes (serotype 7 has yet to be examined in detail).

Coding analysis

The RNA contains three sequential open reading frames (ORFs), termed 1a, 1b and 2 (Fig. 2). ORF 1a commences at the initiation codon at position 89 and terminates at position 2890. However, this ORF overlaps that termed 1b by 71 bases. ORF 1b lacks a suitable initiation codon at its 5′ end, the first possible

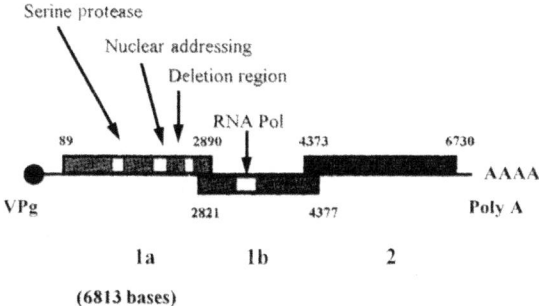

Fig. 2. Astrovirus genome structure. Motifs indicated are described in the text

```
PV-1  VAILPTHASP (26) LEITIIITLK (66) GQCGGVITCT-GK-VIGMHVG

RV14  VCVLPTHAQP (26) LELTVLTLD  (65) GQCGGVLCAT-GK-IFGIHVG

BWY   ALMTATHVLR (30) GDVTLLRGP  (58) GHSGSPYFN--GKTILGVHSG

HAst  DIVTAAHVVG (23) KDIAFLTCP  (52) GMSGAPVCDKYGR-VLAVHQT

      **   x        x* **         * x*      *  **#*#
```

Fig. 3. Comparison of the astrovirus protease with similar enzymes from other viruses: PV-1 (poliovirus type 1); RV14 (rhinovirus type 14); BWY (Barley western yellows virus, a luteovirus); HAst (human astrovirus serotype 1). Catalytic triad residues are marked (X), Residues conserved in type are marked (*), Binding cleft residues are marked (#)

AUG is in a poor context according to the Kozak consensus, and is not reached until position 3274, 383 bases downstream from the stop codon of ORF 1a. Sequence analysis in these two ORFs has identified motifs indicative of enzymic functions, as expected for virus non-structural proteins. Astroviruses may be expected to encode a protease for polyprotein maturation, a RNA-dependent RNA polymerase, and a helicase. Analysis has identified both the protease and RNA-dependent RNA polymerase, but there appears to be no helicase-like sequence. This may be a consequence of the relatively small size of the astrovirus RNA; genomes of approximately 6000 nucleotides may not require this activity [6].

The virus protease has been identified in ORF 1a, centered on the catalytic triad histidine (461), serine (551) aspartic acid (489). A fourth residue (histidine 566) is also conserved and thought to lie in the substrate binding pocket of the enzyme (Fig. 3). This sequence resembles most closely the picornavirus proteases, which are chymotrypsin-like, with the exception that picornaviruses have replaced the active serine residue with cysteine. A potential bipartite nuclear addressing signal has also been detected in ORF1a, and although virus proteins do appear to be moved to the nucleus during replication, the significance of this signal is not known [1]. Sequences towards the carboxy terminal of ORF1a may also determine host cell range (see below). The RNA-dependent RNA polymerase motif has been detected in ORF1b. The sequence YGDD which is thought to lie at the active site of the enzyme is encoded at position 3940. However, this frame is not expressed from its own initiation codon (see below).

ORF 2 encodes a protein of approximately Mr 90000 [15, 19, 25]. This is in agreement with a presumptive precursor for the virion proteins identified in the infected cell [14]. Sequence comparisons between human astroviruses show that this protein has a well-conserved amino terminus, which is relatively basic. This is reminiscent of some plant viruses in which such a sequence is probably involved in RNA binding. Similar basic regions have been observed in some coronavirus capsids and in hepatitis B core antigen, in which they may perform a similar role.

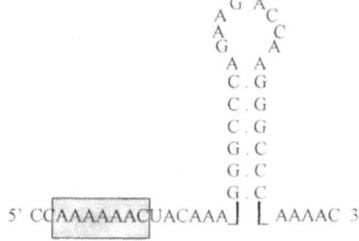

Fig. 4. Predicted secondary structure of the astrovirus frameshifting signal. Residues of the heptanucleotide "slippery" sequence are boxed

Genome expression

Only two species of RNA have been detected intracellularly, the largest, which corresponds to genomic RNA, could act as mRNA for the proteins encoded in ORF 1a. The smaller, subgenomic RNA, is approximately the same size as ORF 2, but there is no candidate mRNA for ORF 1b. However, ORF 1a overlaps ORF 1b by 71 nucleotides, and examination of the region of overlap reveals a heptanucleotide sequence which agrees with the consensus deduced for sites of ribosomal translation frameshift; XXXYYYN, (where X is A, U or G; Y is A or U and N is A, U or C). Such frame-shifts have been detected in a variety of viruses such as corona-, arteri- and retroviruses, but these so called "slippery" sequences act in conjunction with elements of secondary structure; a minimum of a stem-loop and, in many instances, a pseudoknot is also required. Similar features can be identified in this region of the astrovirus RNA (Fig. 4) and mapping studies support the existence of the predicted secondary structure [13]. This region of the RNA has been tested by translation *in vitro* and shown to direct -1 ribosome translational frameshifting at the ORF1a/1b junction. Approximately 5% of ribosomes translating this RNA would slip frame and thus avoid termination at the end of ORF 1a. This would form a 1a/1b fusion protein including sequences from the RNA-dependent RNA polymerase. This system was used to investigate the minimum requirements for ribosomal frameshifting by site-directed mutagenesis; the heptanucleotide sequence is essential; deletion or modification from the XXXYYYN consensus abolishes frame shifting entirely. A stable base-paired stem is also required. Mutations that weaken this secondary structure reduce the efficiency of the frameshift, but those that maintain the strength of the base pairs are tolerated. Thus, a stable stem is needed but its sequence is not critical. Finally, the sequence of the loop is unimportant [13]. This suggests that pseudoknot formation is not involved because these sequences would be required to maintain downstream base pairing interactions. Frameshifting in the astroviruses is thus similar to that seen in animal viruses such as in HTLV-1, but is unique among the non-enveloped animal viruses. The closest similarities in terms of genome organization, presence of a serine protease and expression of the polymerase by means of a translational frame-shift are found in non-enveloped plant viruses, luteoviruses.

The subgenomic RNA corresponds to the 3′-terminus of the RNA and includes ORF2 in its entirety [15, 19]. That this RNA could encode the virion structural proteins was confirmed by N-terminal sequencing of virion proteolytic fragments and by expression of this gene in recombinant baculoviruses [15, 19]. ORF 2 from the serotype 1 virus was found to be 300 bases longer than that of the serotype 2 virus sequenced. This raised the possibility that the structural protein made by type 1 virus could be larger than that produced by other serotypes. However, 5′ end mapping of the subgenomic RNA revealed that the mRNAs formed by these two viruses are in fact the same, and the potential extra information is not represented in the mRNA [19]. The actual proteins formed are thus of very similar size between these two serotypes of human astrovirus. Baculovirus-expressed ORF2 protein reacted strongly with antiserum to purified astrovirions [19]. However, although astroviruses can form empty particles [3, 20], virus-like particle assembly was not detected in insect cells. In contrast, most of the protein formed a high molecular weight aggregate. This presumably implies that unlike Norwalk virus, post-translational processing is required before particles can be assembled.

Comparative sequence analysis

Three complete astrovirus genomic sequences have now been reported: two strains of serotype 1 have been sequenced (one isolated by laboratory adaptation in LLCMK$_2$ cells, and the other several years later in CaCo-2 cells), and a single strain of serotype 2 (isolated by laboratory adaptation in LLCMK$_2$ cells). Non-structural genes from all three were very similar except in one region: both isolates adapted to growth in LLCMK$_2$ cells lacked 15 amino acids that were present in virus isolated in CaCo-2 cells. This suggests that the process of host range adaptation could involve selection of deletion mutants from the viruses originally present in stool. This would be best tested by repetition of the adaptation procedure, but this has not been feasible due to the difficulties in working with human fecal material. An alternative would have been to examine the original stool material from which the type 1 laboratory-adapted virus had been derived. These experiments had taken place in Oxford (UK) some 15 years ago, and no such original material remained. However, the stool in question had also been used for volunteer studies. One such volunteer had become ill and a sample of this stool, containing virus removed by a single passage in the human gut from that used for cell culture adaptation, was still available. This was tested by PCR and was found not to contain deletion mutations, even though the adapted virus derived from it did (Fig. 5). Subsequent sequence studies showed that all new isolates, whether grown in CaCo-2 cells or sequenced directly from stool samples, possess 15 amino acids not present in the five laboratory adapted strains. Furthermore, deletion is not selected even after 10 serial passages in CaCo-2 cells (Fig. 6) [24]. It is therefore highly probable that the process of laboratory adaptation to a different host cell can select for this mutation and thus

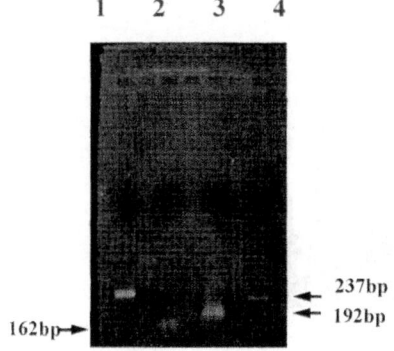

1 2 3 4

237bp

192bp

162bp→

Fig. 5. PCR analysis of deletion in serotype 1 viruses: Track 1, Newcastle CaCo-2 cell isolate A2/88; track 2 HPV PCR size marker 162 bp; Track 3, HAst-1 lab adapted; Track 4, HAst-1 from volunteer stool (ie related by one passage in the human gut from the progenitor virus used to obtain the sample analysed in track 3)

```
PEPEVESQPLDLSQKKEKQSEYEQQVVKSTKPQQLEHEQQVVKPIKPQKSEPQPYSQ
HAst-1 (A2/88 Newcastle)
PEPEVESQPLDLSQKKEKQSEY--------------EQQVVKSIKPQKSEPQPYSQ
HAst-1 (Lab adapted)

PEPEVESQPLDLCQKKEKQSEYEQQVVKSIKPQQLEHEQQVVKSIKPQKSEPQPYSQ
HAst-1 (Volunteer)

PEPETETQPLDLSQKKEKQPE-------------HEQQVVKSTKPQKNEPQPYSQ
HAst-2 (Lab adapted)
PELEAEAQPLDLSQKKEKQPE-------------HEQQVMKPTKPQKSEPQPYSQ
HAst-3 (Lab adapted)
PEPEAEAQPLDLSQKKEKQPE--------------HEQQVMKPTKPQKSEPQPYSQ
HAst-4 (Lab adapted)
PEPEAETQPLDLSQKKEKQPE--------------HEQQVVKSTKPQKNDPQPYSH
HAst-5 (Lab adapted)

PEPEAESQPLDLSQKKEKQSEYEQQVVKSIKPQQLEHEQQVVKPTKPQKSEPQPYSQ
Serotype 4 (CaCo-2 cellp4)
PEPEAESQPLDLSQKKEKQSEYEQQVVKSIKPQQLEHEQQVVKPTKPQKSEPQPYSQ
Serotype 4 (CaCo-2 cellp10)
```

Fig. 6. Sequence analysis of laboratory adapted and CaCo-2 cell grown astrovirus isolates. Isolate designation is given in the Figure. The deletion is indicated as amino acid sequence and removes one copy of a partially repetitious sequence EQQVVK ··· KPQ

this region of the ORF1a gene may influence the host cell range of the virus. The basis of this effect is not known.

Structural proteins must be responsible for the antigenic variation observed between different astrovirus serotypes. The available sequences were compared, and a variation plot was constructed (Fig. 7). Sequences towards the 5′ end appear well conserved between serotypes, but a region between residues 649 to 707 (positions refer to HAstV-1) is highly variable, and this could contribute to antigenic differences between the viruses. There are numerous insertions and deletions between strains in this area. These could be responsible for the size variation observed between serotypes in the smallest virion polypeptide. This small protein is also thought to be located on the outside of the virus, since it is stripped from the virus by detergent treatment [20]. These observations indicate that the smallest virion protein may be derived from this region and could be responsible for most, if not all, of the antigenic variation between serotypes.

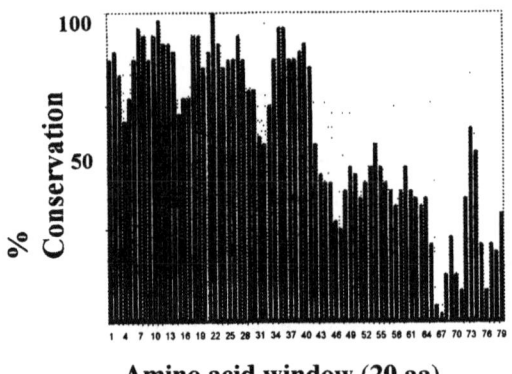

Amino acid window (20 aa)

Fig. 7. Comparison of astrovirus structural protein sequences. All available sequences for proteins encoded in ORF2 were aligned using the CLUSTAL program of the PC Gene software package. Alignments were divided into windows of 20 residues, overlapping by 10. Each was scored with respect to positions where substitutions/deletions were permissible. If all amino acids at that position are identical score = 1; if maintained in character score = 0.5; if differing in character or deleted in some viruses score = 0. Maximum score per window = 20, minimum score = 0

Conclusion

Astrovirus is the only genus in a recently recognized family of animal viruses, the *Astroviridae* [16]. Although much remains to be elucidated, concerning their replication strategy, these agents of disease have already revealed features that make them unique among the non-enveloped animal viruses. Similarly progress in the molecular biology of these agents has allowed the production of novel reagents for diagnosis and has challenged our view of the significance of these agents in human disease.

References

1. Carter MJ (1994) Genomic organization and expression of astroviruses and caliciviruses. In: Proceedings of the Third International Symposium on Positive Strand RNA Viruses. Arch Virol [Suppl 9] 429–439
2. Centers for Disease Control (1990) Viral agents of gastroenteritis: public health importance and outbreak management. Morbidity and Mortality Weekly Report 39 (No RR-5): 5–6
3. Herring AJ, Gray EW, Snodgrass DR (1981) Purification and characterisation of ovine astrovirus. J Gen Virol 53: 47–55
4. Herrmann JE, Taylor DN, Echeverria P, Blacklow NR (1991) Astroviruses as a cause of gastroenteritis in children. N Engl J Med 324: 1757–1760
5. Jiang BM, Monroe SS, Koonin EV, Stine SE, Glass RI (1993) RNA sequence of astrovirus: distinctive genomic organization and a putative retrovirus-like ribosomal frameshifting signal that directs the viral replicase synthesis. Proc Natl Acad Sci USA 90: 10539–10543
6. Koonin EV (1991) Similarities in RNA helicases. Nature 352: 290
7. Kurtz JB, Lee TW (1978) Astrovirus gastroenteritis: Age distribution of antibody. Med Microbiol Immunol 166: 227–230

8. Kurtz JB, Lee TW (1984) Human astrovirus serotypes. Lancet ii: 1405
9. Lee TW, Kurtz JB (1981) Serial propagation of astrovirus in tissue culture with the aid of trypsin. J Gen Virol 57: 421–424
10. Lee TW, Kurtz JB (1982) Human astrovirus serotypes. J Hyg 89: 539–540
11. Lee TW, Kurtz JB (1994) Prevalence of human astrovirus serotypes in the Oxford region 1976–1992, with evidence for two new serotypes. Epidemiol Infect 112: 187–193
12. Madeley CR, Cosgrove BP (1975) 28 nm particles in faeces in infantile gastroenteritis. Lancet ii: 451–452
13. Marczinke B, Bloys AJ, Brown TDK, Willcocks MM, Carter MJ, Brierley I (1994) The human astrovirus RNA-dependent RNA polymerase coding region is expressed by ribosomal frameshifting. J Virol 68: 5588–5595
14. Monroe SS, Stine SE, Gorelkin L, Herrman JE, Blacklow NR, Glass RI (1991) Temporal synthesis of proteins and RNAs during human astrovirus infection in cultured cells. J Virol 65: 641–648
15. Monroe SS, Jiang B, Stine SE, Koopmans M, Glass RI (1993) Subgenomic RNA sequence of human astrovirus supports classification of *Astroviridae* as a new family of RNA viruses. J Virol 65: 67: 3611–3614
16. Monroe SS, Carter MJ, Herrmann JE, Kurtz JB, Matsui SM (1995) Astroviridae. In: Virus Taxonomy. Springer, Wien New York, pp 364–367
17. Sanchez-Fauquier A, Carrasoca AL, Carascosa JL, Overo A, Glass RI, Lopez JA, San Martin C, and Melero JA (1994) Characterisation of a human astrovirus serotype 2 structural protein (VP26) that contains an epitope involved in virus neutralization. Virology 201: 312–320
18. Willcocks MM, Carter MJ (1992) The 3′ terminal sequence of a human astrovirus. Arch Virol 124: 279–289
19. Willcocks MM, Carter MJ (1993) Identification and sequence determination of the capsid protein gene of a human astrovirus serotype 1. FEMS Micro Lett 114: 1–8
20. Willcocks MM, Carter MJ, Laidler F, Madeley CR (1990) Growth and characterisation of a human astrovirus in a continuous cell line. Arch Virol 113: 1–8
21. Willcocks MM, Carter MJ, Silcock JG, Madeley CR (1991) A dot-blot hybridization procedure for the detection of astrovirus in stool samples. Epidemiol Infect 107: 405–410
22. Willcocks MM, Carter MJ, Madeley CR (1992) Astroviruses. Rev Med Virol 2: 97–106
23. Willcocks MM, Brown TDK, Madeley CR, Carter MJ (1994) The complete sequence of a human astrovirus. J Gen Virol 75: 1785–1788
24. Willcocks MM, Ashton N, Kurtz JB, Cubitt WD, Carter MJ (1994) Cell culture adaptation of astrovirus involves a deletion. J Virol 68: 6057–6058
25. Willcocks MM, Kurtz JB, Lee TW, Carter MJ (1995) Prevalence of human astrovirus serotype 4: capsid protein sequence and comparison with other strains. Epidemiol Infect 114: 385–391

Authors' address: Dr. M. J. Carter, School of Biological Sciences, University of Surrey, Surrey GU2 5XH, UK.

Arch Virol (1996) [Suppl] 12: 287–300

© Springer-Verlag 1996

The changing epidemiology of astrovirus-associated gastroenteritis: a review

R. I. Glass[1], J. Noel[1], D. Mitchell[2], J. E. Herrmann[3], N. R. Blacklow[3], L. K. Pickering[2], P. Dennehy[4], G. Ruiz-Palacios[5], M. L. de Guerrero[5], and S. S. Monroe[1]

[1] Viral Gastroenteritis Section, Division of Viral and Rickettsial Diseases, Centers for Disease Control and Prevention, Atlanta, Georgia, U.S.A.
[2] The Center for Pediatric Research, Eastern Virginia Medical School and Children's Hospital of the King's Daughters, Norfolk, Virginia, U.S.A.
[3] The Division of Infectious Diseases and Immunology, University of Massachusetts Medical School, Worcester, Massachusetts, U.S.A.
[4] Division of Pediatric Infectious Disease, Rhode Island Hospital, Providence, Rhode Island, U.S.A. [5] Instituto Nacional de la Nutricion "Salvador Zubiran", Delegacion, Tlalpan, Mexico City, Mexico

Summary. Our understanding of the epidemiology of astrovirus-associated gastroenteritis has changed markedly with each improvement in detection method. In early surveys based on electronmicroscopy (EM), astroviruses appeared to be a rare cause of gastroenteritis, being found in fewer than 1% of children with diarrhea, usually in small outbreaks of disease and primarily during the winter season. The development and use of monoclonal antibodies and enzyme immunoassays (EIA) to detect astroviruses led to reports of a higher prevalence (2.5%–9%) of astrovirus infection among patients hospitalized with diarrhea. Astroviruses appeared second only to rotaviruses as a cause of hospitalization for childhood viral gastroenteritis. Studies based on EIA detection of astroviruses indicate that astroviruses are common causes of diarrhea in children worldwide, and that most children are infected during their first two years of life. The elderly and the immunocompromised represent high-risk groups as well. The observations that newborns monitored prospectively rarely have repeat disease and that the rate of detection decreases with increasing age suggest that immunity to astroviruses, as immunity to rotaviruses, may develop early in life.

The cloning and sequencing of astroviruses have led to more sensitive assays to detect the viruses by reverse transcription, polymerase chain reaction (RT-PCR). Application of RT-PCR for detection of astroviruses in children in day-care centers showed a marked increase in the detected prevalence of astrovirus-associated diarrhea, the rate of asymptomatic infection, and the duration of shedding of virus among those infected, when compared with studies that used other methods. As with rotaviruses, neither the mode of transmission nor the reservoir of astrovirus infection has been identified. Both immune and molecular-based assays to detect astrovirus serotypes indicate that serotype 1 is most common worldwide, although the predominant serotypes may vary by region and time. In the absence of obvious strategies to prevent astrovirus-associated diarrhea, vaccines might be considered if further studies establish that the disease burden would render such a vaccine cost-effective.

Introduction

An Astroviruses were first described in 1975 by Appleton and Higgins [2] and then named by Madeley and Cosgrove who, using electron microscopy (EM), visualized small round viruses with a star-shaped appearance in fecal specimens of children with diarrhea [40]. For the next 15 years, EM remained the only method to detect these viruses in clinical specimens. During this time, our understanding of the epidemiology of astrovirus-associated gastroenteritis was limited to studies at centers where detection was possible and to a single detection method, EM, that we now recognize was relatively insensitive. Consequently, astroviruses were viewed as uncommon causes of gastroenteritis found primarily in small outbreaks of disease.

More recently, however, major advances in our understanding of these viruses have led to diagnostic breakthroughs that permit reassessment of the true role that astroviruses play as causes of gastroenteritis in humans. The viruses have been adapted to culture in continuous cell lines [32, 67] and have been completely sequenced, and information regarding the genomic organization of these viruses [22, 38, 66] allowed for the establishment of its own virus family, the *Astroviridae* [49] comprising a single genus, *Astrovirus*. There now have been seven distinct serotypes identified by immunofluorescence assays (IFA) and characterized by sequence analysis [33]. In addition, the use of monoclonal antibodies has been critical to the development of simple, more sensitive enzyme immuno- and molecular assays (*e.g.*, probes and RT-PCR) to detect strains with high sensitivity and to characterize isolates. Applying these advances in diagnostics, astroviruses are no longer considered rare but are known to be common enteric pathogens that, like rotaviruses, infect most children in their first few years of life and infect other high-risk groups (*e.g.*, immunocompromised, the elderly) as well.

This review will trace advances in diagnostic methods that have changed our understanding of the epidemiology of astroviruses. We begin with early EM studies and progress to the abundance of new research and findings made possible by the application of new diagnostic methods. We conclude by defining research issues that need to be addressed in order to outline future approaches to understanding the disease burden caused by astroviruses and to consider strategies for the control and prevention of astrovirus-associated gastroenteritis in humans.

New diagnostic advances

The evolution of our understanding of the epidemiology of astroviruses has paralleled advances in our ability to detect and work with this virus (Table 1). The first astrovirus was discovered by EM and early surveys underscored the importance of this detection method, even if the epidemiologic picture proved to be incomplete [3–5, 9, 14, 15, 24, 30, 34, 35, 37, 39, 50, 51, 54, 56, 57, 59, 60, 62, 71]. Astroviruses were difficult to distinguish by direct EM because they could often be confused with other small round viruses in the 26–30 nm range [6]. Moreover, the distinct star-shaped structures were evident in only a small portion of the particles. In 1977, Lee and Kurtz managed to adapt astroviruses to replicate in human embryonic kidney (HEK) cells and later to monkey (LLCMK$_2$) cells [31, 32]. This permitted viruses to be amplified and typed by IFA [27, 33] and used to raise antisera to individual serotypes and prepare monoclonal antibodies [18]. More recent improvements in culture techniques have made it possible to culture virus from more than 80% of antigen-positive fecal specimens [67].

The success of Herrmann *et al.* in preparing monoclonal antibodies provided the key ingredient that permitted development of a simple enzyme immunoassay (EIA) that proved to be more sensitive than EM [18, 19, 48]. For example, in a survey of 275 specimens, only one astrovirus was detected by EM, but 13 were found by EIA and all could be confirmed by RT-PCR or cultivation (Judy Lew, pers. comm.). Moreover, EIA could be used to screen field-collected and clinical specimens on a large scale, opening a diagnostic window that has led to major new findings [8, 10, 16, 17, 20, 25, 36, 43, 44, 47, 48, 55, 64].

While EIA was considerably simpler and more sensitive than EM, a large proportion of specimens gave ambiguous results, and, therefore, other confirmatory tests were required. With the sequencing of the astrovirus genome [22, 38, 66], probes for both hybridization [42, 48, 68] and RT-PCR [1, 16, 23] were immediately adapted and applied. Small riboprobes proved to be no more sensitive than EIA, but RT-PCR was extremely sensitive, detecting as few as 10–100 virus particles [58]. Unfortunately, methods have not been simplified to work with the large numbers of specimens collected in field studies.

Serotyping of astrovirus strains has been facilitated by the preparation of seven reference strains and matched, paired rabbit antisera by the Oxford Laboratory [33]. Only one classification scheme has been established based on these Oxford reference reagents, and serotyping performed by immune EM (IEM), IFA, and EIA, all rely on these same reagents [52, 53]. RT-PCR for typing has been developed on the basis of knowledge of the sequences of the reference strains [33, 53, 69].

Epidemiology during the EM period

The EM period of astrovirus studies has been reviewed several times [9, 14, 28]. EM surveys of viral agents of gastroenteritis in Australia [15], the United Kingdom [50], and the United States [35], have sporadically identified astroviruses, but the epidemiologic context was often not well characterized. In

Table 1. Advances in diagnostic methods to detect astrovirus

Method	Year/Investigator	Principle	Sensitivity (virus/gm stool)
Virus Detection			
Electron microscopy	1975 Madeley and Cosgrove	28 nm SRVs with star-shaped morphology	10^{6-7}
Culture	1977 Lee and Kurtz 1981 Lee and Kurtz 1990 Willcocks et al.	adaptation to growth in HEK, LLCMK$_2$, & CaCo$_2$ cells	~10^2
EIA	1990 Herrmann et al. 1991 Moe et al.	MAbs that detected all serotypes	10^{5-6}
RNA probes	1991 Moe et al. 1991 Willcocks et al.	sequence used to make probe for hybridization	10^{5-6}
RT-PCR	1993 Grohmann et al. 1993 Jonassen et al.	amplification of portion of genome shared by all strains	~10^2
Serology			
IFT/ISEM	1978 Kurtz and Lee	high seroprevalence, early acquisition	
IEM	1982 Konno et al.	seroconversion to outbreak strain	
IgM/IgG	1988 Wilson and Cubitt	antibodies in gamma globulin	
RIA	1993 Midthun et al.	few outbreaks with seroconversion	

these studies, astroviruses were rarely detected (< 1% of specimens examined) and were not found at all reporting centers [35, 50]. Astroviruses were detected primarily in diarrheal specimens from children younger than 4 years of age and were more often identified in winter than in summer months.

Seroepidemiology

While astroviruses were rarely detected in fecal specimens by EM, a number of limited serosurveys using various detection methods provided intriguing and contrasting results. In the United Kingdom, antibodies to astroviruses measured by IFA, immunosorbent EM, or RIA were acquired early in life, and more than 70% of older children and young adults had detectable antibody [26, 45, 70]. Konno et al. using IEM, identified antibody responses to the infecting virus among children who became ill in a kindergarten outbreak in Sapporo [24]. Midthun and colleagues, working with reagents to the Marin County agent (serotype 5), found a seroprevalence of 40% in a small group of adults [45]. Gamma globulin tested from a number of settings worldwide had high-titered antibodies to all five astrovirus serotypes to which reference reagents were then available [70]. In both volunteer studies and in small outbreak investigations, some but not all infected people mounted an antibody response to astroviruses [24, 29, 45]. Moreover, the assessment of this response differed depending upon the method of detection used. The early acquisition and high prevalence of antibodies raised the question of whether astroviruses were causing an abundance of asymptomatic infections and were not pathogens, or if the method of virus detection, EM, was inadequate for detecting infections.

Surveys based on EIAs

The preparation of an EIA brought a rapid advance and change in the understanding of the epidemiology of astroviruses (Table 2). The EIA was 10–100

Table 2. Prevalence of astrovirus among children with sporadic episodes of diarrhea

Setting	Investigator/yr	% EIA Positive Cases (N)	Controls (N)	P value
Thailand–outpatient	Herrmann 1991	8.6 (1111)	2.1 (947)	.001
Arizona–day care	Lew 1991	4.0 (524)	0.7 (138)	.04
Guatemala–cohort	Cruz 1992	7.3 (1396)	2.4 (830)	.001
Maryland	Kotloff 1992	2.5 (424)	0.7 (265)	.13
Japan–outbreaks	Utagawa 1994	6–10%	ND	
Mexico–cohort*	Guerrero 1995	5.0 (626)	ND	<.05
Rhode Island*	Dennehy 1992	4.3% (875)	0.8% (118)	.07
Aborigines*	Herrmann 1993	5.5 (1343)	ND	NA

*Abstracts

times more sensitive than EM and easier to perform on a large scale. EIAs were used in studies conducted using various populations and settings, such as children in hospitals for sporadic episodes of diarrhea [8, 10, 20, 25, 48], children with diarrhea in day-care centers [36, 47], birth cohorts [8, 17], outbreak investigations [47, 55, 63], and groups at high risk for diarrhea for which the etiology was often unknown [7, 16, 52]. In every setting, astroviruses have been found, and in most settings, these viruses have been present with a prevalence greater than expected.

The first such study by Herrmann *et al.* was conducted among Thai children admitted to a hospital for diarrhea (Table 2) [20]. Astroviruses were identified in 8.6% of 1111 fecal specimens from cases, but in only 2.1% of 947 controls (P < 0.001), making them the second most common enteric pathogen identified after rotaviruses. In several studies, astroviruses have been found second only to rotavirus as a cause of diarrhea [8, 20, 36]. Lew *et al.* identified astroviruses in 4% of 524 children with diarrhea in a day-care setting; infection was significantly associated with diarrhea as compared with infection in children who were well (< 1%) [36]. Cruz *et al.* detected astroviruses in 7.3% of Guatemalan children with diarrhea but in only 2.4% of controls without diarrhea (P < 0.001) [8]. Most studies focused on astroviruses as causes of childhood diarrhea and found more astroviruses in children with diarrhea than in controls, providing statistical evidence that astroviruses were pathogens in these settings. Nonetheless, the prevalence of astroviruses in children hospitalized with diarrhea has ranged from 2.5% to 9%. These differences could be due more to issues related to the assay itself or to factors related to study design (*e.g.*, age of patient, seasonality) than with the prevalence of the viruses. Hence, the true disease burden caused by astroviruses as detected with EIA diagnostics has not yet been assessed.

Preliminary results from surveys of diarrhea etiology indicate that while astroviruses may cause diarrhea with severe dehydration, its clinical presentation is usually less severe than diarrhea caused by rotaviruses [36, 47]. Cohort studies suggest that repeat disease is uncommon and that astrovirus infection in older children may be associated with a second serotype. The rapid decrease in disease prevalence with age and the apparent lack of repeat infections suggest that some group-specific immunity may develop against astrovirus diarrhea.

RT-PCR

RT-PCR is more sensitive than EIA for the detection of astroviruses and has been used for confirmation of results with specimens that gave ambiguous results by another test [16]. It has not been applied for screening because of the tediousness of performing this method en masse. Nonetheless, the first study to use RT-PCR on selected specimens indicated that our understanding of the epidemiology of astroviruses may change again with this diagnostic advance [46] (Table 3). In their study of diarrhea among children in day-care centers, Mitchell *et al.* used both EIA and RT-PCR to screen fecal specimens in an outbreak and to determine the prevalence of infection and the duration of

Table 3. Prevalence of astrovirus infection by types of Assay*

| | Positive for Astrovirus | | |
	EIA (%)	PCR (%)	P value
Specimens			
with diarrhea (N = 104)	13	40	<.001
well (N = 264)	9	24	<.001
Children's Rooms			
A (N = 12) infants	75	100	.25
B (N = 14) toddlers	50	79	.13
C (N = 10) older toddlers	20	90	<.02
Total (N = 36)	50	89	<.02
Virus shedding-days (median range)			
Episodes	1.5 (1–9)	4 (1–35)	.06

*From Mitchell DK *et al.*, Ref. [46, 47]

shedding of virus during and after illness. By EIA, the prevalence of astroviruses among children in affected day-care rooms was 50%, but was as low as 20% in some rooms of older children. When RT-PCR was used, almost all children, regardless of age, were shown to have had a documented infection and the duration of their shedding was increased from days to weeks. In essence, astroviruses had not selectively infected some children but had indiscriminately infected them all. Furthermore, long periods of asymptomatic shedding, important for secondary transmission, were well documented.

Serotyping studies

Studies in which astrovirus strains have been serotyped have been limited. All investigators have used different assays but the same reference reagents from the Oxford group [23a, 27, 33, 52, 53] (Table 4). From these studies, serotype 1 is predominant, types 2, 3, and 4 are not uncommon, and type 5 is least common. Types 6 and 7 have only recently been identified and appear to be uncommon as well. The distribution of serotype varies greatly by year and location. The usefulness of serotyping for studies of the epidemiology of transmission or the mechanisms of immunity remains to be determined.

High risk groups

As with rotaviruses, astroviruses have not been fully studied with regard to many basic epidemiologic characteristics, such as mode of transmission and reservoirs of infection; however, studies with human volunteers and investigations of outbreaks have shed some light on issues of susceptibility to disease and

Table 4. Distribution of astrovirus serotypes

Country/Yr	Investigator [Ref.]	Method	N	Serotypes (%)						
				1	2	3	4	5	6	7
UK 1974–84	Kurtz and Lee [27]	IFT/ISEM	73	77	5	7	6	5	na	na
UK 1981–93	Noel and Cubitt [52]	IEM	14	75	1	7	5	0	12	na
UK 1976–92	Lee and Kurtz [33]	IFT/ISEM	29	65	11	8	12	2	<1	<1
7 Countries	Noel et al. [53]	EIA/RT-PCR	61	52	11	16	10	2	2	0

IFT = immunofluorescence; ISEM = immunosorbent EM; IEM = immune EM; EIA = enzyme immunoassay, RT-PCR = reverse transcription PCR

Table 5. Epidemic settings in which astroviruses have been associated with gastroenteritis

Setting [Ref.]	Comment
Outbreaks traced to vehicles	
Oysters [28, 72]	
Drinking water [28]	direct epidemiologic link not
School food [55]	well established
Outbreaks in institutions for children, the elderly,	
and the hospitalized	
Day-care centers [36, 46, 47]	
Kindergarten [24]	
School children/staff [55]	high risk patients are the very
Nursing homes [13, 37, 44, 56, 65]	young and the very old
Hospitals (children) [2, 3, 10–12, 30, 41, 51, 59]	
Family contacts [61]	
Immunocompromised patients	
HIV [16]	most common enteropathogen
Bone marrow recipients [7]	most common infectious agent
SCID [52]	prolonged shedding

transmissibility of infection (Table 5). Two studies of healthy adult volunteers were conducted in the period when EM was the sole means to detect virus shedding [29, 45]. Volunteers could be infected by the fecal-oral route, but very few adults who received the astrovirus inoculum developed diarrhea, shed virus in their stool, or produced detectable antibody to the infecting virus. Astrovirus outbreaks have been associated with contaminated vehicles of infection (*i.e.,* food, oysters, and water), but conclusions concerning transmission have not been based on full epidemiologic investigations and the mode of transmission remains unknown (Table 5) [14, 28, 55]. Astrovirus diarrhea is usually confined to children, but in outbreaks in day-care centers and schools, teachers, parents, or caretakers of ill children have also become ill, suggesting that their exposure to astroviruses may be much greater and by a different route than is usual [55]. This observation is very similar to the experience with rotaviruses in that caretakers of children with rotaviruses may themselves become infected and develop symptoms despite having preexisting antibody [21]. In a survey of paired sera collected from more than 20 outbreaks in the United States, Midthun *et al.* suggested that outbreaks of astrovirus gastroenteritis in adults are uncommon [45]. No outbreak could be identified in which a majority of patients experienced seroconversion to an astrovirus. Consequently, while astroviruses can be transmitted by the fecal-oral route, as with volunteers, transmission by contaminated food or water remains in question.

People at greatest risk for astrovirus infection are the very young and the very old, and outbreaks have been documented in both groups (Table 5). In the young, outbreaks have been documented among children in day-care centers, schools, and hospital wards [11, 12, 30]. In nursing homes and among groups of the elderly, similar outbreaks have been well reported, but with no well-defined mode of transmission [9, 13, 37, 44, 56, 65]. Immunocompromised patients who are in general at particular risk for diarrhea are also at increased risk of infection with astroviruses. In a survey of the etiology of diarrhea in a cohort of human immunodeficiency virus- (HIV) infected people, astroviruses were the most common agents associated with their illnesses [16]. Astroviruses were also identified in bone marrow recipients [7] and in patients with severe combined immunodeficiency syndrome (SCIDS), both of whom demonstrated prolonged shedding of virus [52]. It is not clear whether the high prevalence of astrovirus infection in these patients is due to their increased risk of infection, an inability to terminate that infection caused by their immune deficit or both. The problem is associated with disease for which no specific treatment has yet been recommended.

Conclusions

Advances in our ability to detect and characterize astroviruses from fecal specimens have led to a major change in the thinking about the epidemiology of these viruses. Astroviruses are no longer considered to be pathogens found in rare outbreaks of disease, but common agents of childhood diarrhea affecting all children in the first few years of life. Infections tend to have a winter seasonality and small children and the elderly are at greatest risk of disease. The disease can be severe, but is generally less severe than rotavirus diarrhea. Groups at particular risk include those with increased exposure to infection, such as those in day-care centers, hospitals, schools, or retirement homes, and those immunocompromised or otherwise unable to combat or clear infection.

To consider specific interventions for the prevention and treatment of disease caused by astroviruses, the burden of astrovirus-associated gastroenteritis needs to be established. This effort will involve the application of new diagnostic techniques to epidemiologic studies of patients hospitalized for gastroenteritis, as well as to longitudinal studies of diarrhea in children. If the disease burden caused by astroviruses approaches that caused by rotaviruses, it may be worthwhile to consider development of a vaccine. Given that astroviruses can replicate in vitro and have been characterized at the molecular level, and that solid immunity appears to develop following natural infection, this strategy is worth considering if the cost of disease in terms of morbidity and mortality in children appears sufficient. At the same time, if the disease burden is limited to certain high-risk groups, such as the immunocompromised, treatment interventions that might include the use of immune globulin could be considered. True epidemics of astrovirus disease have not been well documented, and if virus transmission can be better traced, other public health interventions might be used to interrupt the spread of this pathogen. Finally, if after application of the best available diagnostic methods, astroviruses remain a frequent cause of less severe childhood diarrhea with a limited disease burden, merely understanding the disease prevalence will encourage physicians and public health workers to increase their use of oral rehydration solution and discourage the use of antibiotics for this problem. All these prospects will require further understanding of the epidemiology of the viruses in diverse settings.

On the research agenda, simple detection methods that are more sensitive and more specific than the current EIA, if made available on a widespread basis, would improve our understanding of the epidemiology and clinical features of this disease. The availability of expressed antigens should permit serologic testing and serosurveys on a larger scale. Such testing could help respond to the enigma of why seropositivity does not correlate with virus detection rates. While infection with an astrovirus appears to confer protection against subsequent infection with that serotype, further information regarding immunity to astroviruses is needed. At the same time, and considering the seven astrovirus serotypes, it would be worthwhile to determine if that immunity is serotype-specific. As with rotaviruses, it may be that determination of transmission of astroviruses will be less feasible in the long run than will be the identification of other means to control this disease and prevent its spread.

References

1. Shi M, Sikotra S, Lee T, Kurtz JB, Getty B, Hart CA, Myint SH (1994) Use of a nested PCR method for the detection of astrovirus serotype 1 in human faecal material. Molecular and Cells for Probes 8: 481–486
2. Appleton H, Higgins PG (1975) Viruses and gastroenteritis in infants (letter). Lancet i: 1297
3. Ashley CR, Caul EO, Paver WK (1978) Astrovirus-associated gastroenteritis in children. J Clin Pathol 31: 939–943
4. Avery RM, Shelton AP, Beards GM, Omotade OO, Oyejjide OC, Olaleye DO (1992) Viral agents associated with infantile gastroenteritis in Nigeria: relative prevalence of adenovirus serotypes 40 and 41, astrovirus, and rotavirus serotypes 1 to 4. J Diarrhoeal Dis Res 10: 105–108

5. Bates PR, Bailey AS, Wood DJ, Morris DJ, Couriel JM (1993) Comparative epidemiology of rotavirus, subgenus F (Types 40 and 41) adenovirus, and astrovirus gastroenteritis in children. J Med Virol 39: 224–228

6. Caul EO, Ashley CR, Egglestone SI (1978) An improved method for the routine identification of faecal viruses using ammonium sulfate precipitation. FEMS Microbiol Lett 4: 1–4

7. Cox GJ, Matsui SM, Lo RS, Hinds M, Bowden RA, Hackman RC, Meyer WG, Mori M, Tarr PI, Oshiro LS, Ludert JE, Meyers JD, McDonald GB (1994) Etiology and outcome of diarrhea after marrow transplantation: A prospective study. Gastroenterology 107: 1398–1407

8. Cruz JR, Bartlett AV, Herrmann JE, Caceres P, Blacklow NR, Cano F (1992) Astrovirus-associated diarrhea among Guatemalan ambulatory rural children. J Clin Microbiol 30: 1140–1144

9. Cubitt WD (1990) Small round structured viruses, caliciviruses and astroviruses. Balliere's Clinical Gastroenterology 4: 643–656

10. Dennehy PH, Martin PA, Spangenberger S: The role of astrovirus in acute diarrhea in hospitalized young children (1992)

11. Esahli H, Breback K, Bennet R, Ehrnst A, Eriksson M, Hedlund Kjell-O (1991) Astroviruses as a cause of nosocomial outbreaks of infant diarrhea. Pediatr Infect Dis J 10: 511–515

12. Ford-Jones EL, Mindorff CM, Gold R, Petric M (1990) The incidence of viral-associated diarrhea after admission to a pediatric hospital. Am J Epidemiol 131: 711–718

13. Gray JJ, Wreghitt TG, Cubitt WD, Elliot PR (1987) An outbreak of gastroenteritis in a home for the elderly associated with astrovirus type 1 and human calicivirus. J Med Virol 23: 377–381

14. Greenberg HB, Matsui SM (1992) Astroviruses and caliciviruses: emerging enteric pathogens. Infectious Agents Dis 1: 71–91

15. Grohmann G (1985) Viral diarrhoea in children in Australia. In: Tzipori S (ed) Infectious diarrhoea in the young. Elsevier, Amsterdam, pp 25–28

16. Grohmann GS, Glass RI, Pereira HG, Monroe SS, Hightower AW, Weber R, Bryan RT (1993) Enteric viruses and diarrhea in HIV-infected patients. N Engl J Med 329: 14–20

17. Guerrero ML, Martinez J, Noel J, Rosales G, Calva JJ, Pickering LK, Glass RI, Ruiz-Palacios GM: (1995) Astrovirus diarrhea in a prospective cohort of young Mexican children (Abstract)

18. Herrmann JE, Hudson RW, Perron-Henry DM, Kurtz JB, Blacklow NR (1988) Antigenic characterization of cell-cultivated astrovirus serotypes and development of astrovirus-specific monoclonal antibodies. J Infect Dis 158: 182–185

19. Herrmann JE, Nowak NA, Perron-Henry DM, Hudson RW, Cubitt WD, Blacklow NR (1990) Diagnosis of astrovirus gastroenteritis by antigen detection with monoclonal antibodies. J Infect Dis 161: 226–229

20. Herrmann JE, Taylor DN, Echeverria P, Blacklow NR (1991) Astroviruses as a cause of gastroenteritis in children. N Engl J Med 324: 1757–1760

21. Hrdy D (1987) Epidemiology of rotaviral infection in adults. Rev Infect Dis 9: 461–469

22. Jiang B, Monroe SS, Koonin EV, Stine SE, Glass RI (1993) RNA sequence of astrovirus: Distinctive genomic organization and a putative retrovirus-like ribosomal frameshifting signal that directs the viral replicase synthesis. Proc Natl Acad Sci USA 90: 10539–10543

23. Jonasson TO, Kjeldsberg E, Grinde B (1993) Detection of human astrovirus serotype 1 by the polymerase chain reaction. J Virol Methods 44: 83–88

23a. Jonasson TO, Monceyron C, Lee TW, Kurtz JB, Grinde B (1995) Detection of all serotypes of human astrovirus by the polymerase chain reaction. J Virol Methods 52: 327–334

298   R. I. Glass et al.

24. Konno T, Suzuki H, Ishida N, Chiba R, Mochizuki K, Tsunoda A (1982) Astrovirus-associated epidemic gastroenteritis in Japan. J Med Virol 9: 11–17
25. Kotloff KL, Herrmann JE, Blacklow NR, Hudson RW, Wasserman SS, Morris JG Jr, Levine MM (1992) The frequency of astrovirus as a cause of diarrhea in Baltimore children. Pediatr Infect Dis J 11: 587–589
26. Kurtz J, Lee T (1978) Astrovirus gastroenteritis: age distribution of antibody. Med Microbiol Immunol 166: 227–230
27. Kurtz JB, Lee TW (1984) Human astrovirus serotypes (letter). Lancet ii: 1405
28. Kurtz JB, Lee TW (1987) Astroviruses: human and animal. In: Bock G, Whelan J (eds) Ciba Foundation Symposium 128. Novel diarrhoea viruses. John Wiley & Sons, New York, pp 92–107
29. Kurtz JB, Lee TW, Craig JW, Reed SE (1979) Astrovirus infection in volunteers. J Med Virol 3: 221–230
30. Kurtz JB, Lee TW, Pickering D (1977) Astrovirus associated gastroenteritis in a children's ward. J Clin Pathol 30: 948–952
31. Lee TW, Kurtz JB (1977) Astroviruses detected by immunofluorescence (letter). Lancet ii: 406
32. Lee TW, Kurtz JB (1981) Serial propagation of astrovirus in tissue culture with the aid of trypsin. J Gen Virol 57: 421–424
33. Lee TW, Kurtz JB (1994) Prevalence of human astrovirus serotypes in the Oxford region 1976–1992, with evidence for two new serotypes. Epidemiol Infect 112: 187–193
34. Leite JPG, Barth OM, Schatzmayr HG (1991) Astrovirus in faeces of children with acute gastroenteritis in Rio de Janeiro, Brazil. Mem Inst Oswaldo Cruz 86: 489–490
35. Lew JF, Glass RI, Petric M, LeBaron CW, Hammond GW, Miller SE, Robinson C, Boutilier J, Riepenhoff-Talty M, Payne CM, Franklin R, Oshiro LS, Jaqua M-J (1990) Six year retrospective surveillance of gastroenteritis viruses identified at ten electron microscopy centers in the United States and Canada. Pediatr Infect Dis J 9: 709–714
36. Lew JF, Moe CL, Monroe SS, Allen JR, Harrison BM, Forrester BD, Stine SE, Woods PA, Hierholzer JC, Herrmann JE, Blacklow NR, Bartlett AV, Glass RI (1991) Astrovirus and adenovirus associated with diarrhea in children in day care settings. J Infect Dis 164: 673–678
37. Lewis DC, Lightfoot NF, Cubitt WD, Wilson SA (1989) Outbreaks of astrovirus type 1 and rotavirus gastroenteritis in a geriatric in-patient population. J Hosp Infect 14: 9–14
38. Lewis TL, Greenberg HB, Herrmann JE, Smith LS, Matsui SM (1994) Analysis of astrovirus serotype 1 RNA, identification of the viral RNA-dependent RNA polymerase motif, and expression of a viral structural protein. J Virol 68: 77–83
39. Madeley CR (1979) Comparison of the features of astroviruses and caliciviruses seen in samples of feces by electron microscopy. J Infect Dis 139: 519–523
40. Madeley CR, Cosgrove BP (1975) 28 nm particles in faeces in infantile gastroenteritis. Lancet ii: 451–452
41. Madeley CR, Cosgrove BP (1975) Viruses in infantile gastroenteritis (letter). Lancet ii: 124
42. Major ME, Eglin RP, Easton AJ (1992) 3′ Terminal nucleotide sequence of human astrovirus type 1 and routine detection of astrovirus nucleic acid and antigens. J Virol Methods 39: 217–225
43. Maldonado Y, Logan L, Sanchez M, Chmyz M, Millan F, Valdespino J, Matsui S, Greenberg H: Population-based prevalence of astrovirus gastroenteritis in rural Mayan children (Abstract)

44. Matsui SM, Lewis TL, Chiu E, Smith LS, Dupuis K, Cahill CK, Oshiro LS: An outbreak of astrovirus gastroenteritis in a nursing home and molecular characterization of the virus (Abstract)
45. Midthun K, Greenberg HB, Kurtz JB, Gary GW, Lin F-Y, Kapikian AZ (1993) Characterization and seroepidemiology of a type 5 astrovirus associated with an outbreak of gastroenteritis in Marin County, California. J Clin Microbiol 31: 955–962
46. Mitchell DK, Monroe SS, Jiang X, Matson DO, Glass RI, Pickering LK (1995) Virologic features of an astrovirus diarrhea outbreak in a day care center revealed by reverse transcription-polymerase chain reaction. J Infect Dis 172: 1437–1444
47. Mitchell DK, Van R, Morrow AL, Monroe SS, Glass RI, Pickering LK (1993) Outbreaks of astrovirus gastroenteritis in day care centers. J Pediatr 123: 725–732
48. Moe CL, Allen JR, Monroe SS, Gary HE Jr, Humphrey CD, Herrmann JE, Blacklow NR, Carcamo C, Koch M, Kin K-H, Glass RI (1991) Detection of astrovirus in pediatric stool samples by immunoassay and RNA probe. J Clin Microbiol 29: 2390–2395
49. Monroe SS, Carter MJ, Herrmann JE, Kurtz JB, Matsui SM (1995) Astroviridae. In: Murphy FA, Fauquet CM, Bishop DHL, Ghabrial SA, Jarvis AW, Martelli GP, Mayo MA, Summers MD (eds) Virus taxonomy: classification and nomenclature of viruses. Springer, Wien New York, pp 364–367
50. Monroe SS, Glass RI, Noah N, Flewett TH, Caul EO, Ashton CL, Curry A, Field AM, Madeley R, Pead PJ (1991) Electronmicroscopic reporting of gastrointestinal viruses in the United Kingdom, 1985–87. J Med Virol 33: 193–198
51. Nazar H, Rice S, Walker-Smith JA (1982) Clinical associations of stool astrovirus in childhood. J Pediatr Gastroenterol Nutr 1: 555–558
52. Noel J, Cubitt D (1994) Identification of astrovirus serotypes from children treated at the Hospitals for Sick Children, London 1981–93. Epidemiol Infect 113: 153–159
53. Noel JS, Lee TW, Kurtz JB, Glass RI, Monroe SS (1995) Typing of human astroviruses from clinical isolates by enzyme immunoassay and nucleotide sequencing. J Clin Microbiol 33: 797–801
54. Nozawa CM, Vaz MG, Guimaraes MA (1985) Detection of astrovirus-like viruses in diarrheic stool and its coexistence with rotavirus. Rev Inst Med Trop Sao Paulo 27: 238–241
55. Oishi I, Yamazaki K, Kimoto T, Minekawa Y, Utagawa E, Yamazaki S, Inouye S, Grohmann GS, Monroe SS, Stine S, Carcamo C, Ando T, Glass RI (1994) A large outbreak of acute gastroenteritis associated with astrovirus among students and teachers in Osaka, Japan. J Infect Dis 170: 439–443
56. Oshiro LS, Haley CE, Roberto RR, Riggs JL, Croughan M, Greenberg H, Kapikian A (1981) A 27-nm virus isolated during an outbreak of acute infectious nonbacterialgastroenteritis in a convalescent hospital: A possible new serotype. J Infect Dis 143: 791–795
57. Pavone R, Schinaia N, Hart CA, Getty B, Molyneux M, Borgstein A (1990) Viral gastro-enteritis in children in Malawi. Ann Trop Paediatr 10: 15–20
58. Saito K, Ushijima H, Nishio O, Oseto M, Motohiro H, Ueda Y, Takegi M, Nakaya S, Ando T, Glass RI, Zaiman K (1995) Detection of astrovirus from stool samples in Japan using reverse transcription and polymerase chain reaction amplification. Microbiology and Immunology (Japan) 39: 825–828
59. Scott TM, Madeley CR, Cosgrove BP, Stanfield JP (1979) Stool viruses in babies in Glasgow 3. Community studies. J Hyg Camb 83: 469–485
60. Stewien KE, Durigon EL, Tanaka H, Gilio AE, Baldacci ER (1991) Ocorrencia de astrovirus humanos na cidade de Sao Paulo, Brasil. Rev Saude Publicz 25: 157–158
61. Tanaka H, Kisielius JJ, Ueda M, Glass RI, Joazeiro PP (1994) Intrafamilial outbreak of astrovirus gastroenteritis in Sao Paulo, Brazil. J Diarrhoeal Dis Res 12: 219–221

62. Timenetsky M, Kisielius J, Grisi S, Escobar A, Ueda M, Tanaka H (1993) Rotavirus, adenovirus, astrovirus, calicivirus, e "small round virus particles" em feces de criancas, com e sem diarreia aguda, no periodo de 1987 a 1988, na Grande Sao Paulo. Rev Inst Med Trop Sao Paulo 35: 275–280

63. Utagawa ET, Nishizawa S, Sekins S, Hayashi Y, Ishihara Y, Oishi I, Iwasaki A, Yamashita I, Miyamura K, Yamazaki S, Inouye S, Glass RI (1994) Astrovirus as a cause of gastroenteritis in Japan. J Clin Microbiol 32: 1841–1845

64. Utagawa ET, Yamazaki S, Inouye S, Glass RI, Nishizawa S, Sekine S, Hayashi Y, Yamashita T, Oishi I, Yamazaki K, Minekawa K, Yamashita I, Iwasaki A (1992) Astrovirus epidemiology in Japan (1982–92) as studied by an ELISA kit (Abstract). US-Japan Viral Diseases Meeting

65. Watkins JS (1984) Astrovirus gastroenteritis on a geriatric ward. Communicable Diseases Report 84: 16

66. Willcocks MM, Brown TDK, Madeley CR, Carter MJ (1994) The complete sequence of a human astrovirus. J Gen Virol 75: 1785–1788

67. Willcocks MM, Carter MJ, Laidler FR, Madeley CR (1990) Growth and characterisation of human faecal astrovirus in a continuous cell line. Arch Virol 113: 73–81

68. Willcocks MM, Carter MJ, Silcock JG, Madeley CR (1991) A dot-blot hybridization procedure for the detection of astrovirus in stool samples. Epidemiol Infect 107: 405–410

69. Willcocks MM, Silcock JG, Carter MJ (1993) Detection of Norwalk virus in the UK by the polymerase chain reaction. FEMS Microbiol Lett 112: 7–12

70. Wilson SA, Cubitt WD (1988) The development and evaluation of radioimmune assays for the detection of immuneglobulins M and G against astrovirus. J Virol Methods 19: 151–159

71. Xu AY, Wan XB, Qiu FX, Pang QF (1981) Astrovirus in autumn infantile gastroenteritis. Chin Med J (Engl) 94: 659–662

72. Yamashita T et al. (1991) Isolation of cytopathic small round viruses with BS-C-1 cells from patients with gastroenteritis. J Infect Dis 164: 954–957

Author's address: Dr. R. I. Glass, Viral Gastroenteritis Section, Mailstop G04, Center for Disease Control and Prevention, 1600 Clifton Road, N.E., Atlanta, Georgia 30333, U.S.A.

Arch Virol (1996) [Suppl] 12: 301–307

Structural features unique to enteric adenoviruses

M. Brown[1], **J. D. Grydsuk**[1], **E. Fortsas**[1], and **M. Petric**[2]

[1]Department of Microbiology, University of Toronto, Toronto, Ontario, Canada,
[2]Department of Microbiology, The Hospital for Sick Children, Toronto, Ontario, Canada, and Department of Microbiology, University of Toronto, Toronto, Ontario, Canada

Summary. Enteric adenoviruses are important agents of pediatric gastroenteritis. Characterization of monoclonal antibodies against human adenovirus 41 (h-41) identified an epitope of interest on protein VI, an internal virion protein. The epitope is common to enteric adenoviruses (subgenus A: h-12, h-18, h-31 and subgenus F: h-40, h-41) but is not shared by non-enteric serotypes (subgenera B, C, D or E). By expressing random oligonucleotide fragments of the protein VI gene as T7 gene 10 fusion proteins in the pTope vector (Novagen), the epitope was mapped within the central domain of protein VI, to the region corresponding to aa 114–125 of the Ad2 protein. Identification of this epitope reflects the close evolutionary relationship of subgenus A and subgenus F adenoviruses and draws attention to structural features of enteric adenoviruses as potential determinants of tropism. Furthermore, this epitope may be valuable for identification of enteric adenoviruses in clinical specimens.

Brief report

Adenoviruses are important etiologic agents of pediatric gastroenteritis around the world, having been identified in association with diarrheal disease in both developed and developing countries [5, 41]. At least 49 serotypes of human adenoviruses have thus far been identified and classified into six subgenera [17, 32]. Serotypes within subgenus A and subgenus F display a selective tropism for the gastrointestinal (GI) tract and rarely infect tissues outside the gastrointestinal (GI) tract; these are referred to as enteric adenoviruses. Other serotypes, including those of subgenus C that can infect multiple sites in the body, are infrequently associated with gastroenteritis.

Serotypes predominantly associated with pediatric gastroenteritis include h-40 and h-41 as well as h-31 [6, 19, 23, 41]. In the early 1980's, h-40 and h-41 were identified with comparable frequency but since the mid-1980's, the relative incidence of h-40 has decreased worldwide [6, 11, 13, 19, 35]. In recent years, h-40 has rarely been found in association with disease, although Mickan and Kok [29] reported that h-40 accounted for 20% of adenoviruses identified in association with gastroenteritis in South Australia from July, 1991-June 1992. Both h-40 and h-41 are classified within subgenus F; h-31, the other serotype associated with pediatric gastroenteritis, is classified within subgenus A, along with h-12 and h-18. Both h-12 and h-18 were orginally isolated from stool specimens but are rarely found in association with human disease [6, 7, 19]. In a retrospective study done in this laboratory, only one h-12 isolate was identified among 140 adenovirus isolates from stool specimens of patients with gastroenteritis [6].

In the natural host, subgenus F adenoviruses exhibit a more restricted tropism than do subgenus A adenoviruses. Specifically, h-31 can cause disseminated disease in immunocompromised hosts [16, 20, 31, 33] but there is no evidence that h-40 or h-41 spreads to cause disease outside the GI tract even in immunocompromised hosts [34, M. Brown, unpubl. results]. In cell culture, h-40 and h-41 grow poorly, relative to other serotypes, and for this reason, they are referred to as "fastidious" enteric adenoviruses. The absence of specific host cell factors, as yet unidentified, result in low level expression of the E1B 55K protein [2, 4, 27] as well as inefficient encapsidation of the genome [37, M. Brown, unpubl. results] and inefficient release of progeny virus from host cells [8]. Recent reviews by Mautner et al. [28] and by Tiemessen and Kidd [38] describe the fastidious growth characteristics of h-40 and h-41 in detail.

The fact that enteric adenoviruses do not normally cause disease outside the GI tract may be related to the host immune response, in the case of subgenus A adenoviruses, or to the absence of specific host cell factors, in the case of subgenus F adenoviruses. However, the fact that other serotypes do not usually cause gastroenteritis, even though they have the ability to replicate in a variety of tissues, suggests that the selective tropism of the enteric adenoviruses for the GI tract cannot be explained solely by the presence of host cell factors required for virus replication. It is possible that other factors, including structural features of the virion, may contribute to the selective tropism of enteric adenoviruses for the GI tract.

Structural proteins that affect early virus/cell interactions. The most obvious structural feature likely to affect adenovirus tropism is the knob of the fibre that serves to mediate initial attachment of the virion to the host cell receptor. Subgenus F adenoviruses (h-40 and h-41) are unique among human adenoviruses in that virions contain both long and short fibres [21, 46]. Alignment of the amino acid sequences shows that the short fibres of h-40 and h-41 are closely related, as are the long fibres, but that the short and long fibres of the same serotype are not closely related to each other (Fig. 1). Given that the knob domains of the two short fibres are virtually identical, as are the knob domains of

Fig. 1. Amino acid sequence aligment of the long and short fibres of h-40 and h-41. Sequences were aligned using DNAsis 2.0 software. GenBank accession numbers are L19443 [h-40 long (L40F) and short (S40F) fibres], X16583 [h-41 long fibre (L41F)] and X17016 [h-41 short fibre (S41F)]

```
              10         20         30         40         50
L40F    1 MKRARPEDDF NPVYPYPHYN PLDIPFITPP FASSNGLQEK PPGVISLKYT   50
L41F    1 MKRARIEDDF NPVYPYPHYN PLDIPFITPP FASSNGLQEK PPGVISLKYT   50
S40F    1 MKRTRIEDDF NPVYPYDTSS TPSIPYVAPP FVSSDGLQEN PPGVLALKYT   50
S41F    1 MKRTRIEDDF NPVYPYDTFS TPSIPYVAPP FVSSDGLQEK PPGVLALKYT   50
              60         70         80         90        100
L40F   51 DPLTTK-NGA LTLKLGTGLN IDKNGDLSSD ASVEVSAPIT KTNKIVGLNY  100
L41F   51 DPLTTK-NGA LTLKLGTGLN IDBNGDLSSD ASVEVSAPIT KTNKIVGLNY  100
S40F   51 DPITTNAKHE LTLKLGSNIT LQN--GLLSA TVPTVSPPLT NSNNSLGLAT  100
S41F   51 DPITTNAKHE LTLKLGSNIT LBN--GLLSA TVPTVSPPIT NSNNSLGLAT  100
             110        120        130        140        150
L40F  101 TKPLALQNNA LTLSYNAPFN VVNNNLALNM SKPVTINANN ELSLLIDAPL  150
L41F  101 TKPLALRSNA LTLSYNAPLN VVNNNLALNI SKPVTVNANN ELSLLIDAPL  150
S40F  101 SAPIAVSANS LTLATAAPLT VSNNQLSINT GRGLVITNN- ----------  150
S41F  101 SAPIAVSANS LTLATAAPLT VSNNQLSINA GRGLVITNN- ----------  150
             160        170        180        190        200
L40F  151 NADTGTLRLR SDAPLGLVDK TLKVLFSSPL YLDNNFLTLA IERPLALSSN  200
L41F  151 NADTGTLRLQ SAAPLGLVDK TLKVLFSSPL YLDNNFLTLA IERPLALSSS  200
S40F  151 -----AVAVN PTGALGFNNT ---------- ---------- ----------  200
S41F  151 -----ALTVN PTGALGFNNT ---------- ---------- ----------  200
             210        220        230        240        250
L40F  201 RAVALKYSPP LKIENENLTL STGGPFTVSG GNLNLATSAP LSVQNNSLSL  250
L41F  201 RAVILKYSPP LKIENENLTL STGGPFTVSG GNLNLTTSAP LSVQNNSLSL  250
S40F  201 ---------- ---------- ---------- GALQLNAACG MRVDGANLIL  250
S41F  201 ---------- ---------- ---------- GALQLNAACG MRVDGANLIL  250
             260        270        280        290        300
L40F  251 ---------- -----GVNPP FLITDSGLAM DLGDGLALGG SKLIINLGPG  300
L41F  251 VITSPLKVIN SMLAVGVNPP FTITDSGLAM DLGDGLALGG SKLIINLGPG  300
S40F  251 HVAYPFEAIN ---------- ---------- ---------- -QLTLRLENG  300
S41F  251 HVAYPFEAIN ---------- ---------- ---------- -QLTLRLENG  300
             310        320        330        340        350
L40F  301 LQMSNGAITL ALDAALPLQY KNNQLQLRIG SASALIMSGV TQTLNVNANT  350
L41F  301 LQMSNGAITL ALDAALPLQY RDNQLQLRIG STSGLIMSGV TQTLNVNANT  350
S40F  301 LEVTNGG--- ---------- ---------- ---------- ----------  350
S41F  301 LEVTSGG--- ---------- ---------- ---------- ----------  350
             360        370        380        390        400
L40F  351 SKGLAIENNS LVVKLGNGLR FDSWGSIAVS PTTTTP---- -TTLWTTADP  400
L41F  351 GKGLAVENNS LVVKLGNGLR FDSWGSITVS PTTTTP---- -TTLWTTADP  400
S40F  351 ---------K LNVKLGSGLQ FDMNGRITIS NRIQTRSVTS LTTIWSIS-P  400
S41F  351 ---------K LNVKLGSGLQ FDSNGRIAIS NSNRTRSVPS LTTIWSIS-P  400
             410        420        430        440        450
L40F  401 SPNATFYESL DAKVWLVLVK CNGMVNGTIS IKAQKGTLLK PTASFISFVM  450
L41F  401 SPNATFYESL DAKVWLVLVK CNGMVNGTIS IKAQKGILLR PTASFISFVM  450
S40F  401 TPNCSIYETQ DANLFLCLTK NGAHVLGTIT IKGLKGALRE MNDNALSVKL  450
S41F  401 TPNCSIYETQ DANLFLCLTK NGAHVLGTIT IKGLKGALRE MHDNALSLKL  450
             460        470        480        490        500
L40F  451 YFYSDGTWRK NYPVFDNEGI LAN----SAT WGYRQGQSAN TNVSNAVEFM  500
L41F  451 YFYSDGTWRK NYPVFDNEGI LAN----SAT WGYRQGQSAN TNVSNAVEFM  500
S40F  451 ---------- ---PFDNQGN LLNCALESST WRY---QEIN AVASNALIFM  500
S41F  451 ---------- ---PFDNQGN LLNCALESST WRY---QEIN AVASNALIFM  500
             510        520        530        540        550
L40F  501 PSSKRYPNEK GSEVQNMALT YTFLQGDPNM AISFQSIYNH AIEGYSLKFT  550
L41F  501 PSSKRYPNQK GSEVQNMALT YTFLQGDPNM AISFQSIYNH ALEGYSLKFT  550
S40F  501 PNSTVYPRNK TADPGNM--- --LICISPN- -IIFSVVYNE INSGYAFIFK  550
S41F  501 PNSTVYPRNK TAHPGNM--- --LICISPN- -IIFSVVYNE INSGYAFIFK  550
             560        570        580        590        600
L40F  551 WRVRNBRFD IPCGSFSYVT EC---..... ..........  600
L41F  551 WRVRNBRFD IPCGSFSYVT EC---..... ..........  600
S40F  551 NSAEPGKPFH PPTAVFCYIT ECWAL..... ..........  600
S41F  551 NSAEPGKPFH PPTAVFCYIT ECIAL..... ..........  600
```

the two long fibres, it appears that h-40 and h-41 recognize the same host cell receptors but that the two fibres recognize different receptors, possibly on different cells. Studies with h-2 (a non-enteric adenovirus) have shown that internalization, which occurs by receptor mediated endocytosis, is distinct from the attachment process and is faciliated by specific interaction of the RGD (arg-gly-asp) motif in the penton base of the virion with $\alpha_v\beta_3$ and $\alpha_v\beta_5$ integrins in the host cell membrane [12, 43, 44]. This RGD sequence is found in representative serotypes of subgenera A, B, C and E (25) but is notably absent in h-40; the corresponding h-40 sequence is RGAD [10].

Internal protein VI. Characterization of monoclonal antibodies prepared against h-41 led to the identification of two epitopes on protein VI, an internal protein of the virion. One epitope was common only to h-40 and h-41 (subgenus F); the other epitope was common to all serotypes within subgenus A (h-12, h-18, h-31) and subgenus F (h-40, h-41) but was not shared by representative serotypes of subgenera B, C, D or E. The DNA sequence of the h-40, h-41 and h-31 protein VI genes was determined for comparison with published sequences of the corresponding h-2, h-5 and h-12 genes. Alignment of the deduced amino acid sequences shows a high degree of conservation in three domains of the protein VI precursor (corresponding to aa 1–108, 128–158 and 228–250 in the h-2 protein) among all six serotypes. The two epitopes were mapped within protein VI by expression of randomly generated oligonucleotide fragments of the protein VI gene, as T7 gene 10 fusion proteins, in a pTope expression vector (Novagen). Both epitopes map to contiguous sites within the region corresponding to aa 114–125 of the h-2 protein. Details of the nucleotide and amino acid sequence alignments, as well as the epitope mapping, are published elsewhere (Grydsuk *et al.*, 1996).

Protein VI is an important structural component of the virion; it is located beneath the vertices, holding the peripentonal hexons together [36]. Deletion mutagenesis has identified the sequences that are important for hexon binding as those corresponding to aa 48–74 and aa 233–239 of h-2 protein VI [26]. These sequences are highly conserved between enteric and non-enteric adenoviruses [14]. Protein VI also has DNA binding properties and the C-terminal 11 aa peptide of the precursor protein has been shown to activate the virion protease responsible for cleavage of precursor proteins IIIa, VI, VII, VIII, μ and terminal protein during virion maturation [42]. This "activation domain" is also conserved in enteric and non-enteric adenoviruses [14].

The epitope common to enteric adenoviruses may simply reflect a close evolutionary relationship between subgenus A and subgenus F serotypes. Comparison of nucleotide sequence data for the ITR E1, E2a, E3b, major late promoter, hexon, protease and fibre regions of the genome [3] as well as VA-RNA sequences [22] indeed shows a close relationship between subgenus A and subgenus F adenoviruses. However, a close evolutionary relationship between subgenus A and subgenus F adenoviruses does not negate the possibility that the conserved epitope may have a functional significance with respect to the selective GI tropism displayed by these serotypes. Furthermore, this epitope is potentially valuable for the laboratory identification of clinically relevant adenoviruses (both subgenus A and subgenus F) in stool specimens of patients with gastroenteritis, using a single test. Such a test would represent an improvement over existing immunoassays which detect only subgenus F adenoviruses (h-40/h-41) [1, 9, 11, 13, 15, 18, 24, 30, 39, 40, 45].

References

1. Ahluwalia GS, Scott-Taylor TH, Klisko B, Hammond GW (1994) Comparison of detection methods for adenovirus from enteric clinical specimens. Diag Microbiol Infec Dis 18: 161–166
2. Bailey AC, MacKay N, Mautner V (1993) Enteric adenovirus type 40: Expression of E1B proteins *in vitro* and *in vivo*. Virol 193: 631–641
3. Bailey A, Mautner V (1994) Phylogenetic relationships among adenovirus serotypes. Virol. 205: 438–452
4. Bailey A, Ullah R, Mautner V (1994) Cell type specific regulation of expression from the Ad40 E1B promoter in recombinant Ad5/Ad40 viruses. Virol 202: 695–706
5. Bern C, Glass RI (1994) Impact of diarrheal diseases worldwide. In: Kapikian AZ (ed) Viral infections of the gastrointestinal tract. Marcel Dekker, New York, pp 1–26
6. Brown M (1990) Laboratory identification of adenoviruses associated with gastroenteritis in Canada from 1983 to 1986. J Clin Microbiol 28: 1525–1529
7. Brown M, Petric M, Middleton PJ (1984) Diagnosis of fastidious enteric adenoviruses 40 and 41 in stool specimens. J Clin Microbiol 20: 334–338
8. Brown M, Wilson-Friesen HL, Doane F (1992) A block in release of progeny virus and a high particle-to-infectious unit ratio contribute to poor growth of enteric adenovirus types 40 and 41 in cell culture. J Virol 66: 3198–3205
9. Cruz JR, Caceres P, Cano F, Flores J, Barlett A, Torun B (1990) Adenovirus types 40 and 41 and rotaviruses associated with diarrhea in children from Guatemala. J Clin Microbiol 28: 1780–1784
10. Davison AJ, Telford EA, Watson MS, McBride K, Mautner V (1993) The DNA sequence of adenovirus type 40. J Mol Biol 234: 1308–1316
11. de Jong J, Bijlsma K, Wermenbol AG, Verweij-Uijterwaal MW, van der Avoort HGAM, Wood DJ, Bailey AS, Osterhaus ADME (1993) Detection, typing, and subtyping of enteric adenoviruses 40 and 41 from fecal samples and observation of changing incidences of infections with these types and subtypes. J Clin Microbiol 31: 1562–1569
12. Goldman MJ, Wilson JM (1995) Expression of $\alpha_v\beta_5$ integrin is necessary for efficient adenovirus-mediated gene transfer in the human airway. J Virol 69: 5951–5958
13. Grimwood K, Carzino R, Barnes GL, Bishop RF (1995) Patients with enteric adenovirus gastroenteritis admitted to an Australian pediatric teaching hospital from 1981 to 1992. J Clin Microbiol 33: 131–136
14. Grydsuk JD, Fortsas E, Petric M, Brown M (1996) Common epitope on protein VI of enteric adenoviruses from subgenus A and F. J Gen Virol 77: 1811–1819
15. Herrmann JE, Perron-Henry DM, Blacklow NR (1987) Antigen detection with monoclonal antibodies for the diagnosis of adenovirus gastroenteritis. J Infect Dis 155: 1167–1171
16. Hierholzer JC (1992) Adenoviruses in the immunocompromised host. Clin Microbiol Rev 5: 262–274
17. Hierholzer JC, Stone YO, Broderson JR (1991) Antigenic relationships among the 47 human adenoviruses determined in reference horse antisera. Arch Virol 121: 179–197

18. Jarecki-Khan K, Tsipori SR, Unicomb LE (1993) Enteric adenovirus infection among infants with diarrhea in rural Bangladesh. J Clin Microbiol 31: 484–489

19. Johansson ME, Andersson MA, Thorner, PA (1994) Adenoviruses isolated in the Stockholm area during 1987–1992: restriction endonuclease analysis and molecular epidemiology. Arch Virol 137: 101–115

20. Johansson ME, Brown M, Hierholzer JC, Thorner A, Ushijma H, Wadell G (1991) Genome analysis of adenovirus type 31 strains from immunocompromised and immunocompetent patients. J Infect Dis 163: 293–299

21. Kidd AH, Chroboczek J, Cusack S, Ruigrok RWH (1993) Adenovirus type 40 virions contain two distinct fibers. Virol 192: 73–84

22. Kidd AH, Garwicz D, Oberg M (1995) Human and simian adenoviruses: Phylogenetic inferences from analysis of VA RNA genes. Virol 207: 32–45

23. Krajden M, Brown M, Petrasek A, Middleton PJ (1990) Clinical features of adenovirus enteritis: a review of 127 cases. Pediatr Infect Dis J 9: 636–641

24. Lew JF, Moe CL, Monroe SS, Allen JR, Harrison BM, Forrester BD, Stine SE, Woods PA, Hierholzer JC, Herrmann JE, Blacklow NR, Bartlett AV, Glass RI (1991) Astrovirus and adenovirus associated with diarrhea in children in day care settings. J Infect Dis 164: 673–678

25. Mathias P, Wickham T, Moore M, Nemerow G (1994) Multiple adenovirus serotypes use αv integrins for infection. J Virol 68: 6811–6814

26. Matthews DA, Russell WC (1995) Adenovirus protein-protein interactions: molecular parameters governing the binding of protein VI to hexon and the activation of the adenovirus 23K protease. J Gen Virol 76: 1959–1969

27. Mautner V, MacKay N, Steinthorsdottir V (1989) Complementation of enteric adenovirus type 40 for lytic growth in tissue culture by E1B 55K function of adenovirus types 5 and 12. Virol 171: 619–622

28. Mautner V, Steinthorsdottir V, Bailey A (1995) Enteric adenoviruses. In: Doerfler W, Bohm P (ed) The Molecular Repertoire of Adenoviruses. Springer, Heidelberg

29. Mickan LD, Kok TW (1994) Recognition of adenovirus types in faecal samples by Southern hybridization in South Australia. Epidemiol Infect 112: 603–613

30. Noel J, Mansoor A, Thaker U, Herrmann J, Perron-Henry D, Cubitt WD (1994) Identification of adenoviruses in faeces from patients with diarrhoea at the Hospitals for Sick Children, London, 1989–1992. J Med Virol 43: 84–90

31. Rodriguez FH, Liuzza GE, Gohd RH (1984) Disseminated adenovirus 31 infection in an immunocompromised host. Am J Clin Pathol 82: 615–618

32. Schnurr D, Dondero ME (1993) Two new candidate adenovirus serotypes. Intervirology 36: 79–83

33. Schnurr D, Bollen A, Crawford-Miksza L, Dondero ME, Yagi S (1995) Adenovirus mixture isolated from the brain of an AIDS patient with encephalitis. J Med Virol 47: 168–171

34. Schofield KP, Morris DJ, Bailey AS, de Jong JC, Corbitt G (1994) Gastroenteritis due to adenovirus type 41 in an adult with chronic lymphocytic leukemia. Clin Infect Dis 19: 311–312

35. Scott-Taylor TH, Hammond GW (1995) Local succession of adenovirus strains in pediatric gastroenteritis. J Med Virol 45: 331–338

36. Stewart PL, Fuller SD, Burnett RM (1993) Difference imaging of adenovirus: bridging the resolution gap between X-ray crystallography and electron microscopy. EMBO J 12: 2589–2593

37. Tiemessen CT, Kidd AH (1994) Adenovirus type 40 and 41 growth in vitro: host range diversity reflected by differences in patterns of DNA replication. J Virol 68: 1239–1244

38. Tiemessen CT, Kidd AH (1995) The subgroup F adenoviruses. J Gen Virol 76: 481–497
39. Van R, Wun C-C, O'Ryan M, Matson DO, Jackson L, Pickering LK (1992) Outbreaks of human enteric adenovirus types 40 and 41 in Houston day care centers. J Pediatr 120: 516–521
40. van der Avoort HGAM, Wermenbol AG, Zomerdijk TPL, Kleijne JAFW, van Asten JAAM, Jensma P, Osterhaus ADME, Kidd AH, de Jong JC (1989) Characterization of fastidious adenovirus types 40 and 41 by DNA restriction enzyme analysis and by neutralizing monoclonal antibodies. Vir Res 12: 139–158
41. Wadell G, Allard A, Johansson M, Svensson L, Uhnoo I (1994) Enteric adenoviruses. In: Kapikian AZ (ed) Viral infections of the gastrointestinal tract. Marcel Dekker, New York, pp 519–547
42. Webster A, Hay RT, Kemp G (1993) The adenovirus protease is activated by a virus-coded disulphide-linked peptide. Cell 72: 97–104
43. White JM (1993) Integrins as virus receptors. Current Biology 3: 596–599
44. Wickham TJ, Mathias P, Cheresh DA, Nemerow GR (1993) Integrins $\alpha v \beta 3$ and $\alpha v \beta 5$ promote adenovirus internalization but not virus attachment. Cell 73: 309–319
45. Wood DJ, Bijlsma K, de Jong JC, Tonkin C (1989) Evaluation of a commercial monoclonal antibody-based enzyme immunoassay for detection of adenovirus types 40 and 41 in stool specimens. J Clin Microbiol 27: 1155–1158
46. Yeh HY, Pieniazek N, Pieniazek D, Gelderblom H, Luftig RB (1994) Human adenovirus type 41 contains two fibers. Virus Res 33: 179–198

Authors' address: Dr. M. Brown, Department of Medical Genetics and Microbiology, University of Toronto, Fitz Gerald Building, 150 College St., Toronto, Ontario M5S 1A8, Canada.

Arch Virol (1996) [Suppl] 12: 309–311

Closing remarks

M. K. Estes

These Proceedings present a record of fundamental and practical scientific advances representing the latest progress aimed towards understanding and conquering the devastating international effects of viral gastroenteritis. During the three days between June 28 and June 30, 1995, scientists from around the world gathered in the outskirts of Sapporo, Japan, to become a relatively secluded group of people in the forest, or "Nidom" as described by the Ainu or early native settlers of this beautiful region of the world. This retreat from daily research activities in a place where Leonard Bernstein came to rejuvenate his musical creativity resulted in the presentation of outstanding new research achievements, and a renewal of scientific creativity. Artistic creativity also emerged as illustrated in a painting of the meeting site which is striking as a record of the uniqueness and beauty of this Northern hideaway in Japan (Fig. 1).

The visualization of the small viruses (the Norwalk virus) in 1972 in stool filtrates from adults which could be passaged and cause illness in adult volunteers provided the first proof that a viral agent can cause gastroenteritis. This was rapidly followed by visualization of the larger rotaviruses in intestinal biopsies of children with gastroentertis in 1973. These early discoveries opened a new field of investigation which has culminated in the diverse presentations given at this meeting. More than twenty years after the initial discoveries of gastroenteritis viruses, we now know about additional such viruses (astroviruses, caliciviruses, enteric adenoviruses) although the natural history and epidemiology of infections with these more recently recognized agents remains to be completely defined. Such needed information is now being gathered rather quickly due to breakthroughs in understanding the basic properties of these viruses, and the development of new sensitive diagnostic assays in formats amenable for use in large scale epidemiologic studies. While caliciviruses and astroviruses are being found more frequently, the rotaviruses remain the single most important viral agent causing severe, life-threatening diarrheal disease in children. Thus, rotaviruses were the emphasis of this meeting.

The presentations on rotaviruses exemplify the complexity and diversity of these viruses, and the payoff obtained from investigation of these agents from a broad range of approaches. Directed studies to develop a vaccine have proceeded with greater confidence and ease as fundamental information about the genes and proteins of the rotaviruses have become known. Structural studies about these

Fig. 1. Painting of the forest area at the Hotel Nidom meeting site of the Sapporo
International Symposium on Viral Gastroenteritis. Painting by Miyoko Yamazaki

large, complex viruses have not only opened new arenas in the discipline of
structural biology, but are leading to a new level of understanding of how these
pathogens assemble and function. Such information coupled with new insights
presented at this meeting about the signals needed to regulate how the rotavirus
genes replicate can be expected to lead to use of these viruses as delivery systems or
vectors for gene therapy in the future.

Vaccine progress remains encouraging with two vaccine candidates moving
towards possible licensure by the turn of the century. This achievement is expected
to have a major impact on the health of children. The vaccine efforts have raised
many key questions such as what is the basis of protective immunity, and will
vaccine efficacy already observed in developed countries be as potent in developing countries where living conditions may vary quite extensively. Information
about immunity from studies of the natural history of rotavirus infections in
children in developed and developing countries and studies in animal models
have raised questions about whether the immune responses to an attenuated live
oral vaccine (with the viruses being of nonhuman origin) differ from those
following natural infection with the more virulent human rotaviruses. Vaccination

using nonreplicating virus-like particles as parenteral or oral immunogens and DNA vaccines offer new strategies not even anticipated when the rotaviruses were first discovered. Combinations of such new vaccine approaches and a greater understanding of the basis of protective immunity, which may not be simply based on the induction of neutralizing antibodies as once thought, may help improve vaccine efficacy even further or lead to new approaches of treatment and therapy. The new information presented in these Proceedings is often based on collaborations established at previous meetings, and it is hoped that reading this volume will update the reader and also stimulate new ideas and foster continued progress towards reducing the daily global toll of diarrheal and other viral disease.

Author's address: Prof. Dr. Mary K. Estes, Division of Molecular Virology, Baylor College of Medicine, Houston, Texas 77030, U.S.A.

SpringerVirology

T.F. Schwarz,
G. Siegl (eds.)

F.A. Murphy, C.M. Fauquet,
D.H.L. Bishop, S.A. Ghabrial,
A.W. Jarvis, G.P. Martelli,
M.A. Mayo, M.D. Summers (eds.)

M.A. Brinton, C.H. Calisher,
R. Rueckert (eds.)

Imported Virus Infections

1996. 68 figures. VIII, 202 pages.
Cloth DM 210,–, approx. US $ 143.00
ISBN 3-211-82870-2
Special edition of
Archives of Virology / Supplement 11, 1996
(Soft cover edition of Suppl. 11 only available
for subscribers to "Archives of Virology")

This book deals both with the epidemiologic background and the specific characteristics of vector-borne and emerging viral infections which may be spread all over the world due to today's rapid transport of infected individuals or animal vectors. Detailed description of the situation with e.g. Dengue, Japanese encephalitis, Lassa, hepatitis, HIV and filoviruses helps to plan diagnostic approaches and to develop scenarios for the handling of patients suspected of carrying high hazard viruses.

Virus Taxonomy

Classification and
Nomenclature of Viruses
Sixth Report of the International
Committee on Taxonomy of Viruses

1995. 185 figures. IX, 586 pages.
Soft cover DM 160,–, approx. US $ 98.00*
ISBN 3-211-82594-0
Archives of Virology / Supplement 10

The Committee's Sixth Report includes one order, 71 families, 11 subfamilies, and 175 genera and more than 4,000 member viruses. On 600 printed pages large amounts of molecular biologic data, illustrated by micrographs and virion diagrams, gene maps and tables give a comprehensive overview and prove helpful in teaching, in diagnostics, in scholarly research, and in the practical areas of medicine, veterinary medicine, plant pathology, insect pest management, and biotechnology.

10 % price reduction for subscribers to the journal "Archives of Virology"

Positive-Strand RNA Viruses

1994. 182 figures. X, 558 pages.
Soft cover DM 380,–, approx. US $ 224.00*
ISBN 3-211-82522-3
Archives of Virology / Supplement 9

Positive-strand RNA viruses include the majority of the plant viruses, a number of insect viruses, and animal viruses, such as coronaviruses, togaviruses, flaviviruses, poliovirus, hepatitis C, and rhinoviruses. Works from more than 50 leading laboratories represent latest research on strategies for the control of virus diseases molecular aspects of pathogenesis and virulence; genome replication and transcription; RNA recombination; RNA-protein interactions and host-virus interactions; protein expression and virion maturation; RNA replication; virus receptors; and virus structure and assembly. Highlights include analysis of the picorna-virus IRES element, evidence for long term persistence of viral RNA in host cells, acquisition of new genes from the host and other viruses via copy-choice recombination, identification of molecular targets and use of structural and molecular biological studies for development of novel antiviral agents.

SpringerWienNewYork

P.O.Box 89, A-1201 Wien • New York, NY 10010, 175 Fifth Avenue
Heidelberger Platz 3, D-14197 Berlin • Tokyo 113, 3-13, Hongo 3-chome, Bunkyo-ku